RADIOELEMENT ANALYSIS
PROGRESS AND PROBLEMS

Proceedings of the Twenty-Third Conference
on Analytical Chemistry in Energy Technology

RADIOELEMENT ANALYSIS
PROGRESS AND PROBLEMS

Proceedings of the Twenty-Third Conference
on Analytical Chemistry in Energy Technology

Gatlinburg, Tennessee, October 9-11, 1979, sponsored
by the Analytical Chemistry Division, Oak Ridge
National Laboratory, operated by Union Carbide
Corporation for the Department of Energy under
Contract No. W-7405-eng-26.

W.S. LYON, Editor

Analytical Chemistry Division
Oak Ridge National Laboratory
Oak Ridge, Tennessee

Published 1980 by Ann Arbor Science Publishers, Inc.
230 Collingwood, P.O. Box 1425, Ann Arbor, Michigan 48106

Library of Congress Catalog Card Number 79-55145
ISBN 0-250-40343-9

Manufactured in the United States of America

PREFACE

Themes for the annual conferences on Analytical Chemistry in Energy Technology are selected months, sometimes a year, in advance. In times of rapidly changing problems and priorities, it is not easy to forecast a theme that will attract the future interest of analytical chemists within the energy technology community.

"Radioelement Analysis: Progress and Problems" was selected this year because of the growing interest in radio-element analysis, spurred by concerns of environmental questions and waste management problems. While we recognized that much progress had been made, it was obvious that significant problems were as yet unsolved. In the interim between theme selection and the conference, the Three Mile Island accident dramatically underscored the importance of analytical chemistry in general and radioelement analysis in particular.

We are pleased with the success of the conference. Attendance was good, new personal contacts were made and old friendships renewed. Impromptu meetings were held to consider specific problems during which much useful information was transmitted.

Through this volume we hope to enlarge and preserve the value of this 23rd Conference.

W. D. Shults
Director
Analytical Chemistry Division
Oak Ridge National Laboratory

W. S. Lyon is Section Head of Analytical Methodology in the
Analytical Chemistry Division at Oak Ridge National Laboratory,
Oak Ridge, Tennessee. He holds a BS in Chemistry degree from
the University of Virginia (1943), a MS from the University of
Tennessee (1967), and has numerous graduate hours in chemistry
from the latter institution. Since 1947 he has been associated
with ORNL. His main research interests have been radiochemistry,
nuclear measurements, activation analysis and, more recently,
sampling and analysis of coal and coal combustion products.

Lyon is the author of one book, *Trace Element Measurements
at the Coal Fired Steam Plant* (CRC Press, 1977), editor of
three others, *Guide to Activation Analysis* (Van Nostrand, 1964),
Nuclear and Atomic Activation Analysis (with T. Braun and
E. Bujdoso, Akademia Kiado, 1976), and *Analytical Chemistry in
Nuclear Fuel Reprocessing* (Science Press, 1978). He is on the
editorial board of two international journals, has coauthored
the biennial review "Nucleonics" for *Analytical Chemistry* since
1966, and has edited the ORNL Analytical Chemistry Division
Annual Report for ten years. He has over 150 journal publica-
tions, is active in the American Nuclear Society and on the
organizing committees of several international conferences, and
has given numerous invited lectures both in the USA and abroad.

CONFERENCE COMMITTEE

A. L. Harrod, General Chairman

W. S. Lyon, Technical Program Chairman

C. B. Brooks, Treasurer

G. Long and P. Mullins,
Conference Secretaries

J. A. Carter	T. G. Scott
L. T. Corbin	W. D. Shults
L. M. Jenkins	J. R. Stokely

CONTENTS

PLENARY LECTURE

GAMMA SPECTROMETRY AND ACTIVATION ANALYSIS

MASS SPECTROMETRY

QUALITY ASSURANCE AND STANDARDS

PLENARY LECTURE

A SPECIAL RADIOELEMENT PROBLEM: ORNL ASSISTANCE TO
THREE MILE ISLAND IN HANDLING CONTAMINATED AIR AND WATER

R. E. Brooksbank, Chemical Technology Division,
Oak Ridge National Laboratory, Oak Ridge, Tennessee

ABSTRACT

Oak Ridge National Laboratory has played a significant
role in the recovery of Three Mile Island. Work has been
performed in waste handling, analytical service, flowsheet
development, and decontamination. These efforts in assistance to TMI are really the result of the close working
relationship between chemical engineers and analytical chemists. The two major areas of involvement are handling of
contaminated air and contaminated water that resulted from
the accident. These are discussed and an outline is given
of remedial actions proposed and taken.

INTRODUCTION

Before launching into the technical aspects of this
presentation, I would like to make a few personal observations. First, I would like to thank Dub Shults for the
opportunity to present this lecture. I do consider it a
great honor. My rapid affirmative response to Dub to present this paper was not based on the technical subject to
be presented, but rather the fact that this paper would be
presented as the first one in a conference. If you have
ever had the opportunity to present the _last_ paper in the
last session on the _last_ day of a conference as I have done
on occasion, you would also jump at the chance to be first.

As the investigative work concerning the post-accident
period comes to light, it will be apparent that ORNL will
have played a significant role in the recovery of TMI. Work
in the area of waste handling, analytical service, flowsheet
development, and decontamination will all be regarded as

significant. Efforts in this specific area of assistance
to TMI are really the results of the close working relationship between chemical engineers and analytical chemists.

The two major areas of involvement, and the subject of
this discussion, are contaminated air and water handling
which resulted from the accident. In order to give you an
appreciation for the specific problems involved in this area,
it will be necessary to present the status of the various
systems as they existed shortly following the accident. The
outline to be followed in this presentation is presented on
Fig. 1.

An original photograph of the Three Mile Island site
is presented in Fig. 2. The site is located in a predominantly farming community with two major population commun-
ities, namely Middletown and Goldsboro. Goldsboro is located to the west of the site and may be seen on the lower
portion of the photograph. Middletown, located north of the
site, is just off of the photograph on the left-hand side.
It was estimated that approximately 2 million people reside
within a 50-mile radius of the site.

The accident occurred in the TMI-2 reactor system which
may be observed in the lower left hand corner of the photograph presented as Fig. 3. The significant items in this
photograph include the auxiliary buildings and the fuel
handling structure. The stack, through which minor releases
of radioactivity emanated early in the accident, is the
black structure adjacent to TMI-2 and the Auxiliary Building-
2. The headquarters for the on-site ORNL team was one of
the trailers located in the upper right-hand corner of the
photograph.

CONTAMINATED AIR HANDLING

The release of radioactive material to the environment
as the result of the accident will be an item of concern for
a long period. Insignificant quantities of radioactive
material were released to the Susquehanna River via the
liquid pathway; however, these releases were below the established normal plant technical specifications for discharge
under normal operations. The major insult to the population
surrounding the TMI site resulted from the release of the
^{133}Xe isotope via the air pathway which emanated from the
stack in the days following the accident. Figure 4 presents
the average dose to the population as a function of distance
from the site during the period from March 28 through April
7, 1979. As is recognized, the most hazardous isotope present in the air pathway was ^{131}I, because of its adverse

- INTRODUCTION
- STATUS OF OFF-GAS SYSTEM FOLLOWING ACCIDENT
- MODIFICATIONS TO OFF-GAS SYSTEM
- STATUS OF TMI WATER FOLLOWING ACCIDENT
- LOW ACTIVITY LEVEL WATER PROCESSING
- INTERMEDIATE ACTIVITY LEVEL WATER PROCESSING
- HIGH ACTIVITY LEVEL WATER PROCESSING
- TECHNICAL AND POLITICAL STATUS

Figure 1. Presentation Outline

Figure 2. TMI Site

Figure 3.   TMI Reactor System

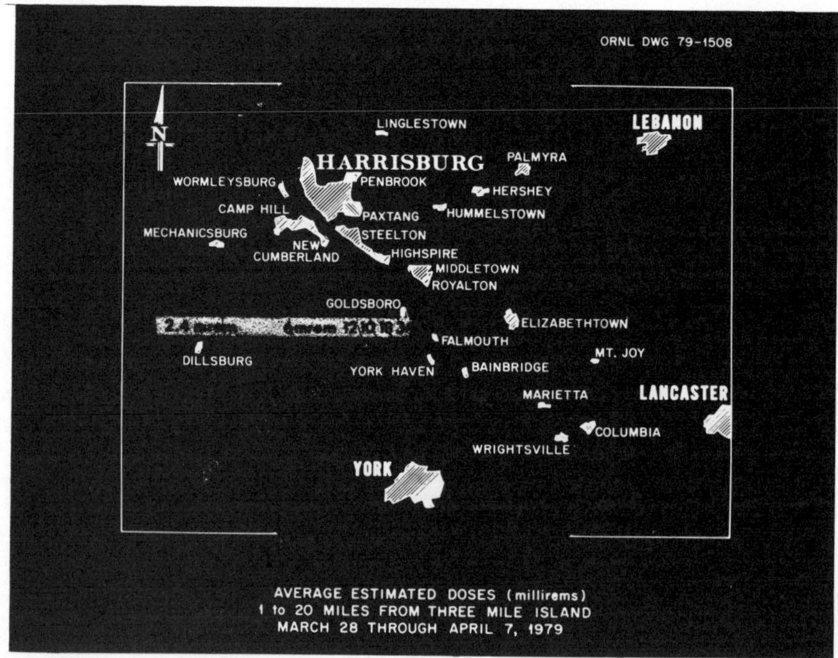

Figure 4.   Average Estimated Dose

effect on the human environment through the food chain. In
all, a total of 13-16 Ci of this isotope was released over
the accident period. When compared to the Windscale incident
of 1957, where 2200 Ci of iodine was released in a single
burst, the TMI release is small. Many things were done
on the site to maintain this release at a minimal level.

Status of Off-Gas System Following the Accident

     An assessment of the condition of the off-gas handling
and treatment system and support buildings was begun shortly
after the accident and is still in progress. The immediate
problem following the accident was the release of iodine
and the noble gases in excess of release specifications for
normal operations. Because iodine has a more pronounced effect on the health and welfare of the downstream population,
serious attention was given to the effectiveness of the charcoal traps designed to remove this isotope. Both downstream
and upstream samples of the charcoal traps contained in the
Auxiliary 2 and Fuel Handling Buildings, through which all
gaseous releases from TMI-2 emanated, indicated that the
traps were ineffective for the removal of iodine. Problems
inherent in establishing the effectiveness of the off-gas
removal systems involved high radiation levels surrounding
both the monitoring equipment and the traps themselves.
Figure 5 shows a schematic representation of the off-gas
system immediately following the accident.

Modifications to the Off-Gas System

     Results of the tests conducted on the iodine trapping
efficiency of the charcoal units within the Auxiliary 2
and Fuel Handling Buildings indicated that all the traps
should be replaced. Therefore, a total of 300 traps were
changed (180 in the Auxiliary Building and 120 in the Fuel
Handling Building) throughout the period of April 20-May 3.

     Because the reactor system was not yet stabilized from
the standpoint of the natural convection cooling mode and the
primary loop contained an estimated 6 million Ci of iodine,
the decision was made to provide the existing off-gas trains
with a supplemental system. This system, which contained
four trains totaling a treatment capacity of 100,000 cfm,
was located in Pasco, Washington. It was flown to the TMI-
2 site for installation on the Auxiliary 2 Building roof
and was placed onstream on May 3, 1979. Currently, three
of the four trains are in operation. Figure 6 summarizes,
in schematic fashion, the overall modifications made to the
off-gas system. An additional modification, shown on the
figure, is the capping of the stack vent; this provided an
added margin of safety.

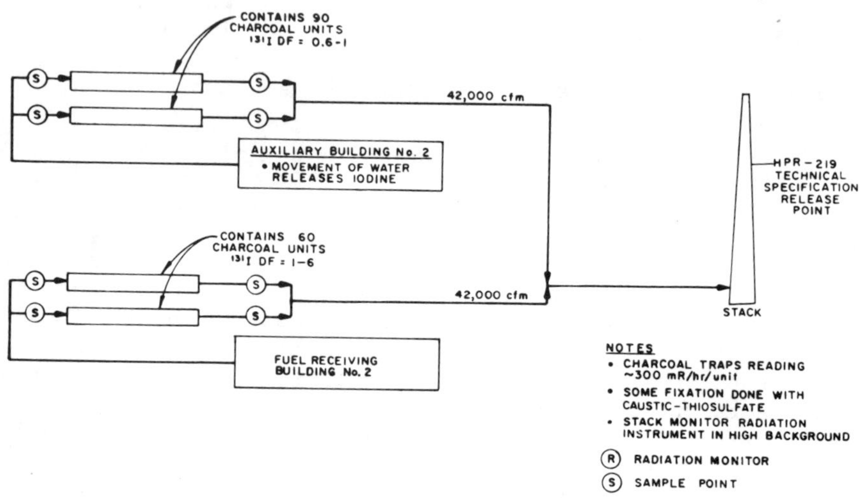

Figure 5. Off-Gas System Following Accident

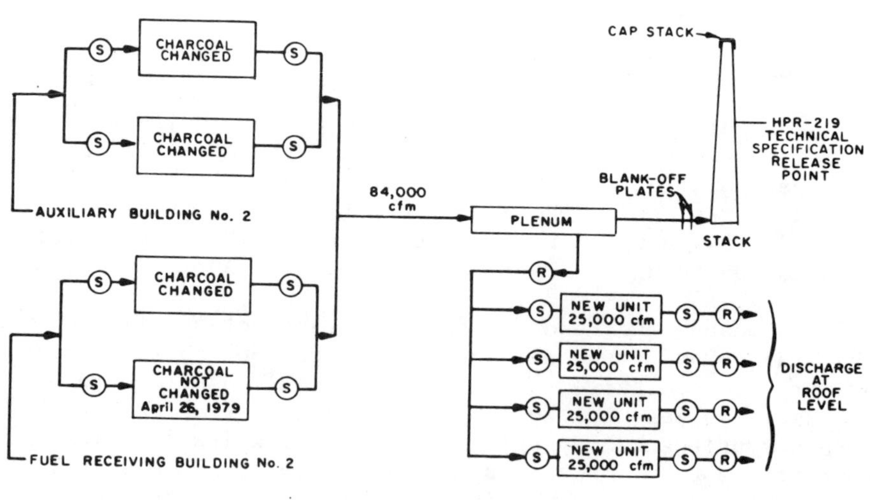

Figure 6. Off-Gas System Malfunction

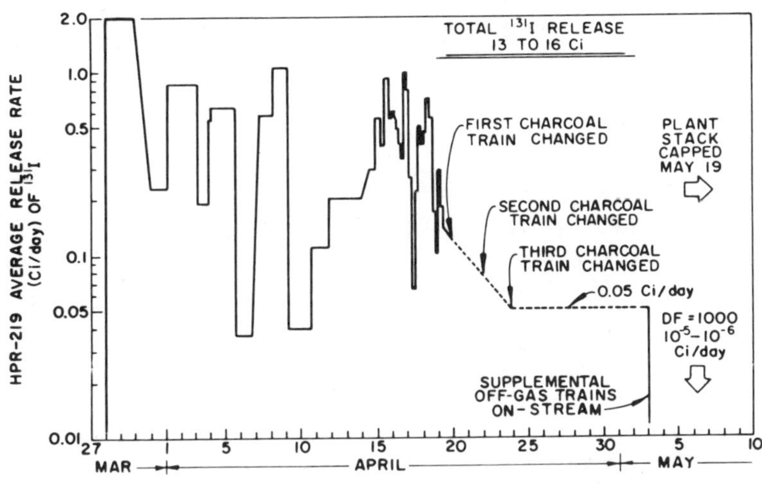

Figure 7. ^{131}I Release Date

Figure 8. New Charcoal Adsorbers During Installation

In addition to the modifications outlined, the floor areas suspected to be contaminated by iodine-containing solution were frequently wet with sodium thiosulfate in an attempt to decrease the level of iodine activity in the Auxiliary Building atmosphere (and thus reduce the iodine release).

Results

The actions discussed above, along with a major effort to minimize transfer and/or leakage of solution containing iodine and noble gases, led to a steady decrease in iodine release (Fig. 7). The most significant reductions were achieved when the existing charcoal adsorbers were changed (Fig. 8), when the new charcoal treatment system became operative, and when the plant stack was capped. The level of ^{131}I release varied from 0.05 to 2 Ci/da, which exceeded the technical specification quarterly average release rate limit of 0.002 Ci/day. When the new set of adsorbers was put into operation on May 3, the ^{131}I release rate dropped to approximately 1 μCi/day.

CONTAMINATED WATER TREATMENT

Contaminated water was continually being generated at TMI following the accident because of leakage through pump seals, flushing of sampling systems, and flushing of contaminated floor areas. The major concern relative to this water was that the quantity to be accumulated might exceed the storage capacity. There was also concern that the water level in the Containment Building might rise high enough to render inoperative some vital instruments. The eventual need to treat all of the liquids, including the primary coolant and all decontamination solutions, was considered throughout the planning for water handling.

Status of TMI Water Following the Accident

The status of the liquid handling systems as of April 1, 1979, is shown in Fig. 9. The primary reactor coolant loop contained 87,000 gal of highly radioactive coolant with an ^{131}I inventory of about 6,000,000 Ci. In addition, the Reactor Containment Building was estimated to contain about 225,000 gal of water which had been contaminated by a large volume of the radioactive reactor coolant. Some instruments were inoperative, probably because they were becoming full, and floor areas had become flooded with water that had over-flowed or leaked from the tanks. Portions of this water

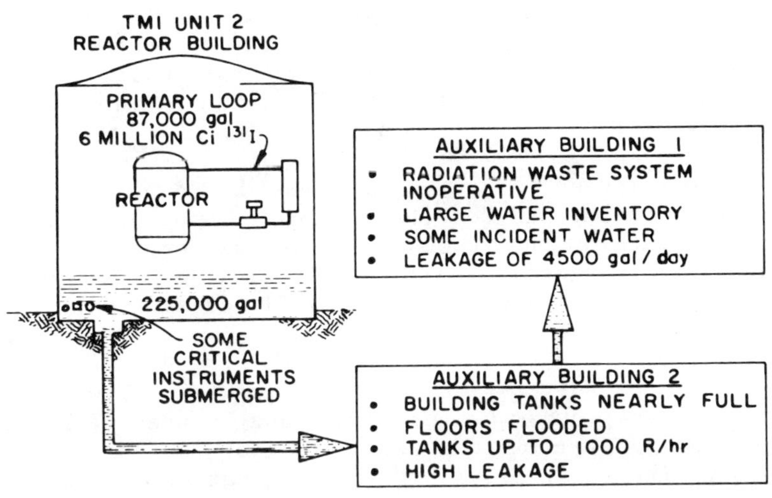

Figure 9.   Post-Accident Liquid Status

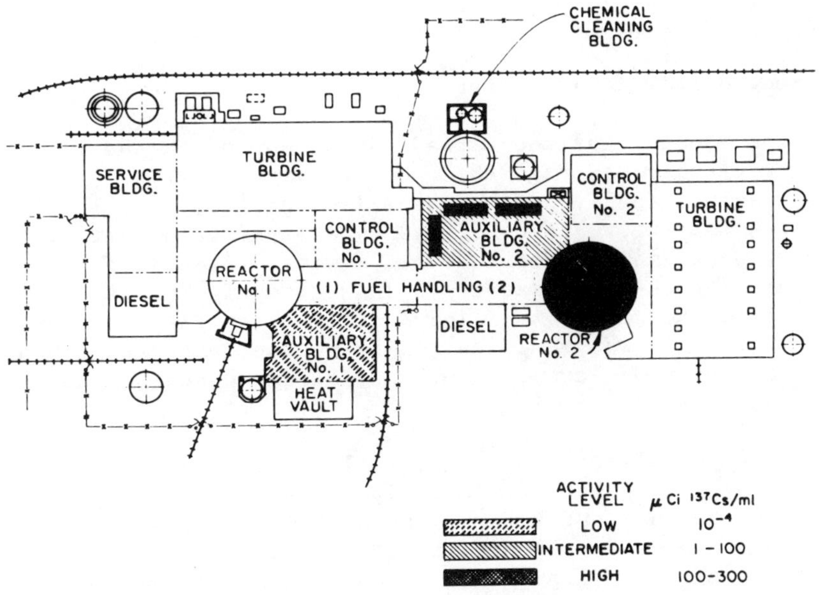

Figure 10.   Contaminated Liquid Distribution

were contaminated to varying degrees by water that had been transferred from the Containment Building sump before the Containment Building had been isolated, and subsequently had been transferred within the Auxiliary Building system during post-accident operations. The Unit 2 Reactor Building went into containment approximately 4 hr after the accident and has remained in this state ever since.

The Unit 1 Reactor, which had been shut down for refueling prior to the accident, was being brought up to operating temperature by the reactor coolant pump energy input prior to going critical. The available tankage within the Unit 1 Auxiliary Building was becoming filled with water due to normal leaks.

None of the Unit 2 water could be treated. The Unit 2 reactor coolant letdown stream could not be treated because of mechanical problems in the Unit 2 reactor coolant letdown evaporator. The other liquid wastes originating in Unit 2 are normally treated in the Unit 1 miscellaneous waste evaporator, which was out of service because a demineralizer bed was being changed. In any case, the transfer of Unit 2 post-event water to the Unit 1 Auxiliary Building was considered to be undesirable.

Water inventories in both Auxiliary Buildings were increasing. There was an urgent need for additional storage and/or water treatment facilities.

Treatment of Low-Activity-Level Water

The location of the varied water sources on the TMI site is presented in Fig. 10. Low-activity-level water (LALW) was originally defined as all water from Unit 1 and any pre-event water, as confirmed by analysis, in Unit 2. However, when analysis of the Unit 1 water revealed that some Unit 2 post-event water had inadvertently been transferred into the Unit 1 Auxiliary Building vessels, the definition of LALW was modified to include any water that had an ^{131}I activity of less than 0.1 µCi/ml and contained no actinides.

A demineralizer system (Fig. 11) consisting of a filter followed by a mixed-bed demineralizer for activity removal was set up on the west side of the Unit 1 Fuel Handling Building to process Unit 1 LALW. This system was referred to as Epicor-1 (initially Cap-Gun-1) because it was being operated by Capolupo & Gundal, Inc. Two 20,000-gal Haliburton tanks were available for the decontaminated water. The first-pass decontaminated water went into one Haliburton tank and was

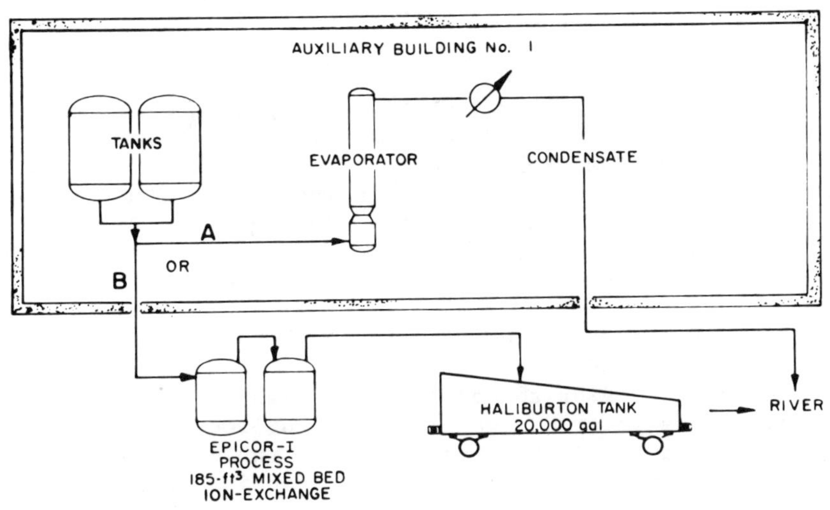

Figure 11.   Epicor-1 Demineralizer System

TANK	VOLUME (gal)	CONC. OF PRINCIPAL NUCLIDES[a] (μ Ci/ml)					
		^{131}I	^{134}Cs	^{136}Cs	^{137}Cs	^{140}Ba	^{3}H
REACTOR COOLANT BLEED TANK A	77,250	1.9	6.5	0.28	28	0.09	0.23
REACTOR COOLANT BLEED TANK B	77,250	2.8	7.6	0.29	35	0.3	0.27
REACTOR COOLANT BLEED TANK C	77,250	3.0	7.7	0.28	35	0.29	0.29
NEUTRALIZER TANK A	8,780	0.15	0.56	0.01	2.5	0.01	—
NEUTRALIZER TANK B	8,780	0.18	0.72	0.02	3.3	0.03	—
MISC. WASTE HOLDUP, AUX. BLDG. SUMP AND TANK, MISC. SUMPS	13,500	1.0	2.4	0.08	10.1	0.80	0.98
WASTE EVAPORATOR COND., CONTAMINATED DRAIN TANKS	16,200	<0.1	<0.1	<0.1	<0.1	<0.1	—
TOTAL	279,010						

a. CORRECTED FOR RADIOACTIVE DECAY TO JUNE 15, 1979

Figure 12.   Volumes of Solution and Concentrations of Principal Nuclides in Auxiliary Building Tanks

sampled. If the water had not been decontaminated sufficiently to permit release in one treatment cycle, the filter
and mixed demineralizer beds were changed and a second decontamination run was made. The second-pass decontaminated
water was routed to the second 20,000-gal Haliburton tank.

The first batch of water treated by this system required two passes to meet the technical specifications for release to the Susquehanna River. All subsequent batches
required two cycles of treatment. The first batch of
water that was successfully treated was released to the
Susquehanna River beginning on the night of April 11, 1979.
By June 6, a total of 103,500 gal of water had been treated
and released to the river.

Treatment of Intermediate-Activity-Level Water

As the result of the accident, a significant quantity
of radioactive water was generated and collected in the
Unit 2 Auxiliary Building tankage. For the most part, this
solution can be characterized as "intermediate-activity-
level" water, which can be defined as water containing ^{131}I
and ^{137}Cs at concentrations greater than 1 µCi/ml but less
than 100 µCi/ml.

Quantity and Characteristics

The wastewater in this category was produced from the
following four sources: (1) an inventory of wastewater
(130,000 gal) that existed in the Unit 2 Auxiliary Building
tankage prior to the accident; (2) contaminated water from
the Reactor Containment Building sump that had been transferred to the Auxiliary Building and collected in various
tanks (4200 gal) during the early phases of the accident;
(3) letdown water from the reactor coolant system, which
resulted in a net increase in the inventory; and (4) normal
leakage from system components in the Auxiliary Building.

As can be seen from Fig. 12, the total volume contained
in the Auxiliary Building tankage is approximately 279,000
gal. Figure 12 also gives the concentrations of the principal radionuclides present in the various solutions contained in the tanks.

Processing Justification

Although the Auxiliary Building is of sufficiently high
integrity that contaminated water can be positively controlled for an indefinite period, there are several significant reasons why the decontamination of this water is

beneficial. Available capacity of the tanks in the
Auxiliary Building is required in the event that the water
has to be pumped from the Reactor Building in order to protect the operability of Reactor Building components which
maintain continued safe shutdown of the facility. The
wastewater in the Auxiliary Building continues to be a
source of exposure to personnel needing entry into the Auxiliary Building. The continued safe shutdown of TMI Unit 2
depends on the operability of original plant equipment located in the Auxiliary Building and the use of additional
equipment being installed in the course of completing the
modifications now in progress. The surveillance and personnel exposures associated with these actions are adversely
affected by radiation levels associated with the stored
liquid.

The removal of the stored contaminated water will also
provide additional benefit to the surface decontamination
effort currently under way in the Auxiliary Building--now
precluded by high radiation levels.

Process Description

The process to be employed for the treatment of intermediate-level water is an extension of that used for low-
level water decontamination (Epicor-I) and is based on existing commercial technology currently in practice at numerous nuclear power plants. This process, designated as
Epicor-II, uses a liquid radwaste processing system supplied
by Epicor, Inc., and is designed to decontaminate radioactive water contained in the Auxiliary Building tanks via
filtration and ion exchange.

A simplified schematic flow diagram of the Epicor-II
system is presented in Fig. 13. Contaminated water is pumped
from the miscellaneous waste holdup tank in the Auxiliary
Building to a prefilter in the process which removes particulate radioactive materials and suspended solids. This
prefilter also contains a cation exchange resin which is
highly effective for removing cesium and other cationic
radionuclides from the water (removal efficiency, approximately 90%). Following the prefilter, the solution is passed
through two demineralizers placed in series. The first
demineralizer also contains cation resin which further
decontaminates the solution from cation-sorbing nuclides.
The second demineralizer contains mixed resins (cation and
anion) which are efficient for both cationic and anionic
radionuclides, including cesium and iodine. After processing,
the water is collected in a clean water-receiving tank which
has a capacity of 133,000 gal. Should analysis indicate that

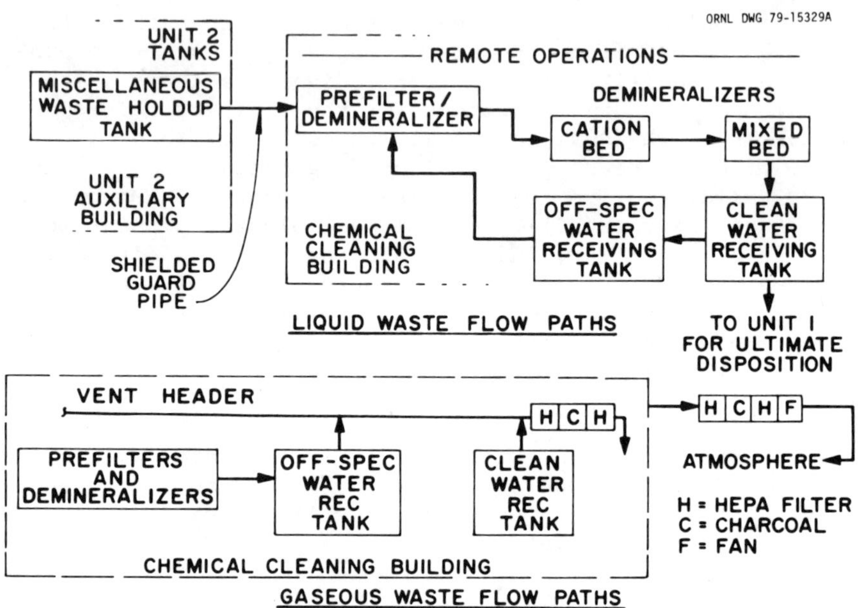

Figure 13. Flow Schematic of Epicor-II Processing System

BORON >500 ppm (2600 ±200 mg/ml)

pH   8.0

U   <10 ppb

Pu   ND*

FISSION PRODUCTS	ACTIVITY (μCi/cc)
^{99}Mo	179
^{131}I	8.2×10^3
^{132}I	<20
^{134}Cs	82
^{137}Cs	330
^{136}Cs	108
^{140}Ba	290
^{140}La	160
^{89}Sr	600
^{90}Sr	50
^{103}Ru	ND*
^{144}Ce	—
^{95}Zr	ND*
^{97}Zr	ND*

*DID NOT BECAUSE OF ph

Figure 14. Preliminary Primary Coolant Analysis
April 11, 1979

the radionuclide content of this tank is above specifications,
the water can be transferred to an off-specification vessel
(95,000 gal) for rework. Product water below the predetermined limits contained in the plant's technical specifications will be transferred to the TMI Unit 1 or 2 liquid
waste management system to be held for ultimate disposition.
Thus far, no decision has been made of the ultimate disposition of this water (to the Susquehanna River, for example) because of political concerns.

The Epicor-II processing of intermediate-activity-level
water will be done in the Chemical Cleaning Building, which
has been modified to ensure the safety of workers and the
general public as the result of more stringent radiation
controls. Basically, the modifications to this building
were made in the general area of radioactive material containment. The Chemical Cleaning Building has been converted
into a low-leakage confinement area and has been equipped
with an exhaust system to maintain the building at a negative pressure. HEPA and charcoal systems have been provided on the ventilation system, which will discharge
through a localized stack. All effluents from the process
will be subjected to both gaseous and liquid release monitoring. The processing system will be operated entirely by
remote means, except for infrequent tasks such as sampling
and chemical additions. All remote system operations are
controlled from the TV Monitor Control Building, which is
located adjacent to the Process Building. The remote transfer of spent filters and resins from this position in the
Processing Building into shielded casks for removal to the
solids staging area can be accomplished and has been incorporated into the design.

Treatment of High-Activity-Level Water

The largest volume of water generated from the accident
may be regarded as "high-activity-level water," which may
be defined as water with a radionuclide concentration in
excess of 100 µCi/ml based on the ^{137}Cs content.

Quantity and Characteristics.

Water in this category is primarily from two major
sources. The first source (90,000 gal) is the water contained in the primary loop coolant circuit which is used to
maintain the reactor in a safe condition by removing heat in
a natural convection mode. The second source (540,000 gal)
includes (1) the water contained in the reactor containment
structure which resulted from the release of water from the
primary coolant circuit during the early phase of the acci-

dent, (2) the volume of liquid transferred into the building
through the containment spray system, and (3) a large volume
of water which was released through miscellaneous equipment
(pump seals, space coolers, etc.) during the post-accident
period. Depending on the fission product removal efficiency
of the Epicor-II treatment system, the water in the three
reactor coolant bleed tanks may be processed later, if
necessary, in this system.

In addition to the increased radioactivity contained in
this water, the ionic contamination mandates that the water
be given special treatment. The major ionic constituent is
boron, present at levels up to approximately 2600 ppm, which
is used as a neutron poison in the coolant circuit. In
addition to this element, sodium is also present as the
result of spray activation with NaOH solution. Figure 14
presents significant information related to the chemical
and ionic characteristics of the high-activity-level waste
solution.

Necessity for Emergency Tankage

Soon after the accident, it was recognized that additional tankage would be required to receive the high-activity-
level water accumulating in the Reactor Containment Building. There was considerable concern that the water level
in the building might rise high enough to flood and thereby
prevent operation of some of the vital instruments that monitor and control the reactor. These instruments had not been
designed for operation in a hostile environment of high radiation fields and submergence in liquids. Subsequently, a
tank farm was designed and installed in the fuel storage
pool A contained in the Unit 2 Fuel Handling Building.

The design of the tank farm incorporated six tanks
with a total capacity of 110,000 gal (Fig. 15). Two
25,000-gal tanks were installed near the bottom of the pool
and connected to each other with a standpipe for which devices were to be designed later for sampling and solution
transfer. Four 15,000-gal tanks, installed above the 25,000-
gal vessels, were connected to a second standpipe. Shielding
for all vessels was provided by installing concrete slabs on
the top of the pool structure. An independent off-gas
treatment system was also installed on top of the tanks to
decontaminate any gaseous effluent that might be evolved
during tank operations. A photograph of the bottom tanks
during installation is presented in Fig. 16.

Fortunately, the tank farm system was not needed for its
original purpose. This eliminated the potential risk of

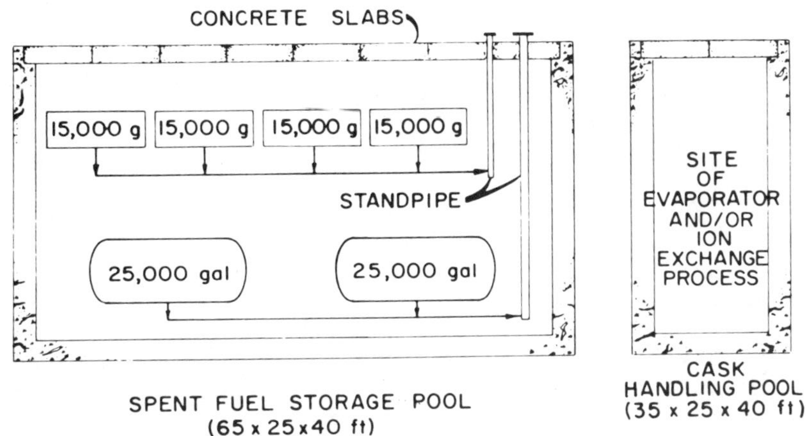

Figure 15.   Tank Farm Schematic

Figure 16.   Tank Farm During Installation

transferring additional liquid out of the Containment Building (into the tank farm) and, thereby, increasing the inventory of iodine in a structure external to the reactor system. The decision had been made not to disturb the solution in containment so the 8-day ^{131}I could decay to safe limits.

The tank farm will be utilized as an integral part of the high-level water treatment system.

Processing Justification

In order to proceed with the recovery of the reactor system, the liquid being held in the Containment Building will require removal and treatment. Currently, this liquid is standing at a depth of approximately 7-1/2 to 8 ft and is covering several components, including instruments. Although the leakage of water into the containment area has been minimized, the possibility for increased leakage continues to exist. In the early phases of the accident, the water in containment had ^{131}I concentrations estimated to be of the order of 10^4 µCi/ml. Because of this factor, every effort was made to avoid disturbing this solution until the radioiodine had been allowed to decay. Figure 17 shows the fission product decay curves, based on the analysis of primary coolant, for the radioactive nuclides that were of greatest concern with regard to treatment of the solution and indicates the most desirable processing period.

Flowsheet Development

Because of the unique nature of the water to be decontaminated, certain phases of the development of a flowsheet were required. The first phase of the flowsheet development work involved the selection of a suitable exchange medium for the processing of this solution. Limited samples of actual TMI primary loop water were tested in an ORNL hot cell to establish the characteristic of sorbents believed to be selective for the predominant $^{134-137}$Cs and $^{89-90}$Sr isotopes. A total of seven materials, both organic and inorganic ion exchange sorbents, were tested (Fig. 18). Results from these tests established that one of the zeolites, AW-500 (an inorganic exchanger), was highly selective for the cesium isotopes under the conditions prevailing at TMI and that Dow HCR-S, an organic resin, would be suitable for selection of the strontium isotopes. Following these studies, a series of small-scale column runs was made to verify sorbents using solutions adjusted to TMI's water conditions. Information resulting from these studies was then applied to a "scale-up" computer program at the Savannah River Laboratory to provide information on design on the ion exchange

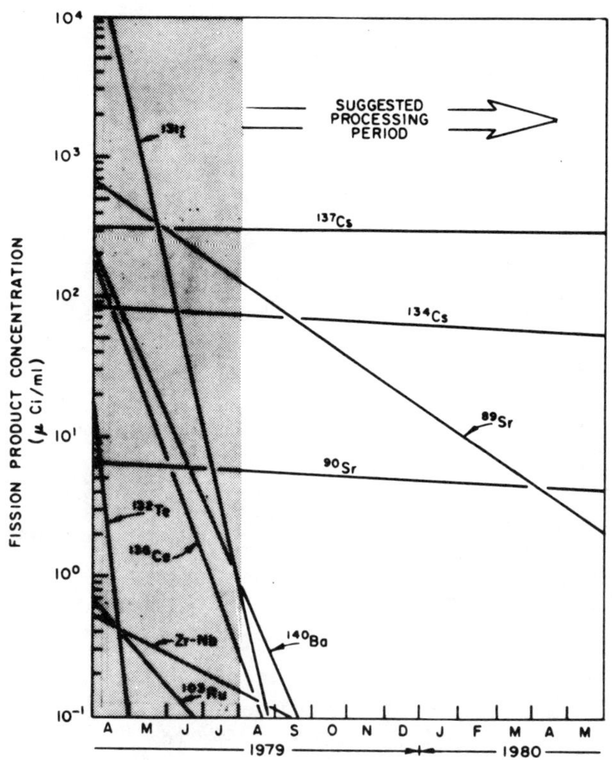

Figure 17. FP Decay in Primary Loop

| EXCHANGER | Sr K_d | | | Cs K_d | | | FINAL pH | PHASE RATIO (ml/g) |
	FIVE MIN	TWO HOUR	SECOND PASS	FIVE MIN	TWO HOUR	SECOND PASS		
HCR-S	80	230	[illegible]	3700	8700	>330	3.6	27
IR-200	120	290	28	85	200	250	8.3	71
Z-900	~0	50	39	280	1800	1400	– –	144
AW-500	~2	100	–	[illegible]	[illegible]	[illegible]	– –	[illegible]
CLINO.	22	96	130	135	990	3000	8.6	67
GLASS	96	250	54	160	440	400	8.6	68
CHARCOAL	20	50	60	0	0.7	2	8.7	42

Figure 18. Distribution Coefficients (K_d) for Primary
Coolant (Primary Water as Received - 960 ppm Na)

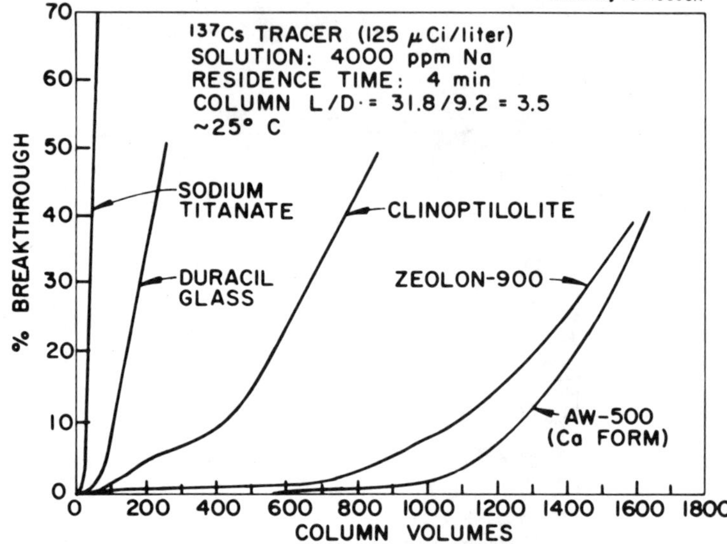

Figure 19.   Small-Column Loading Tests of Ion Exchange
Materials with Synthetic
TMI-2 Primary Loop Water and ^{137}Cs Tracer

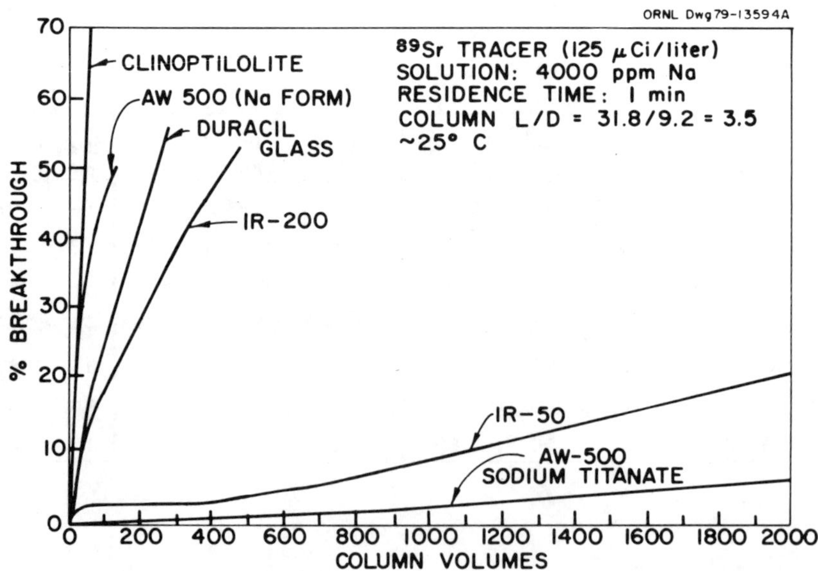

Figure 20.   Small-Volume Loading Tests of Ion Exchange
Materials with Synthetic
TMI-2 Primary Loop Water and ^{89}Sr Tracer

Figure 21.   Samples Taken 28 Aug 79

LABORATORY MEASUREMENT*

ISOTOPE	TOP SAMPLE	MIDDLE SAMPLE	BOTTOM SAMPLE	PREDICTED VALUES
^{137}Cs	176	179	174	200–260
^{134}Cs	40	40	39.6	30–40
^{140}La	0.09	0.078	0.014	0.25–0.49
^{3}H	1.03	1.05	1.01	1.0–1.5
^{131}I	0.012	0.012	0.013	0.015–0.044
^{90}Sr	2.70	2.90	2.83	10–18
^{89}Sr	43.6	40.6	42.1	171–228

* $\mu Ci/ml$

Figure 22.   Analytical Data

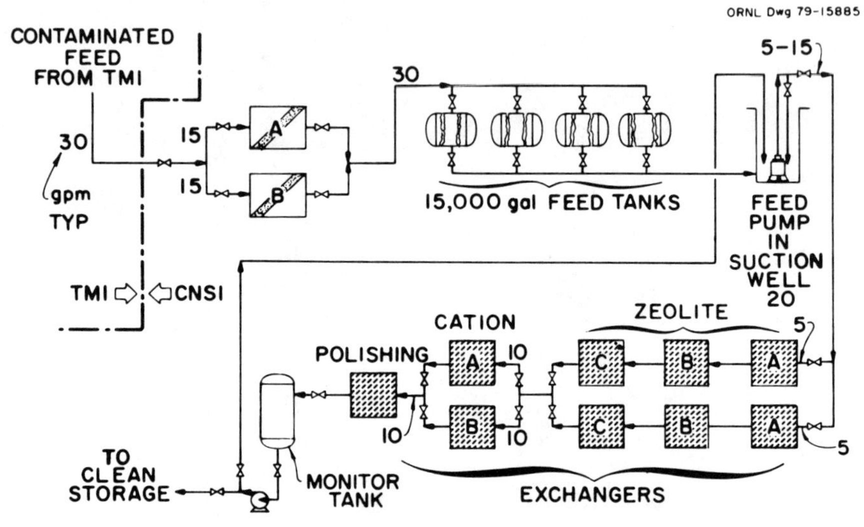

Figure 23.   Liquid Waste Decontamination Flowsheet

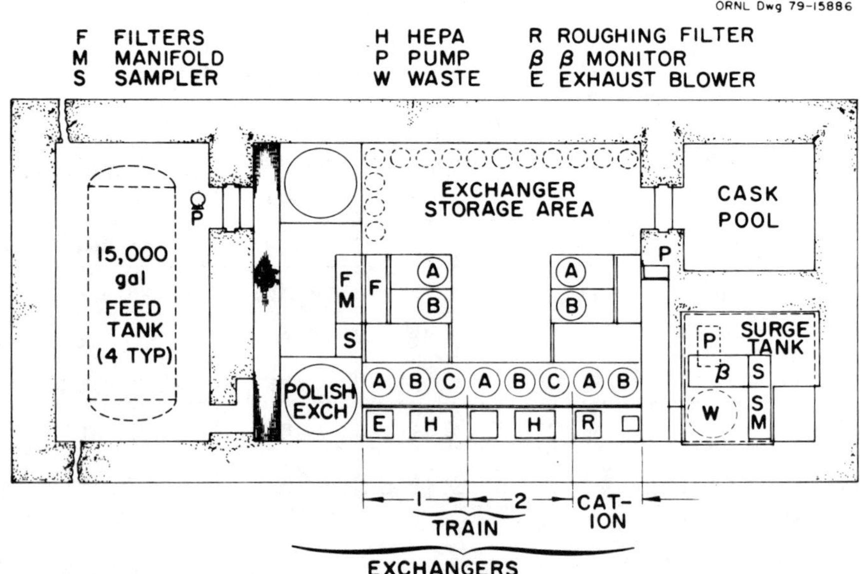

Figure 24.   Liquid Waste Decontamination Layout

columns. Pertinent information yielded by these studies is
presented in Figs. 19 and 20.

Sampling of Containment Water

    The original samples on which the flowsheet development
work was based were withdrawn from the primary coolant circuit and were taken with some difficulty because of personnel exposure and equipment logistic problems. Daily detailed flowsheet work based on small, limited volumes of solution is a risky undertaking at best. It was recognized
that sampling of the larger volume of water within the Reactor Containment Building was a necessity. Recommendations
to sample this body of water have been made since the early
days of the accident. Within the past months, a remote
3-in. hole was cut in a blanket reactor building penetration,
located 2 ft from the surface of the liquid, and 30-ml samples were withdrawn from the top, middle, and bottom of the
liquid on August 28. A photograph of these samples is presented in Fig. 21, following transfer from the steel containers. Samples ranged from 400-800 mR/hr at contact and
had a yellow coloration. Significant quantities of solids
were observed and analyzed in the bottom sample. The radiochemical results of these samples were slightly lower than
predicted from estimates based on event chronology and analysis of the primary loop water (Fig. 22). NRC has recently
concluded from these values that the damage to the core is
less than anticipated. The most significant information
revealed by the analysis would suggest that our earlier
work on flowsheet development was valid.

    Based on the early flowsheet development work, Chem
Nuclear Systems Corporation and the Allied Chemical Corporation entered into a contract with the General Public Util-
ities Service Corporation to provide an engineering scale
flowsheet for the treatment of high-activity-level water.
A schematic of the proposed flowsheet is presented in Fig.
23. Basically, the flowsheet consists of a series of filters
to remove solids, tankage in the tank farm mentioned earlier,
followed by 3 stages of zeolite sorption of ^{137}Cs followed
by cation exchange for ^{90}Sr-^{89}Sr sorption. A commercial
mixed bed unit will be placed downstream to serve as a
polisher. The system will be operated by remote means by
installing all equipment under water to provide for shielding.
A layout of the proposed system is presented on Fig. 24.

CONCLUSIONS AND STATUS

    In conclusion, I would like to summarize the status of
TMI at this time (Fig. 25). With regard to the release of

- **AIRBORNE RELEASES**
    - IODINE NO LONGER A CONCERN
    - ^{85}Kr RELEASE FROM CONTAINMENT UNDER STUDY

- **WATER TREATMENT**
    - LOW–ACTIVITY–LEVEL WATER PROCESSING COMPLETED
    - INTERMEDIATE–ACTIVITY–LEVEL WATER PROCESSING SYSTEM AWAITS APPROVAL
    - HIGH–ACTIVITY–LEVEL WATER PROCESSING SYSTEM BEING DESIGNED

- **DECONTAMINATION AND RECOMMISSIONING**
    - AUXILIARY BUILDING 80% COMPLETED
    - PLANS FOR CONTAINMENT BUILDING AND REACTOR UNDER STUDY

Figure 25. TMI Technical Status

radioactivity from TMI via the air pathway, the iodine no longer presents a hazard to the environment as the result of engineered systems and the natural radioactive decay of this isotope. Recently, the Metropolitan Edison Company released information pertaining to the controlled release of the krypton gas presently accumulated in the Reactor Containment Building. Following an assessment of the alternatives that could be used to remove this 10.4 year half-life material, it was concluded that the safest approach would be to release the gas, after filtration, over a period of 51 days. This release is expected to be below the plant's normal operation technical specifications and within NRC legal and federal limits.

In the case of water treatment, the low-activity-level water has been treated without incident, and the product water has been discharged to the Susquehanna River. Product water from this treatment system was well below release limits.

The construction and installation of the intermediate-
activity-level water treatment system have been completed
and awaits NRC approval to proceed. In this instance, the
NRC staff has submitted an environmental assessment and concluded that this system posed no hazard to the human environment. Following the public comment period, the NRC Commis-
sioners will decide whether or not to proceed.

In the case of high-activity-level water treatment, the
system is currently being designed and equipment is being
procured. Some limited flowsheet and unit operations development remain to be done, mostly in the area of flowsheet
improvement. Of course, prior to operation of the system,
the same NRC approval route must be followed as it was
in the intermediate-activity treatment system.

The decontamination of the contaminated surface of the
Auxiliary Building 2 is approximately 80% completed. No
major technological hurdles have been encountered as this
work proceeds.

Preliminary plans to decontaminate the Reactor Containment Building and recommission the reactor have been drafted
by the Bechtel Power Corporation for GPU. Basically, the
plan anticipates an estimated 42 months' period to accomplish
this objective with an estimated expenditure of $300 million.

RADIOCHEMICAL SEPARATIONS
AND MEASUREMENTS

OPTIMIZATION OF ZnS ALPHA COUNTING SENSITIVITY

<u>Dennis P. Bouldin</u>, IBM General Technology Division,
Essex Junction, VT

ABSTRACT

Alpha particles emitted from contaminants in materials associated with semiconductor memories can cause random errors in these devices. In some cases, the reliability requirements and sensitivity are both so high that extremely low-level alpha fluxes $(< 0.01\alpha/cm^2 - h)$ are of concern. Therefore, a program has been undertaken to optimize the alpha flux sensitivity attainable by ZnS scintillation counting.

The nature of the measurements constrains the time, area, and efficiency variables. Thus, the background rate becomes the main point of attack. Studies are discussed relevant to selection of background defining samples, time dependencies of background counts, background-efficiency trade-offs as a function of lower discriminator setting, and identification of the origin of the background.

The ultimate result is that detection limits substantially less than $0.01 \alpha/cm^2 - h$ have been attained for typical samples and for counting times of about one week.

INTRODUCTION

For about 20 years, the ionizations produced by sub-atomic particles in crystalline silicon have been used to measure these particles. However, recently May and Woods pointed out the possible detrimental effects of such reactions on the reliable operation of microelectronic circuitry under normal conditions.[1] They showed that extremely low-level fluxes of alpha particles emitted by trace contaminants in packaging materials produced "soft errors" in memory circuits, i.e., random changes in information content without lasting physical damage.

The problem was found to be most acute for newer and projected circuit designs in which ever decreasing amounts of electronic "charge per bit" are used. Thus, semiconductor manufacturers have developed a great interest in techniques for assaying materials for alpha emissions and in advancing these techniques to their ultimate sensitivity. Our experience has indicated that zinc sulfide (ZnS) scintillation counting offers the best sensitivity attainable with the methods readily implemented.

EQUIPMENT AND PROCEDURES

Investigations were limited to making the most of readily available equipment (Fig. 1). No special provisions were made for "clean" power supplies or for extraordinary controls of the measurement environment.

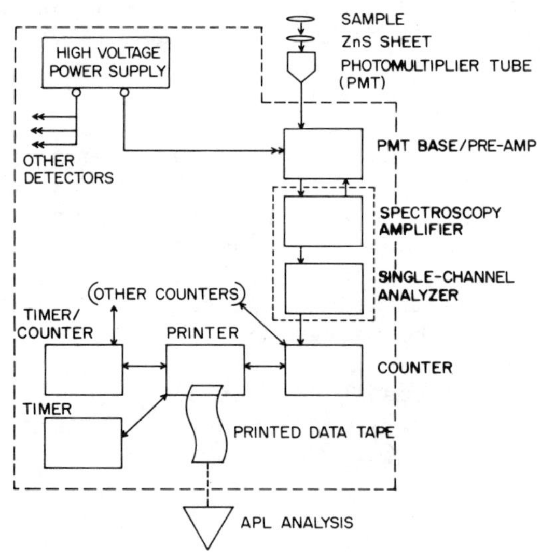

Fig. 1   Schematic of Low-Level ZnS
Alpha Counting System

A particularly useful aspect of our approach is the use of inexpensive ZnS coated plastic. This allows placing the sample in intimate contact with ZnS and phototube with the assurance that should the ZnS surface become contaminated it can be easily replaced. Moreover, the scintillation area can be readily adapted to the shape of samples if necessary, thus minimizing background levels.

Detection Limit

The definition of the flux detection limit is based on Currie's formulation.[2] For our purposes, geometry is essentially 2π; area is restricted to 3-in., 3½-in., or 5-in. photomultiplier-tube faces; and counting times range up to one week. The main point of attack is environmental or detector operating parameters which affect background.

BACKGROUND STUDIES

Background characterization is very time-consuming. For example, background rates associated with a sensitivity limit of 0.005 α/cm^2 - h under our operating conditions may yield only 60 counts per week.

For operational purposes, it is convenient to divide the background rate into components shown in equation 1.

$$b = b_n + b_c + b_e + b_z \tag{1}$$

Where:

b_n = noise background, measured rate for sample without ZnS sheet.

b_c = ZnS reactions with cosmic activity.

b_e = ZnS reactions with (other) environmental radiation.

b_z = ZnS reactions caused by sheet contamination.

Contribution of ZnS Sheets (b_z)

Figure 2 gives the results of one-week readings of silicon wafers on 3-in. detectors. Silicon wafers were chosen because of their availability, anticipated purity, and inherent interest. Moreover, they were to be the support matrix for many subsequent thin-film measurements.

All wafers had some measurable gross activity, with significant variations among samples and no clearcut time dependence in activity.

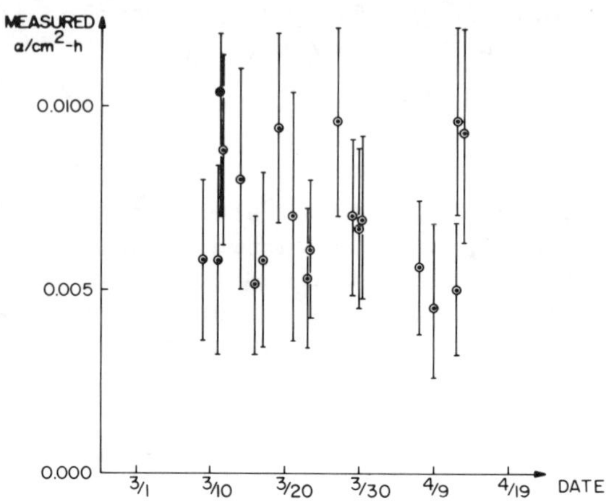

Fig. 2   Background Measurements with Silicon Wafers

Many points plotted in Fig. 2 represent repeat measurements. If the data is reordered according to the wafer
measured, it becomes clear that variations in Fig. 2 are not
genuine differences in the wafers but reflect inconsistencies
in the measurements.

When the results are reordered according to the particular piece of ZnS used (Fig. 3), a meaningful pattern emerges:
the main cause of the variations observed in background are
the variations in ZnS pieces used. These variations arise
in pieces cut from only two 12-in. X 12-in. ZnS sheets;
therefore, they represent either localized differences in the
material itself or differences in contamination from handling.

The data in Fig. 2 and 3 and all other measurements taken
to date also indicate that although variations in readings
exist, there is a lower bound or residual rate of approximately 0.005 α/cm^2 - h below which no samples have been measured.

To further focus on the contribution of the ZnS material,
experiments were run in which two sheets were run consecutive-
ly with silicon wafers and then with one another. The plastic
backing of the sheets is sufficiently thick to assure that
facing sheets are isolated from the environment (except at
the edges).

These experiments showed that the gross readings of two
facing sheets were approximately the sum of those of the
individual sheets with silicon wafers.

If it is assumed that b_z is principally a volume effect
(i.e., caused by α - producing contaminants in the volume of
the ZnS), then this data supports the hypothesis that the
major portion of the background is the b_z component.

However, if b_z arises predominantly from a surface contamination of the ZnS, it would be expected to be multiplied
fourfold in measurements of facing ZnS sheets. Under this
assumption, the roughly twofold increase in counts for facing
ZnS sheets would indicate an inherent ZnS rate of approximately 1/3 of the residual.

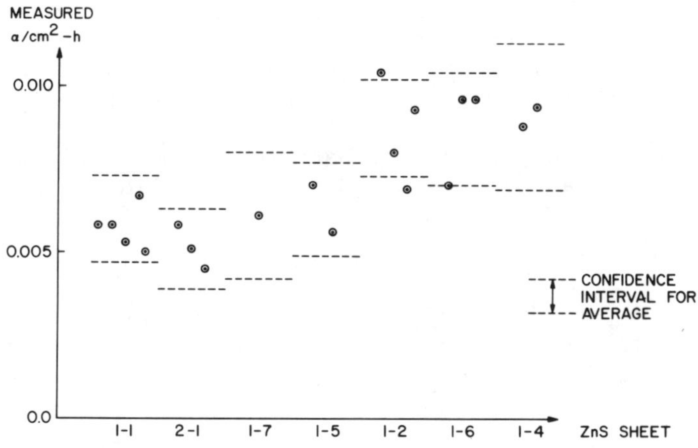

Fig. 3 ZnS Sheet Dependence of Raw Wafer
Flux Measurements

Contributions of Electronic Noise (b_n) and Non-Cosmic
Environmental Radiation (b_e)

Tests of the noise background (b_n) with 3-in. detectors
indicate a contribution of no more than 20% of the residual
rate. Limited tests with larger detectors indicate some
scaling; further studies are under way.

For most samples tested, intimate contact may be attained between the sample and ZnS. This minimizes the influence
of radon, which has been shown to double the residual background for samples separated from the ZnS by only a 2-mm air
gap.

Tests with intense gamma and beta sources in the vicinity
of detectors indicate negligible effects.

Further Studies of Residual Rate

Background counts as a function of time of day have been
measured continuously for a period of over one month. No
trends have been observed. This demonstrates the unimportance
of factors expected to have obvious time dependencies, such
as electrical noise from activities in or near to the lab,
light leakage, or solar cosmic activity.

In many tests of samples other than silicon wafers, no
samples have been found to give lower rates than silicon
wafers. This implies that silicon wafers are valid control
samples and do not contribute to the residual rate.

A dramatic dependence of background upon lower discriminator setting has been observed (Fig. 4). Since efficiency

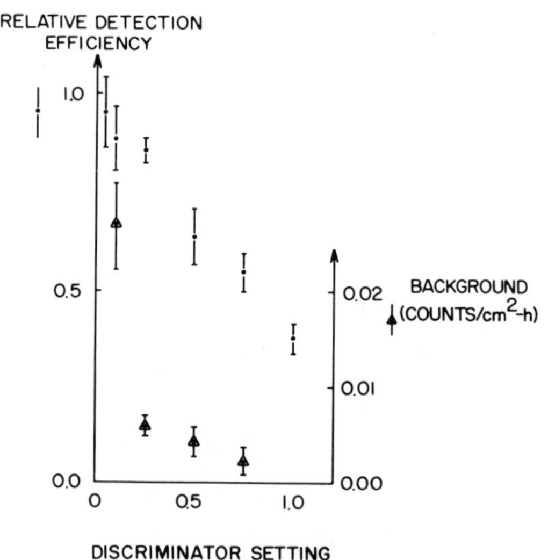

Fig. 4 Efficiency/Background Tradeoffs

has a much more gradual dependence on lower discriminator
level, this indicates the existence of a factor not consistent
with the expected behavior of b_z. However, it is not clear
what role this factor plays under normal operating conditions.
Figure 4 shows the operational importance of recognizing the
background/efficiency trade-off. In our measurements, a
lower discriminator setting of 0.25 V has been selected,
resulting in an efficiency for bulk source spectra of
approximately 70% relative to a comparable surface barrier
detector measurement of particles over 1 Mev from a bulk
source (Th foil).

Background measurements with one 3-in. detector at
200-minute intervals over a 3-month period have shown several
instances of bursts of counts inconsistent with the overall
rate if uncorrelated single counts are assumed. However,
since natural decay chains have several instances of decays
which would be correlated, this data does not negate the
possibility of simple ZnS contamination as a major contributor to the residual.

The combined background rates of several 3-in. detectors
run continuously for one month are shown in Fig. 5.

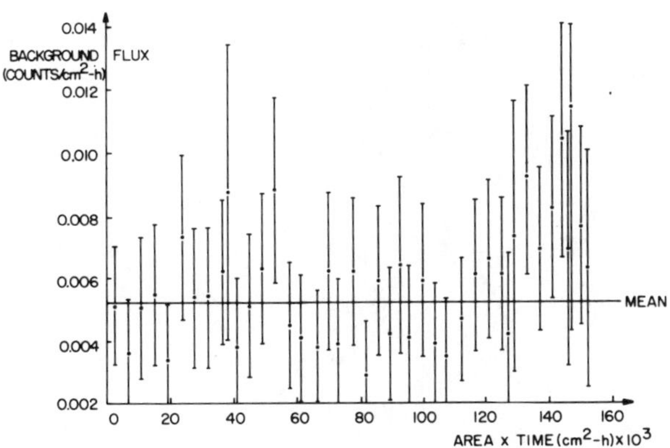

Fig. 5 Long Term Background

This figure shows both the existence of a relatively long-term background stability and an instance of a correlated significant rise in the background level. This rise indicates that the time-dependent behavior of the background is not yet well understood and, thus, that measurements which hope to achieve ultimate sensitivity must be accompanied by constant monitoring of the background.

CONCLUSIONS

From the preceding discussion, it is evident that optimum sensitivity depends upon the choice of a best-case piece of ZnS, a sample which allows for intimate contact with the detector (or steps taken to avoid the presence of radon), careful monitoring of the background for short-term bursts and long-term variations, and optimization of lower discriminator setting.

With these factors in mind, an estimated 90% confidence detection limit of 0.0043 α/cm^2 - h for 1-week counting times has been achieved. Further studies are being conducted to better understand the source of the residual background and the nature of its long-term time variations.

REFERENCES

1. T. C. May and Murray H. Woods, IEEE Trans. Elec. Dev., ED 26, 10 (1979).

2. L. Currie, Anal. Chem. 40, 586 (1968).

POLONIUM-209: A NEW SOURCE OF ANALYTICAL TRACER

P. E. Figgins, V. R. Casella, C. T. Bishop, and A. A.
Glosby. Mound Facility, Miamisburg, Ohio, USA.

ABSTRACT

Polonium-208, currently used as a tracer for polonium-
210 determinations, is scarce and expensive. The nuclear
properties indicate polonium-209 would have advantages as a
tracer. A small amount of polonium-209 is formed during the
neutron irradiation of bismuth-209 to form polonium-210.
Short-lived polonium-210 decays out leaving polonium-209.
Ten-year-old polonium "foils" were examined by GeLi counting
and showed a definite spectrum of polonium-209. Alpha pulse
height analysis of one sample showed polonium-209 to be 99%
of the total count. This tracer was tested in the analysis
of environmental samples.

INTRODUCTION

The two longest-lived isotopes of polonium are
polonium-209 with a half-life of 103 years and polonium-208
with a half-life of 2.90 years. These two isotopes have
found some applications as tracers in the determination of
naturally occurring, 138-day polonium-210. These tracers are
currently in short supply, and are quite expensive.

Mound Facility was involved in the production of polonium
from the late 1940's until 1972. Most of this production was
the isotope polonium-210, produced by irradiating bismuth
metal. However, the isotopes of polonium-208 and polonium-209
were of interest, and one polonium 208/209 product was
recovered from bismuth irradiated with protons at Oak Ridge
National Laboratory, and processed at Mound.[1,2] During the
1960's, it was discovered that some polonium-208 and

polonium-209 was being produced along with the polonium-210,
apparently by (n,2n) and (n,3n) reactions on polonium-210.
Typical nuclear reactions leading to various polonium
isotopes are listed in Table I, and the details of decay
modes are listed in Table II.

Table I

Nuclear Reactions Producing Polonium Isotopes

$$^{209}\text{Bi} \ (\text{n},\gamma) \ ^{210}\text{Bi} \quad \xrightarrow[\text{5.013 d.}]{\beta^-} \quad ^{210}\text{Po}$$

$^{210}\text{Po} \ (\text{n},2\text{n}) \ ^{209}\text{Po}$	$^{210}\text{Po} \ (\text{n},3\text{n}) \ ^{208}\text{Po}$
$^{209}\text{Bi} \ (\text{d},2\text{n}) \ ^{209}\text{Po}$	$^{209}\text{Bi} \ (\text{d},3\text{n}) \ ^{208}\text{Po}$
$^{209}\text{Bi} \ (\text{p},\text{n}) \ ^{209}\text{Po}$	$^{209}\text{Bi} \ (\text{p},2\text{n}) \ ^{208}\text{Po}$

Table II

Decay Modes of Polonium Isotopes[3]

Isotope	Half-life	Decay Modes and Energies (MeV)
^{210}Po	138.4 d	alpha: 5.305 gamma: 0.803
^{209}Po	102.5 yr	alpha: 4.882 (99+%) gamma: 0.2605, 0.2628 electron capture (0.263%) gamma: 0.8966
^{208}Po	2.8976 yr	alpha: 5.116 (99+%) electron capture (0.00181%) gamma: 0.0318, 0.0635, 0.2919, 0.5391, 0.5707, 0.6029, 0.8617

As an example of what might be expected in the way of
isotopic ratios, in their 1966 paper on the nuclear structure
of polonium-208 and polonium-209, Hagee et al.[4] reported
data on the two sources they used. The first source was
produced by proton irradiation and had a 208/209 mass ratio of
14.1 at that time. The second source was produced by neutron
irradiation and had a 208/209 mass ratio of 1.31 at that
time. After a recent request for polonium tracer, some old
polonium "foils" which had been retained for another purpose,
were gamma counted. The spectra showed the characteristic
polonium-209 peaks, and these items were considered for their
potential as a source of polonium tracer.

The polonium-210 recovery process started with the
neutron irradiation of bismuth metal slugs encased in an
aluminum can. The aluminum can was removed, either mechanically or by a caustic dissolution, and the bismuth was
dissolved in a nitric-hydrochloric acid mixture. After
denitrating the solution, the polonium was separated and
concentrated by spontaneous plating on a small amount of
bismuth metal powder. The bismuth was dissolved and the
cycle repeated until the desired concentration of polonium
was achieved. The final polonium purification was electroplating from hydrochloric acid medium onto a platinum gauze
or "foil". The platinum foil was transferred to a nickel
tube and closed with a crimped nickel cap. The nickel tube
was sealed in a glass ampule, and the amount of polonium
determined by a calorimetric measurement.

EXPERIMENTAL

A polonium foil, originally containing 15 curie of
polonium-210, and plated in October 1969, was selected for
study. After crushing the glass ampule, the crimped nickel
cap was removed and the platinum foil, with a visible deposit
was extracted. The platinum foil was leached with hot 4
molar nitric acid containing a little hydrochloric acid, then
transferred to a 50-ml volumetric flask and made up to the
mark with 4 $\underline{M}$ HNO_3. Progress of the dissolution was followed
by gamma counting, and approximately 15% of the count
remained in the nickel tube, where it could be recovered only
by complete dissolution of the nickel tube. Alpha mounts,
prepared by direct evaporation of an aliquot of the main
solution on stainless steel planchets showed a total activity

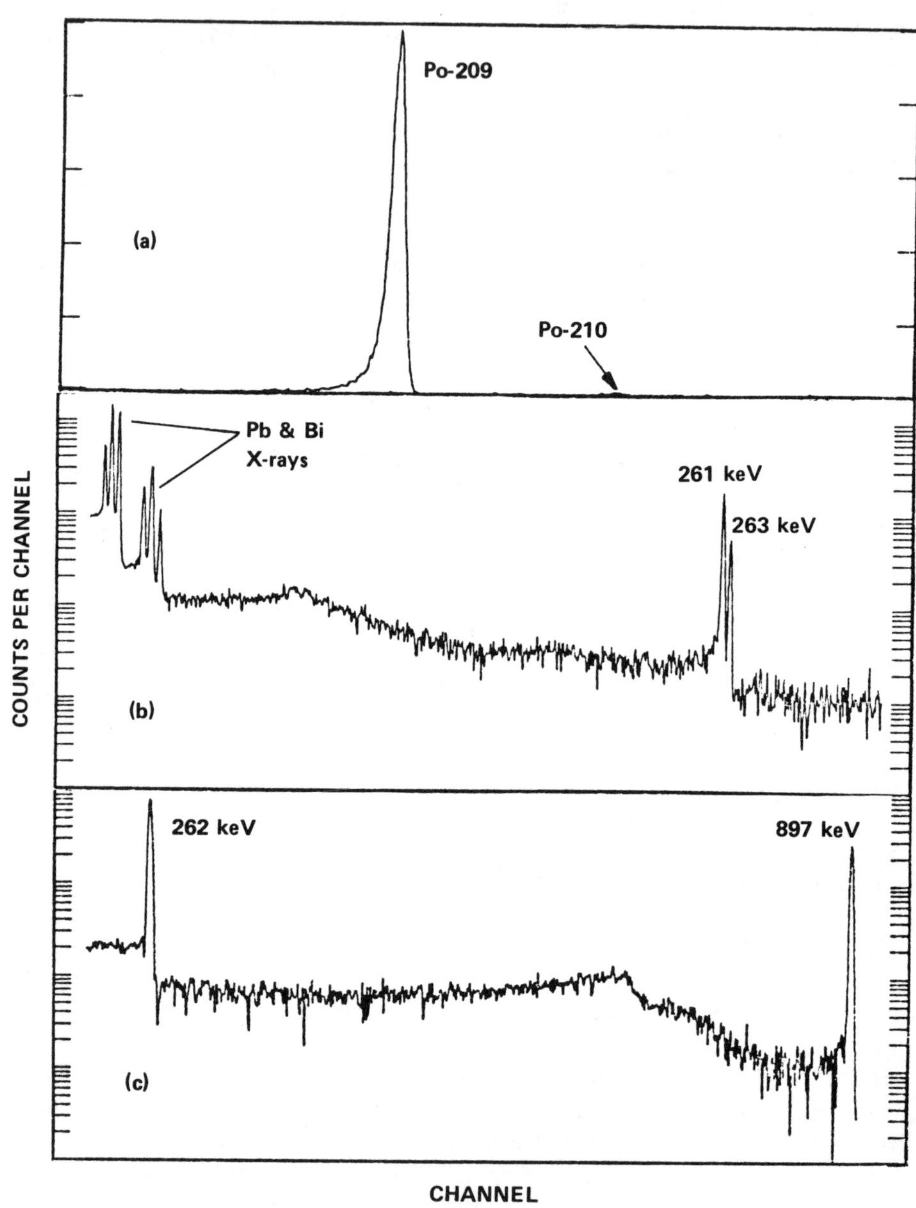

Figure 1. (a) Alpha spectrum of polonium-209 tracer (Po-210 <1.0%); (b) and (c) Ge(Li) spectra from 70-300 keV and 200-900 keV, respectively

of 22 microcuries in the main solution. An alpha pulse
height spectrum (Figure 1a) showed nearly all of the activity
to be in a peak at 4.9 MeV, with approximately 1% of the
activity in a peak at 5.3 MeV, attributed to polonium-210.
No special efforts were made to look for a polonium-208 peak
at 5.11 MeV, and the possibility of small amounts of this
isotope cannot be discounted. The x-ray and gamma spectrum
of a liquid sample over the range of 0 to 300 keV was
examined with a high resolution detector, and is reproduced
in Figure 1b. The triplet peaks at 72.8 to 77.1 keV and at
84.9 to 98.8 keV are attributed to K x-rays from lead and
bismuth. Figure 1b also shows the two polonium-209 peaks at
260.5 and 262.8 keV. Figure 1c shows the high energy gamma
spectrum, where the two peaks around 260 keV show as a single
peak, and the polonium-209 peak at 897 keV. No other gamma
peaks were identified.

Approximately 20 old polonium foils were counted on the
GeLi detector and all showed the typical polonium-209
spectrum shown in Figure 1c. Old records indicated that these
items had been electroplated in the years 1968 to 1970, and
the initial amount of polonium-210 had varied from 3 to over
30 curies. There was an expected variation in the size of the
polonium-209 peak depending on the amount of polonium originally present; however, certain batches seem to contain above
average amounts of polonium-209. Presumably this variation
was caused by differences in the neutron spectrum during the
irradiation, or by a variation of irradiation time. Extrapolating from the single foil to all of the foils currently on
hand, the total polonium-209 is conservatively estimated to
be 100 microcuries.

A portion of the polonium-209 solution was used as a
polonium tracer in the determination of polonium-210 in coal
ash samples. Parallel determinations were made using
polonium-208 tracer. The details of the procedure are
reported in a subsequent paper of this Symposium[5], and will
not be given here; however, some of the alpha spectra are
reproduced in Figure 2. The relative positions of the alpha
peaks from polonium-209, polonium-208, and polonium-210 are
shown in Figure 2a. Figure 2b shows the alpha spectrum from
the coal ash analysis using polonium-209 tracer, and Figure 2c
shows the alpha spectrum using the polonium-208 tracer. In
the case of the polonium-209 tracer, the greater separation
of the two peaks is apparent. This larger separation would

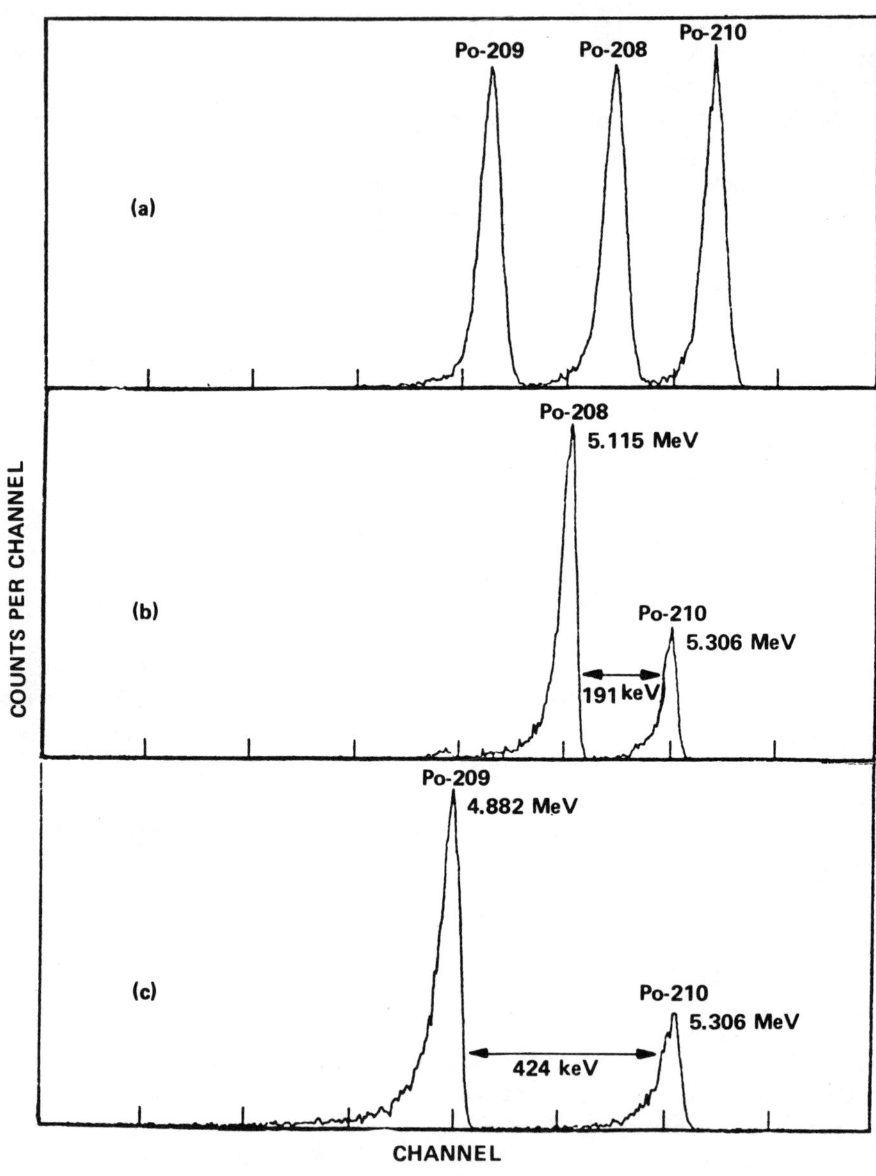

Figure 2.   Alpha spectra of the polonium isotopes. Spectra (b) and (c) are from analysis of coal ash using Po-208 and Po-209 tracers, respectively

44

be quite helpful where low resolution of the counting slide
causes "tailing" of the polonium-210 peak down into the
tracer peak.

Polonium-209, in general, and this source of polonium-209
in particular, seem to have a number of advantages for use as
polonium tracer. The greater degree of separations of the
alpha peaks has been pointed out. A second general advantage
of the polonium-209 relates to its longer half-life. Polonium-
209 will lose only 0.68% of its count by decay in one year,
while polonium-208 decays over 21% in the same time period.
This means that polonium-209 tracer can be stored over fairly
long times, whereas the polonium-208 should be used without
delay. The longer half-life of the polonium-209 also means
that re-standardization of solutions or decay corrections
will not be required as often. This particular source of
polonium-209 has the further advantage that the contribution
from all other polonium isotopes is either small or negligible. If appreciable polonium-208 were present, the total
activity of the tracer would change with time, and it would
be necessary to make corrections for this. In this material,
the amount of polonium-210 is low (less than 1%), and,
therefore, contributions of the tracer to the polonium-210
peak can often be neglected. If the correction is made, it
would normally be quite small compared to polonium-210 in the
sample.

SUMMARY

The recovery of polonium-209 tracer from aged polonium-
210 materials has been demonstrated. The polonium-209 tracer
contained negligible polonium-208 and only 1% polonium-210,
which means fewer complications in its use as a tracer. The
use and advantages of polonium-209 tracer in the analysis of
coal ash samples were demonstrated.

ACKNOWLEDGMENT

Mound Facility is operated by Monsanto Research Corp.
for the U. S. Department of Energy under Contract No.
DE-AC04-76-DP00053.

REFERENCES

1. R. S. Livingston and J. A. Martin, <u>Production of Polonium-208</u>, ORNL-1392 (August 1952).

2. G. L. Fox, G. D. Nelson, and W. H. Powers, <u>Processing and Properties of Polonium-208</u>, MLM-914 (January 1964).

3. C. M. Lederer and V. S. Shirley, editors, <u>Table of Isotopes</u>, Seventh Edition, John Wiley and Sons Inc., 1979.

4. G. R. Hagee, R. C. Lange, and J. T. McCarthy, <u>Nucl. Phys.</u>, 84, 62-80 (1966).
    Also G. R. Hagee, R. C. Lange, and J. T. McCarthy, <u>Weak Alpha and Electron Capture in Polonium-208 and Polonium-209</u>, MLM-1251 (May 1965).

5. A. A. Glosby, V. R. Casella, C. T. Bishop, and L. C. Hopkins, "Measurement of Polonium-210 in Low-Temperature Ashed Coal and Coal Ash," to be presented at 23rd Conference on Analytical Chemistry, Gatlinburg, TN, October 9-11, 1979.

AN IMPROVED ION EXCHANGE PROCEDURE FOR
THE SEPARATION OF BARIUM FROM RADIUM

Geoffrey Gleason, Oak Ridge Associated Universities,
Oak Ridge, Tennessee, USA.

ABSTRACT

Cyclohexylenediaminetetraacetic acid, DCYTA, has been
found to be superior to EDTA for the separation of barium
from radium by cation exchange chromatography. Complete
separation of high concentrations of barium can be rapidly
accomplished on a small column of Dowex-50 by elution with
a 0.05 molar solution of DCYTA buffered to pH 8.5.

INTRODUCTION

The ion exchange separation of radium and barium has
received extensive attention. This is primarily due to the
great similarity of the chemistry of the two elements which
precludes a single step separation. In general, the most
satisfactory separations have been demonstrated with cation
exchange resin columns and chelating solutions such as
ammonium citrate,[1,2] lactate[3] or formate[4] as eluants.
Ethylenediaminetetraacetic acid (EDTA) forms more stable
complexes with radium and barium and allows their elution
with relatively dilute solutions.[5]

Nelson[6] utilized 0.01 M solutions of ammonium-EDTA for
the separation of barium and radium on small columns of
Dowex-50. The procedure requires a small volume application
of low concentrations of both elements, a condition impossible to meet where radium has been originally recovered by
co-precipitation with milligram amounts of barium sulfate.
Greater barium concentrations also require larger columns
and increased concentrations of EDTA eluant to provide
mobility. With such high ratios of barium to radium, complete separation with 0.05 M EDTA eluant under a variety of
column operating conditions could not be achieved.

Cyclohexylenediaminetetraacetic acid, DCYTA, forms complexes with barium and radium similar to those of EDTA. The stability of the barium complex is comparable but that of radium much weaker. DCYTA is thus a more effective eluant than EDTA. Conditions have been established which permit a relatively large volume application and the elution of the barium at milligram concentrations, leaving the radium on the column.

EXPERIMENTAL

Radioactive ^{133}Ba was used as a barium tracer. A small amount of ^{226}Ra was recovered from carnotite solutions by co-precipitation with barium sulfate. The barium sulfate was dissolved by heating with EDTA solution, and sulfate ion was removed by passage through a small column of Dowex-1 anion exchange resin in the chloride form.

Radium and barium contained in the fractions taken from column separation experiments were determined by NaI(T1) scintillation spectrometry. The 186 keV gamma emission from ^{226}Ra was used to measure radium. The 30 keV cesium x-ray was used to measure barium since interference by radium and its daughter products is minimal.

Columns 0.4 cm I.D with a 10-cm depth were used in the experiments. Resin beds were prepared from 100-200 mesh Dowex-50x8 and converted to the sodium form by treatment with saturated sodium chloride solution.

Prior to loading, the column was washed with distilled water and then 5-ml of saturated boric acid was passed into the bed for pH adjustment. In this way, barium-radium solutions could be applied in the presence or absence of complexing agents at a pH of 5. As much as 15 to 20 ml of solution could be passed into the column at a flow rate of about 0.2 ml/min. forming a band limited to the top centimeter of the bed.

After washing with distilled water, the resin bed was adjusted to the desired pH with borate buffer prepared from saturated boric acid. No movement of the radium-barium zone occurred during this step which was continued until the column effluent showed the proper pH.

The column was next developed with a buffered solution of 0.05 M DCYTA prepared by diluting a 0.1 M solution with an equal volume of saturated boric acid and adjusting the pH with sodium hydroxide. A flow rate of 0.2 ml/min was maintained and fractions corresponding to a column volume were taken at 6-minute intervals.

At a pH of 8.5, barium rapidly moved down the column
and was completely removed in the first 10 fractions at which
point the radium was about half way down the column. This
could then be eluted by switching to 0.05 M EDTA at the same
pH which rapidly removed the radium. When radium free from
complexing agent was desired, the column was converted to the
acid form with 0.5 N HCl and the radium eluted with 3 N HCl.

RESULTS

A typical separation of barium and radium is shown in
Figure 1. In this example, 13.8 mg (0.1 mmol) of barium was
added to an EDTA solution containing 0.1 μg of radium to
provide a barium-to-radium ratio in excess of 10^5. The pH
was adjusted to 5 and the entire volume of 15 ml applied to
the prepared column as described.

From the elution data shown in Figure 1, the distribution coefficient for barium (at pH 8.5) in 0.05 M DCYTA
solution (Dowex-50x8) computes to be about 5. From the
position of the radium zone after the passage of 10 column
volumes, the distribution coefficient for radium is estimated
at about 25. The very rapid removal of the radium by 0.05 M
EDTA at the same pH suggests that in this media, the distribution coefficient is less than 1.

SUMMARY

A procedure is described for the cation exchange separation of barium from radium using an eluant of 0.05 M DCYTA at
a controlled pH of 8.5. The system has the advantages of
tolerating a relatively large volume application and high
barium-to-radium ratios.

ACKNOWLEDGEMENTS

This report is based on work performed under Contract
No. DE-AC05-760R00033 between the U.S. Department of Energy
and Oak Ridge Associated Universities.

REFERENCES

1.   E. R. Tompkins, J. Am. Chem. Soc., _70_, 3520 (1948).

2.   W. H. Power, etal, Anal. Chem., _31_, 1077 (1959).

3.   G. M. Milton and W. E. Grummit, Can. J. Chem., _35_,
     541 (1957).

4.   Y. Sugimura and H. Tsubota, J. Mar. Res., _21_, 74
     (1963).

REFERENCES (contd)

5.   G. Duyckarets and R. Lejeune, J. Chromatog., <u>3</u>, 61
      (1960).

6.   F. Nelson, J. Chromatog., <u>16</u>, 403 (1964).

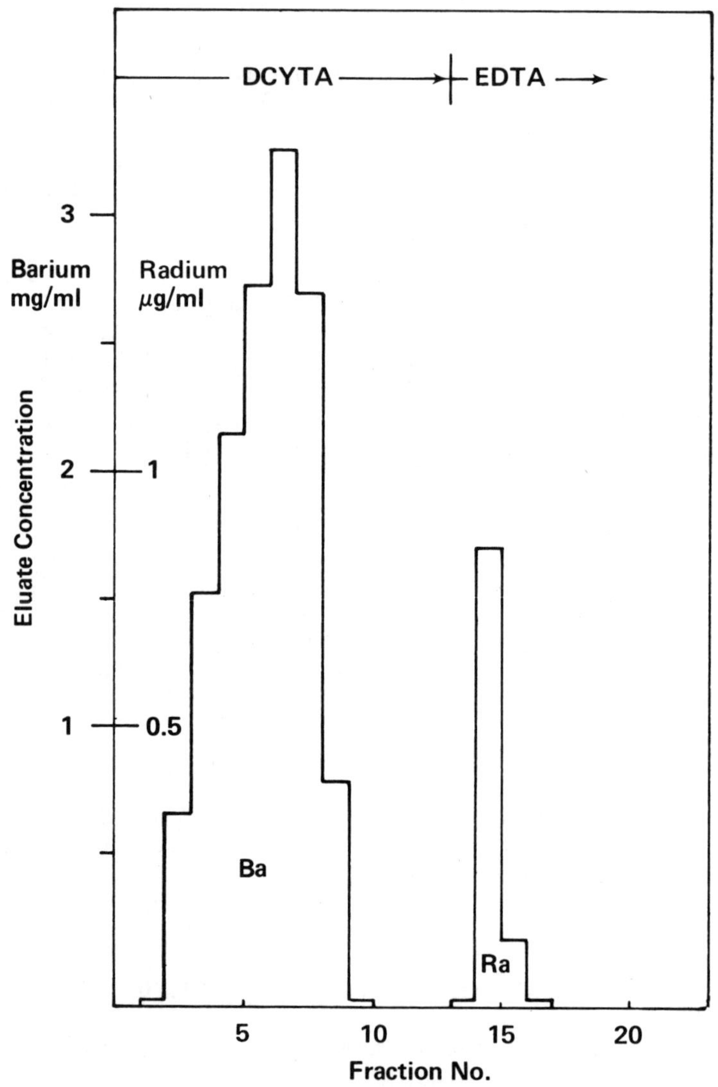

Figure 1.   Elution of barium and radium from
Dowex-50x8 column with 0.05 M reagents at pH 8.5.

GALVANIC DEPOSITION OF ^{210}Po AT A ROTATING NICKEL DISK
ELECTRODE

James W. Dillard, B. B. Hobbs*, Tennessee Valley
Authority, Muscle Shoals, Alabama   35660

ABSTRACT

With increasing environmental concerns over the mining
and milling of uranium for nuclear power production, the
radiochemical determination of ^{210}Po and ^{210}Pb in environmental samples is assuming greater importance. The procedures for both ^{210}Po and ^{210}Pb determinations usually
employ the galvanic deposition of ^{210}Po onto a nickel support
in preparation for alpha spectroscopy. The ^{210}Pb is determined from the ingrowth of ^{210}Po into the deposition solution.
After an appropriate sample preparation, currently accepted
plating procedures call for the deposition solution to be
0.5N in HCl with 2 ml of 40% citric acid and 1 g of hydroxylamine hydrochloride added. This prevents ions such as Fe^{2+}
from interfering in the deposition. The nickel electrode is
either suspended in the deposition solution on a glass hook
or secured in the bottom of a deposition cell with a removable screw cap. Plating is then carried out for one, two,
or more hours at between 50° and 80°C. In this study no
attempt has been made to change the chemical composition of
the deposition solution, only the plating hydrodynamics were
modified. This paper presents a new plating technique using
a rotating nickel disk electrode which achieves uniformly
thin plates at ambient deposition temperatures in one hour
or less, depending on the sample activity.

---

*Dr. Dillard and Mr. Hobbs are employees of the Tennessee
 Valley Authority (TVA). The contents of this paper do not
 necessarily reflect the views and policies of TVA, nor does
 the mention of trade names or commercial reports constitute
 an endorsement or recommendation for use.

INTRODUCTION

With increasing environmental concerns over the mining
and milling of uranium for nuclear power production, the
radiochemical determination of ^{210}Po and ^{210}Pb in environmental samples is assuming greater importance. The procedures for both ^{210}Po and ^{210}Pb determinations usually
employ the galvanic deposition of ^{210}Po onto a nickel support in preparation for alpha spectroscopy. The ^{210}Pb is
determined from the ingrowth of ^{210}Po into the deposition
solution. This paper presents a new plating technique, which
achieves uniformly thin plates at ambient deposition temperatures in 1 h or less, depending on the sample activity.

PROCEDURE

After an appropriate sample preparation, currently
accepted plating procedures call for the deposition solution
to be 0.5$\underline{N}$ in HCl with 2 ml of 40% citric acid and 1 g of
hydroxylamine hydrochloride added. This prevents ions such
as Fe^{2+} from interfering in the deposition. The nickel
electrode is either suspended in the deposition solution on
a glass hook or secured in the bottom of a deposition cell
with a removable screw cap.[1] Plating is then carried out
for 1 to 2 h, or more at 50 to 80°C. In this study no
attempt has been made to change the chemical composition of
the deposition solution; only the plating hydrodynamics were
modified.

The use of a ^{208}Po tracer does not require exhaustive
deposition because the electrochemistry of both ^{210}Po
and ^{208}Po are identical, and proportional quantities of each
will be deposited. For alpha spectrometry, thin uniform
plating of the polonium is desirable to reduce spectra
tailing and to obtain maximum resolution. The peak energy
separation between the two isotopes is only 190 keV, and
optimum detector resolution is normally greater than 25 keV.
Thick or non-uniform plates broaden the spectra and seriously
degrade the resolution, making quantification of ^{210}Po
difficult. Also, non-uniform plating prevents the use of
spectral analysis techniques, such as least-squares resolution, which require consistent spectral shapes between
standards and samples. Therefore, to improve the plating
technique, diffusion and mass transport of the electroactive
species were examined.

The transport mechanism of the electroactive species
from bulk solution to electrode surface is governed by
diffusion across a concentration gradient. Application of

Fick's laws of diffusion and approximation of the Nernst diffusion-layer model results in a generalized formula for the limiting current density:[2]

$$(1) \qquad i_L = \frac{n\bar{F}DC}{\delta},$$

where

i_L = limiting current density,

n = number of electrons,

$\bar{F}$ = Faraday,

D = diffusion coefficient,

C = bulk concentration,

δ = diffusion-layer thickness.

By reducing the diffusion-layer thickness through stirring, the limiting current density will be increased. For a fixed electrode area, the higher current density will result in shorter deposition times for the same amount of material plated. Achieving a reasonably short deposition time is significant, but achieving well-defined hydrodynamic conditions is equally important.

The rotating-disk electrode has well-defined hydrodynamic conditions, which were described mathematically by Levich[3]

$$(2) \qquad i_L = 0.62 \ n\bar{F}D^{2/3}v^{1/6}w^{1/2}C,$$

where

i_L = limiting current density,

v = kinematic solution viscosity,

w = angular disk velocity.

The limiting current density is governed by the rotation rate of the electrode if all other terms are constant. Because rotation rates are easily monitored, use of a rotating disk electrode is an excellent way of controlling the deposition process to ensure uniformity and predictability of plating. A brief description of the plating procedure follows.

The nickel electrode disks used in this study are 1.91
cm in diameter and 0.41 mm thick. After descriptive sample
information is written on the back side of the electrode, it
is sprayed with a light coat of lacquer to ensure permanence
of the sample record and to seal the nickel surface. The
disk is then pressfit into a Teflon electrode holder (Figure
1).

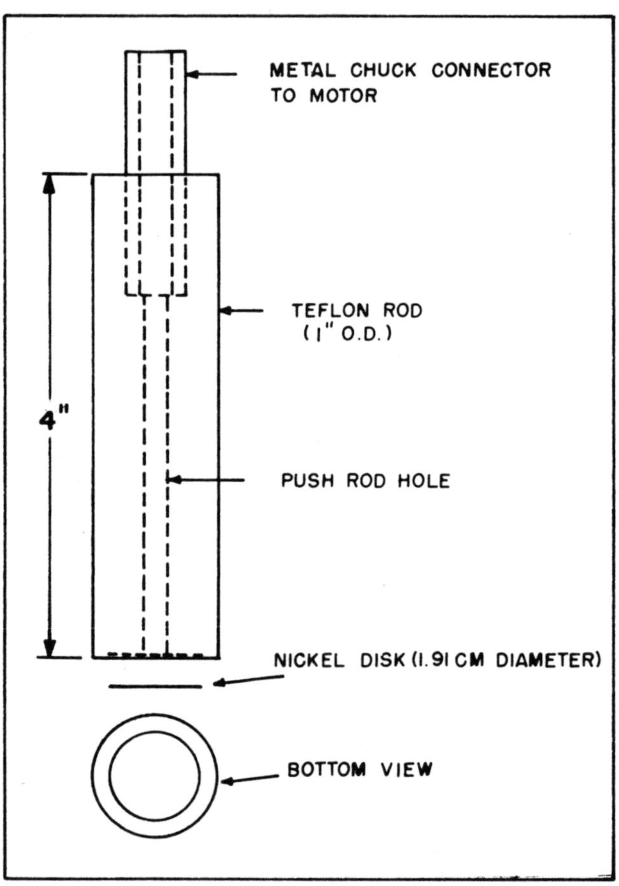

Figure 1.   Teflon Electrode Holder

After the electrode assembly has been secured to a variable-
speed motor, the nickel surface is polished at high rotation
speed with 0.1 μm alumina supported on a polishing cloth.
The electrode is then rinsed in deionized water and air-dried
at a fast rotation rate. Just before sample deposition, the
electrode is rotated in 0.5N HCl for 5 min. During the
deposition step of the sample, the electrode assembly is
rotated at 500 RPM with sufficient immersion into the solution to prevent vortex bubbles from traversing the electrode

surface. Plating times are generally not more than 1 h.
After being rinsed in deionized water, the electrode is air-
dried at a fast rotation rate. Insufficient rinsing and
drying may result in formation of a deposition reagent film
on the electrode surface. This film may cause degradation of
the spectra. A hole passing through the entire length of the
holder allows for removal of the electrode with a pushrod.

RESULTS AND DISCUSSION

     A typical alpha spectrum depicting the well-defined
peaks of ^{208}Po and ^{210}Po is illustrated in Figure 2.

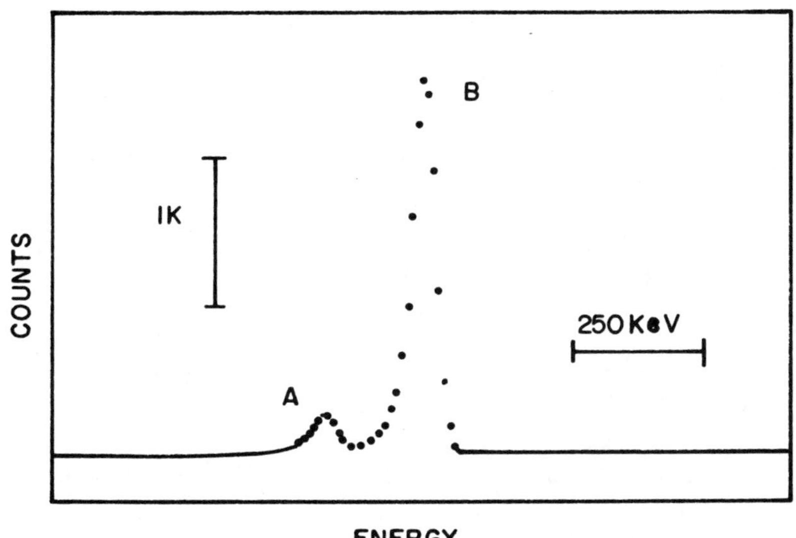

Figure 2. Alpha Spectra. (A) ^{208}Po (3pCi, 5.12 MeV).
(B) ^{210}Po (56pCi, 5.31 MeV)

Even with much higher activities of ^{210}Po, there is adequate
peak separation from the peak of ^{208}Po to quantitate the
activity. Determination of a ^{210}Po–^{210}Pb standard was within
5% uncertainty at the 100 pCi ^{210}Po level. The ^{208}Po tracer
used in the experiment is not known to better than 5% uncertainty.

     Research is also being conducted in our laboratory on
rotating-disk deposition in media other than HCl. Much
simpler and more time-effective procedures for ^{210}Po and
^{210}Pb are being developed. These techniques will be reported
in the future.

REFERENCES

1.   N. A. Talvitie, Anal. Chem., <u>44</u>, 280 (1972).

2.   P. Delahay, <u>New Instrumental Methods in Electrochemistry</u>,
     Interscience Publishers, Inc., New York, NY, 1954, page
     282.

3.   V. G. Levich, Acta Physicochim URSS, <u>17</u>, 257 (1942).

RADIOCHEMICAL SEPARATION OF NEPTUNIUM AND PLUTONIUM FROM
LEACHING OF REACTOR WASTE GLASS IN BRINE SOLUTIONS*

J. Rego, Nuclear Chemistry Division, Lawrence Livermore
Laboratory, Livermore, California, USA

ABSTRACT

The work described in this paper is part of a leaching
study being conducted for the Battelle Pacific Northwest Laboratory's Waste Isolation Safety Assessment Program (WISAP).
Simulated high-level reactor waste glass were leached with
three solutions, one of which was a saturated-salt brine.
Because chemical separation of Np and Pu using organic extraction or anionic exchange is not effective for the brine
samples, a procedure has been developed to first separate neptunium and plutonium from high concentrations of brine before
proceeding with an extraction of neptunium from plutonium.

Samples were equilibrated with tracers, Np and Pu were
co-precipitated with $La(OH)_3$, and interfering ions were removed by washing the hydroxide precipitate with water. Pu and
Np were separated by reducing Pu to Pu^{3+} and extracting the
Np^{4+} into thenoyltrifluoroacetone (TTA); control of oxidation
states and contaminant concentration is critical.

INTRODUCTION

Neptunium and plutonium are of particular interest in
leaching studies because it is a goal of the Waste Isolation
Safety Assessment Program (WISAP) to understand the rate at
which these long-lived radionuclides are released to groundwater which could invade the repositories. The single-pass
leaching of simulated waste glass doped with actinides[1,2]
has been described.[3] The concentration of ^{239}Pu and ^{237}Np in

---

*Work performed under the auspices of the U.S. Department of
Energy by the Lawrence Livermore Laboratory under contract
number W-7405-ENG-48.

the glass is similar to that expected for actual fully radioactive waste glass.[3] Samples were studied over eleven time
intervals within the 420-day experiment. It was necessary to
develop a radiochemical procedure to separate ^{239}Pu and ^{237}Np
in the high-salt samples generated by this leaching study,
because the concentration of the brine hindered their measurement by direct alpha counting.

The composition of the leach brine was based on the composition of a salt rock core from the Waste Isolation Pilot
Plant in New Mexico (WIPP). The major salt component in the
WIPP brine is sodium chloride (Table I). Samples containing
sulfates, fluorides, and phosphates have been shown to interfere with neptunium extraction.[4] However, in these brine solutions there was no measurable fluoride or phosphate, nor was
either used in the procedure as precipitating agents. Sulfate was present, but the massive amount of sodium chloride
probably accounted for the major extraction interference.

TABLE I.

Composition of New Mexico Salt Rock Core

WIPP Brine B	Composition (mg/l)
Na^+	115,000.
K^+	15.
Mg^{+2}	10.
Ca^{+2}	900.
Sr^{+2}	1.5
Cs^+	1.
Cl^-	175,000.
SO_4^{-2}	3,500
BO_3^{-3}	10.
HCO_3^-	10.
Br^-	400.

PH 6.5   0.5

In order to approximate a brine of the composition shown in Table I, we prepared a solution containing the following reagents:

NaCl	287.	g/l
Na_2SO_4	6.2	g/l
$Na_2B_4O_7 \cdot 10\ H_2O$	16.	mg/l
$NaHCO_3$	14.	mg/l
NaBr	520.	mg/l
KCl	29.	mg/l
$MgCl_2 \cdot 6\ H_2O$	85.4	mg/l
$CaCl_2$ anhyd.	2.49	g/l
$SrCl_2 \cdot 6\ H_2O$	4.52	mg/l
CsCl	1.3	mg/l

All attempts to separate neptunium and plutonium before removing the salts had previously failed. Trial separations were attempted using larger dilutions of the brine. At even 90% dilution (0.5 M NaCl), only 50% of the neptunium could be recovered by extraction techniques. Similar losses were experienced using the more time-consuming ion exchange procedures. In addition to the salt interference, the presence of calcium in the sample during a Pu-fluoride precipitation also caused losses of plutonium. Coleman[5] also observed that calcium must be removed because it was one of the elements which he found interfered with LaF_3 co-precipitation.

The goal of the work described here was to separate neptunium and plutonium in a brine medium so each isotope could be measured quantitatively. The procedure developed is described below, and its critical aspects are discussed.

PROCEDURE

Tracer levels of ^{239}Np and ^{242}Pu were added to each 10-milliliter sample in order to measure chemical yield. Ten milligrams of La^{+3} were added. Initially, equilibrium between tracers and samples was attained by using 1 M sodium nitrite in 8 M nitric acid solution. The solution was evaporated by boiling to about 2 ml, then diluted with water to 20 ml. Fifty percent sodium hydroxide was added to pH 12 or greater. The resultant lanthanum hydroxide co-precipitated Np and Pu, and provided an effective separation from the massive concentrations of other ions, which remained in solution. The precipitate was dissolved in 1 ml concentrated HCl and diluted to 10 ml, adjusted to a concentration of 4 M HCl. Two ml of hydroxylamine hydrochloride were added and the solution was heated in a hot water bath for 1-2 minutes. Next, 3/4 ml 1 M SnCl$_2$ and 8 ml 4 M KI were added and the solution was heated for 10 minutes in a hot water bath.[4] The solution, cooled to room temperature, was transferred to a separatory funnel. An equal volume of a 0.4 M solution of thenoyltrifluoroacetone (TTA) in toluene was added. Under these conditions the neptunium 4+ is extracted into TTA and Pu 3+ is not. The neptunium fraction was back-extracted from the organic with 9 M HCl.

The plutonium fraction must undergo further purification. Ten milligrams of La^{3+} and 5 ml HF were added to the Pu solution; LaF$_3$ was co-precipitated with PuF$_3$. The resulting precipitate was dissolved in 10 M HCl and the solution was placed on an anionic exchange column, preconditioned with 10 M HCl. After the column was washed and converted to a NO$_3^-$ column with 8 M HNO$_3$, the column was converted back to a Cl$^-$ column with 10 M HCl. The plutonium was selectively eluted with 10 M HCl$\cdot$0.5 M NH$_4$I.

The final separated and purified samples of neptunium and plutonium were electroplated onto platinum disks from a barely acidic sulfate medium for one hour at 1/2 ampere. The disks were counted on alpha detectors controlled by a Nuclear Data Model 600 Multichannel Analyzer (MCA). Pictures were taken of the alpha spectra using a scope camera attached to the ND600 to document the quality of the spectra (Fig. 1).

The procedure requires approximately two days for eight samples. Typical yields were in the range of 70-80% for neptunium and slightly less for plutonium.

Illustration of Alpha Spectra

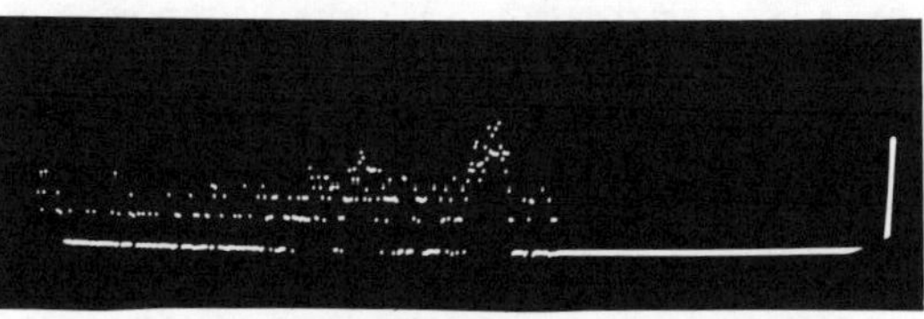

Evaporated brine

Evaporated carbonate

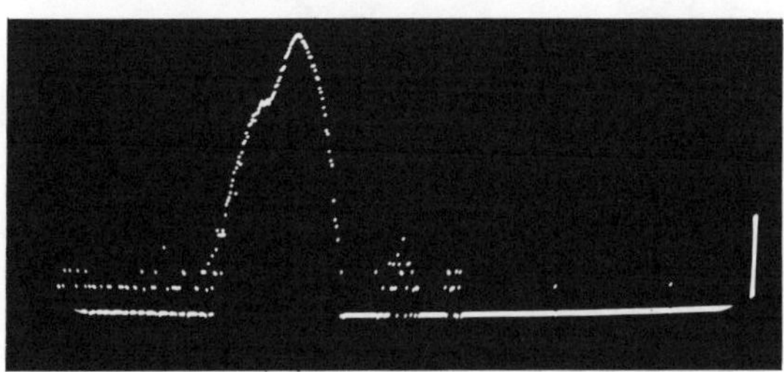

Separated neptunium

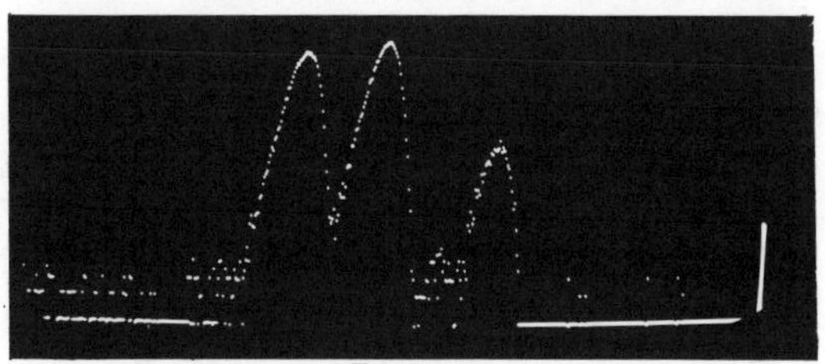

Separated plutonium

DISCUSSION AND CONCLUSIONS

The ^{242}Pu tracer was obtained from the isotopic separation of Pu isotopes. The ^{239}Np tracer was prepared by extraction from a ^{243}Am solution into TTA. Because ^{239}Np has a short half-life ($t_{1/2}$ = 2.35 days), it was necessary to expedite the chemical separation of neptunium and plutonium. Neptunium-239 was selected as a chemical yield tracer for neptunium because it is easily detected in an end-window, lead-shielded NaI crystal counter. Neither ^{239}Pu nor ^{242}Pu emit radiations which interfere with detection of ^{237}Np or ^{239}Pu (respectively). The oxidation states of the neptunium and plutonium during extraction are critical. The presence of interfering salts will jeopardize the separation and reduce the ultimate yields. The presence of small amounts of I_2 in the final plating solution will cause severe losses of plutonium. If these critical factors are controlled, Np and Pu can be effectively separated from highly concentrated brine solutions.

ACKNOWLEDGMENTS

I thank D. Coles and H. Weed for their time and advice.

REFERENCES

1. H. C. Weed and D. D. Jackson, "Design of a Variable Flow Rate, Single-Pass Leaching System, UCRL-52785 (1979).

2. D. G. Coles, H. C. Weed, D. D. Jackson, and J. S. Schweiger, "Single-Pass Leaching of Nuclear Melt Glass by Ground-in Geologic Storage, S. Fried, Editor (1979) pp 93-114.

3. H. C. Weed, D. G. Coles, D.J. Bradley, R. W. Mensing, and J. S. Schweiger, "Leaching Characteristics of Actinides from Simulated Reactor Waste Glass," Lawrence Livermore Lab. Rept. UCRL-81147, Oct. 24, 1978.

4. V. A. Mikhailov, Analytical Chemistry of Neptunium, Halsted Press (1973) 71, 108.

5. G. H. Coleman, "The Radiochemistry of Plutonium," Lawrence Radiation Lab. NAS-NS-3058, Sept. 1, 1965.

AUTOMATIC ION-EXCHANGE CHROMATOGRAPHY IN
CORROSIVE SOLVENT SYSTEMS*

D. W. Hosmer and A. L. Gazlay,
Nuclear Chemistry Division,
Lawrence Livermore Laboratory, Livermore, CA USA

ABSTRACT

An automated system for the performance of ion-exchange
chromatography with corrosive solvents has been built. It is
LSI-11 microcomputer controlled and is capable of selecting
six sample/solvents and collecting six separate fractions.
Preliminary results on the separation of plutonium from 8M
HNO_3 solutions indicate that the system can achieve chemical
yields, analytical accuracy, and precision fully comparable
to manual column operation at a considerable savings of operator time.

INTRODUCTION

The Lawrence Livermore Laboratory maintains a program of
environmental monitoring in the Livermore Valley to detect
changes in the inventory of hazardous materials for which the
Laboratory is a potential source. This Site Environmental
Monitoring Program (SEMP) includes the collection and analysis of air and water supplies, local vegetation and agricul-
tural products, soils, as well as environmental radiation
measurements by thermoluminescent dosimeters. Some of the
analyses include gross alpha and beta measurements, gamma
spectrometry, and the radiochemical purification and determination of several specific nuclides. Each year more than
600 separate analyses are performed for the determination
of plutonium, uranium, and ^{137}Cs on more than 200 samples.

---

*Work performed under the auspices of the U. S. Department
of Energy by the Lawrence Livermore Laboratory under
contract number W-7405-ENG-48.

The generalized scheme for the analyses of these nuclides is presented in Fig. 1.

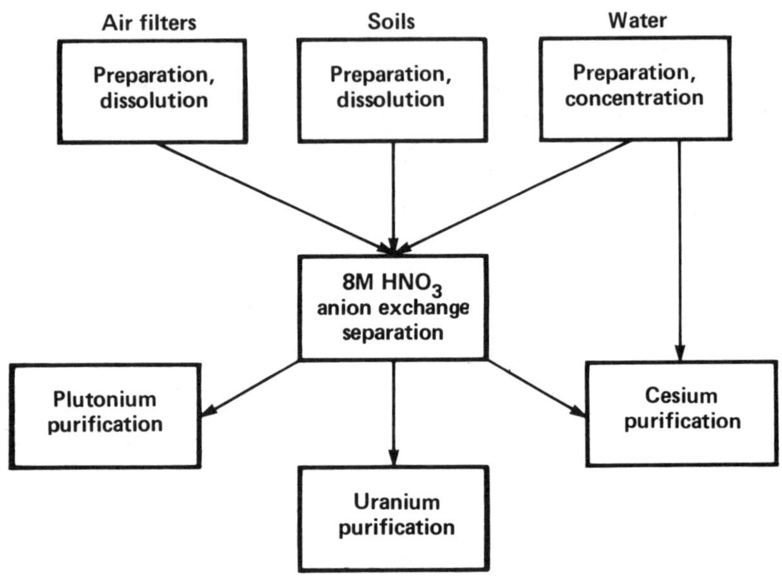

FIG. 1

For all but a few determinations, the anion-exchange step is the critical process upon which the success of all subsequent purifications and determinations depend. The fractions obtained from the original column separation are subsequently purified by techniques that require an additional ion-exchange step for each radionuclide of interest. Final determinations are made by alpha spectrometry for the plutonium isotopes, mass spectrometry for uranium, and gross beta counting for [137]Cs.

Automation of the anion-exchange step should result in an increase in the overall effectiveness of the analytical scheme. Historically, automation of a repetitive process has lead to improvements in both personnel efficiency and sample turnover rate. In addition, the assurance of the quality of the analysis is higher for a process that treats all samples in an identical manner. For these reasons it was decided that an effort to automate the anion-exchange step was justified.

System Requirements

The anion-exchange separation scheme that we employ has
some special characteristics that need to be considered when
designing an automated system. These include the number and
corrosiveness of solvents used. The sample itself is in
8M HNO$_3$, while 8M HNO$_3$, 10M HCl, 0.1M HCl, and freshly prepared 0.1M NH$_4$I in 10M HCl are all used in the column elution
process. Fractions for waste, cesium, plutonium, and uranium
need to be collected. Also, the extremely low levels of the
nuclides we normally measure required that cross-contamination, especially from an occasional "hot" sample, be kept to
a minimum.

In addition to these special considerations, the system
should be able to handle sample sizes up to 1 liter and
should accommodate other types of ion-exchange chemistry
with a minimum of modification. It should also minimize the
amount of human interaction required for operation, but be
capable of manual override if necessary. Liquid volumes
should be metered to within 10% accuracy at low flow rates
and the system capacity should be expandable without major
redesign of the components.

System Description

The chemistry hardware currently consists of six column
assemblies, each with identical components. (See Fig. 2).

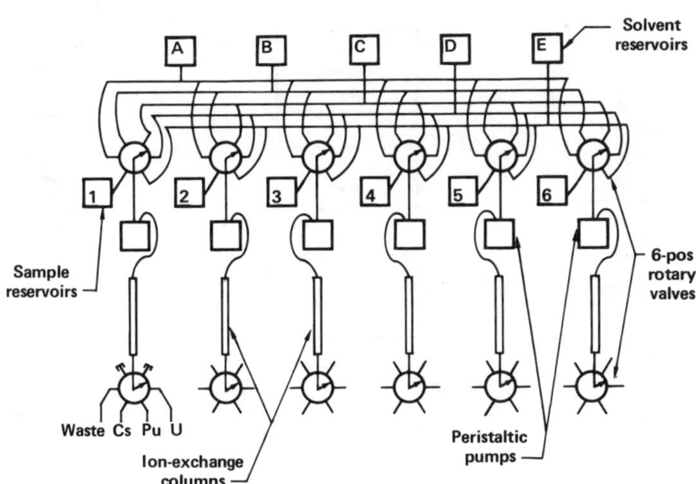

FIG. 2

65

A six-position, pneumatically actuated rotary valve is at
each end of the column assembly. The upper position valve
selects the solvent/sample to be introduced into the column,
while the lower valve selects the port to receive the column
effluent. The column is a 15 mm (id) by 300 mm pyrex glass
tube equipped with quick disconnect tubing attachments at
either end. A peristaltic pump, rated at 2.1 mls/min, meters
the sample and solvent volumes through the column. All
tubing is 1.6 mm (id) with teflon used wherever possible.
Chemically resistant tygon is used for the sample syphon,
peristaltic pump, and fraction collection lines; it is
replaced between sample runs, as is the ion-exchange resin
in each column. All liquid containing and transporting
components are located in a fume hood and are connected to
motors and advancing mechanisms via mechanical couplings.

An LSI-11 microcomputer with 28k memory, dual floppy
disks, and teletype makes up the nucleus of the electronic
controlling hardware (See Fig. 3). Interfaced to the computer are three control chasses which drive the air sole-
noids and peristaltic pumps under program control. There is
also one chassis which monitors position indicators on each
of the rotary valves. According to Stockwell[1] the inclusion
of a feedback mechanism defines an automated system. All of
the electronics have been designed to fulfill the control
requirements of up to 16 column assemblies, the only modifications necessary being to install the additional assemblies
and to change one constant in the controller program.

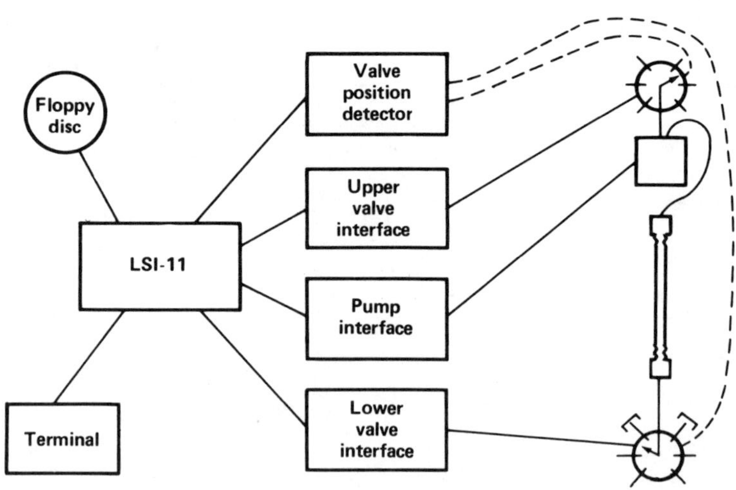

Each column assembly, consisting of two six-position
rotary valves, a peristaltic pump, and one glass column,
costs approximately $1K. The cost of the electronic con-
trolling hardware was about $10K. The development and
fabrication time amounted to nine man-months of effort.
We expect to realize an immediate savings of two to three
man-months per year by using this system in place of our
traditional hand-column techniques. This should increase
substantially as the other ion-exchange chemistries are
adapted to the system, and as the column numbers are expanded to design capacity.

System Evaluation

As a test of the system's effectiveness in separating
plutonium, the following procedure was performed. Pu-242
was added to 5.28 grams of coral sand from Enewetok Atoll
and the resulting mixture dissolved with 8M HNO_3 and H_2O_2.
After filtering and diluting to 100.0 mls with 8M HNO_3, six
5.0 ml aliquots were taken and diluted to approximately 100
mls with 8N HNO_3. Six aliquots of pure Pu-236 tracer (for
use as blanks) were measured and diluted to approximately
100 mls with 8M HNO_3, as were two aliquots from a plutonium
standard solution traced with Pu-242. Approximately one
gram of $NaNO_2$ was added to all the above aliquots.

Three soil aliquots, one standard, and two Pu-236 blanks
were separated on the automatic system according to the chemistry scheme presented in Fig. 4, using DOWEX AG1X8 as the

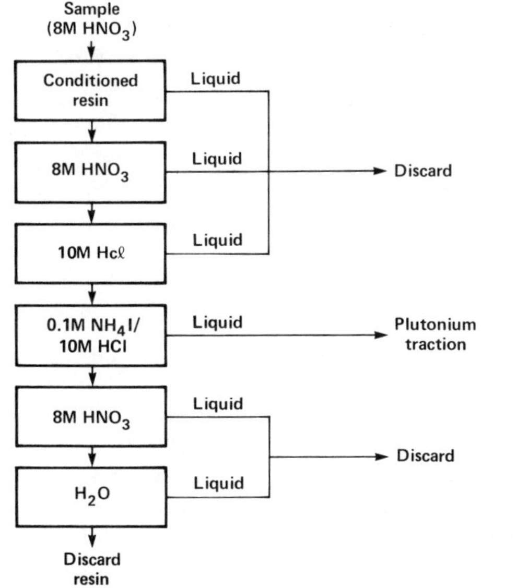

FIG. 4

anion-exchange resin. After the plutonium fractions were
collected, the whole system was washed with approximately
25 mls of 8N HNO₃ and the resin changed. The remaining four
of the Pu-236 traced blanks were run in the same column
positions in which the standard and soil aliquots had previously been run. The three remaining standard and soil
aliquots were separated by the traditional hand-column
techniques. Finally, purification, electroplating, and
alpha spectrometry were performed on all of the plutonium
fractions.

RESULTS

All the plutonium values and tracer recoveries are listed
in Table 1. They show that the average Pu-239 values for the
for the automatically and manually separated aliquots are
indistinguishable. The values for the two plutonium standards are also in good agreement with each other and with
the accepted value of 44 ± 1 pCi.

The recovery averages for the samples separated by the
two techniques show the recoveries to be comparable, though
the auto system samples averages are slightly lower. This
apparent difference is not large enough, however, to noticeably affect the measurement precision of the automatically
separated samples.

The Pu-236 blanks run after the samples are well within
the normal (manual) average blank value range of 0.011 ± 0.014
pCi for environmental samples. However, more studies are in
progress to determine whether or not the Column 1 and 2
results are indicative of cross-contamination, and the continuous monitoring of column backgrounds will be a part of
the regular operating procedure.

FUTURE WORK PLANNED

Future work includes the expansion of the system to
its designed capacity of 16 column assemblies. This will
further reduce the personnel time involved in the anion-
exchange process and will provide sample capacity as LLL's
SEMP continues to increase the yearly load of samples requiring plutonium, uranium, and cesium-137 analysis.

Also desirable is the adaptation to the automatic
system of the ion-exchange purification steps as well as
the separation of different radionuclides. This appears to
be feasible since the anion-exchange separation step is the
worst case in terms of materials compatibility. There are
considerable differences in the sample volumes, however,

TABLE 1

Plutonium Values and Tracer Recoveries for
Automatic and Traditional Hand-Separated Samples

Aliquot	Column Number	Plutonium Values (pCi)	Tracer Recovery %
Auto System			
Soil #1	1	10.2 ± 0.5 g^{-1}	60
#2	2	9.9 ± 0.4 g^{-1}	75
#3	3	9.7 ± 0.4 g^{-1}	68
		ave. 9.9 ± 0.3 g^{-1}	
Standard #1	4	43 ± 1	67
			ave. 68 ± 6
Traditional Method (manual)			
Soil #4	–	9.3 ± 0.5 g^{-1}	84
#5	–	9.8 ± 0.5 g^{-1}	75
#6	–	9.6 ± 0.4 g^{-1}	77
		ave. 9.6 ± 0.3 g^{-1}	
Standard #2	–	43 ± 1	75
			ave. 78 ± 4
Before samples (auto)			
Blank A	5	0.002 ± 0.002	69
Blank B	6	0.003 ± 0.002	83
		ave. 0.002 ± 0.003	ave. 76 ± 10
After samples (auto)			
Blank C	1	0.011 ± 0.004	77
Blank D	2	0.006 ± 0.003	76
Blank E	3	0.002 ± 0.002	53
Blank F	4	0.003 ± 0.003	79
		ave. 0.006 ± 0.004	ave. 71 ± 12

and it appears that a large part of this adaptation effort
will be devoted to the elimination of mass interferences.

A detailed filing system for sample information and
system run parameters is also being developed. It is possible to have the controlling program automatically write such
a file as part of its execution, and this function would be
in line with with the SEMP's increasing emphasis on quality
assurance.

Finally, the inclusion of additional feedback devices
to develop a more complete process sensing capability is
being considered. This may include devices to monitor
sample delivery, and flow sensing devices to detect possible leaks and blockage. All additional feedback would make
the system more reliable and even less time-consuming to
operate.

CONCLUSION

An automated system for performing ion exchange chromatography with corrosive solutions has been built. It is
controlled by a LSI-11 microcomputer and has flexibility to
accomodate up to 16-column assemblies without modification of
the electronic hardware. Time savings estimates for performing anion-exchange separations on LLL's SEMP samples are
from two to three months per year.

Plutonium determinations on solution aliquots separated by the automatic system are comparable to values
obtained from the traditional hand-column method. Tracer
recoveries are not significantly different for the two
methds. There does not appear to be any problem of sample
intercontamination, although the normal column operation
will continuously monitor for that parameter.

Future work planned includes expanding the number of
columns and the types of chemistries that can be accommodated, as well as developing sample filing system that
will be controlled automatically by the operating program.
There are also plans to add additional feedback mechanisms
to the system to allow for more complete process control.

REFERENCES

1. P. B. Stockwell, The Journal of Automatic Chemistry,
    1, 10 (1978).

RADIOCHEMICAL DETERMINATIONS
AND MEASUREMENTS

RADIOMETRIC DETERMINATION OF PLUTONIUM
IN CAUSTIC SCRUBBER SOLUTIONS

B. P. Freeman, J. R. Weiss, and R. A. Mason, Jr.,
U.S. Department of Energy, New Brunswick Laboratory,
Argonne, Illinois, USA

ABSTRACT

The New Brunswick Laboratory incorporates a caustic
scrubber exhaust system to absorb and neutralize acid fumes
from plutonium gloved box operations. The sodium salt
solution resulting from this acid neutralization reaction is
retained in a holding tank along with traces of entrained
plutonium. Safety guidelines require that a reliable
determination of plutonium in this solution be made periodically. Simple alpha radiocounting procedures cannot be
applied to these solutions since the sample cannot be deposited on a planchet in a relatively "solids-free" condition
to avoid self-absorption of alpha particles. Other techniques
involving separations, such as precipitation and solvent
extraction, are time-consuming and not quantitative mainly
because of the high salt content in the sample. To minimize
these problems, a procedure consisting of an ion exchange
separation followed by direct alpha counting of the purified
plutonium was adapted for use. This method is relatively
accurate and precise, and requires less than 16 hours to
complete five determinations.

INTRODUCTION

The New Brunswick Laboratory scrubber system absorbs
and neutralizes HNO_3-H_2SO_4-HCl-HF exhaust fumes from
plutonium gloved box operations with NaOH solution. The
sodium salts of these acids are retained in a holding tank
along with any entrained plutonium. It is necessary to
reliably determine the amount of plutonium in the scrubber
wash solution on a regular basis for purposes of radiation

safety, contamination control, and containment. Simple
alpha radiocounting procedures require a relatively "solids-
free" deposit on the counting planchet for highest
reliability. (The presence of solids cause "masking" or low
results due to absorption of alpha particles.) In order to
determine the concentration of plutonium in the holding tank,
it is necessary to convert the plutonium to a form which can
readily be assayed. Since the level of plutonium in the
holding tank is typically 5,000-50,000 d/m/ml, it is also
necessary to modify existing analytical procedures to accommodate such trace levels. This paper describes a
procedure for the removal of solids and subsequent alpha
counting of scrubber solutions.

EXPERIMENTAL

Reagents and Apparatus

Plutonium Standard Solution, NBS SRM 944, Plutonium Sulfate
Tetrahydrate, $Pu(SO_4)_2 \cdot 4H_2O$, diluted to a concentration of
about 3,000 d/m/ml in 8N HNO_3.

Nuclear Measurement Corporation, Model PC-4 Windowless Gas
Flow Alpha Proportional Counter.

Nitric Acid, 8N, 1N, and concentrated solutions.

Sulfuric Acid, concentrated.

Stripping Solution, 0.35N HCl-0.01N HF.

Bio-Rad Disposable Econo-Columns, No. 739-1110.

Dowex-1, X-2, 400 mesh Anion Exchange Resin.

Salt Solution, 100 g/l $NaNO_3$, 20 g/l $Na_2(SO_4)$, 5 g/l NaCl,
and 1 g/l NaF in 8N HNO_3, to simulate actual scrubber
solution matrix.

Variable temperature hotplate.

Polished stainless steel planchets, 2" diameter, flat.

P-10 gas, Matheson, 90% Argon - 10% Methane.

Burner, propane gas.

Alpha counting standard, Eberline Instrument Corporation,
Serial No. S-1023, Pu-239 principle nuclide, calibrated May
29, 1975; 1617 $\pm$ 320 d/m.

Procedure

<u>Calibration of the alpha proportional counter</u>: The
alpha proportional counter's sample chambers are flushed
with P-10 gas for at least 100 seconds before taking an alpha
count. The efficiency of the counter is determined using an
alpha counting standard disk.

<u>Sample Preparation</u>: A 200-ml scrubber solution sample
containing approximately 100-200 mg of undissolved material
is digested in an equal volume of concentrated HNO_3 with a
trace of HF for several hours, until it is free of all undissolved matter. This digestion will convert existing
plutonium as $Pu(OH)_x$ to $Pu(NO_3)_x$; a form which can be readily
separated from interfering salts by a modification of the
NBL ion exchange procedure,[1,2] described below. If necessary, a portion of the digested scrubber solution is diluted
with 8N HNO_3 such that a level of 3-5,000 d/m/ml is achieved.
Several 1-ml aliquots of the solution are prepared by adding
3 drops of concentrated H_2SO_4, and fuming each aliquot to
incipient dryness on a hotplate. This fuming removes
volatile interferences such as HF and HCl. Na_2SO_4 is not
removed by this procedure, but it has been determined that
the ion exchange separation of plutonium is satisfactory
even when relatively high concentrations of the salt are
present.[3] Aliquots are then processed through the ion
exchange procedure.

<u>Ion Exchange Separation</u>: The ion exchange separation
method used is based on the adsorption of a plutonium complex, $Pu(NO_3)_6^{=}$, formed in 8N HNO_3. This complex is strongly
adsorbed onto anion exchange resins such as Dowex 1X-2, and
is quantitatively eluted with dilute HCl-HF solutions. A
sample is dissolved in 30 ml of 1N HNO_3, and the plutonium is
reduced to Pu^{+3} by addition of 1 ml of 1.0 M $Fe(SO_4)$, followed by oxidation to Pu^{+4} with 30 ml of concentrated HNO_3.
The acid concentration is thus adjusted to 8N HNO_3, and
$Pu(NO_3)_6^{=}$ is formed in this step. An ion exchange column is
washed with several ml of 8N HNO_3, and then the sample is
loaded. Afterwards, the column is washed with at least
100 ml of 8N HNO_3, and the plutonium is eluted with 40 ml of
0.35N HCl-0.01N HF solution into a clean 50 ml beaker.

After the separation is completed, the eluates are
evaporated to approximately 1 ml, and quantitatively transferred to a planchet with 1N HNO_3 and dried on a hotplate.
Each planchet is flamed to red heat over a gas burner,
cooled and stored.

Characterization of the $Pu(SO_4)_2 \cdot 4H_2O$ Standard Solution:
The standard solution aliquots for ion exchange separation
are processed in the same manner as mentioned in the Ion
Exchange Separation section above. Each standard plachet is
counted for ten minutes, according to the calibration procedure above, and is corrected for [241]Am content.

Analysis of Synthetic and Actual Process Scrubber
Solutions: Synthetic scrubber solutions were processed and
counted according to the above procedures. Actual process
solutions were spiked with a known amount of standard solution and treated similarly.

RESULTS AND DISCUSSION

Initially, the ion exchange procedure was tested at a
low level of plutonium (approximately 3,253 d/m/ml), using
the plutonium standard solution. The plutonium level was
chosen to approximate the concentration expected in diluted
scrubber solutions. After ion exchange separation, the eluted
plutonium was counted and the recovery was calculated. The
results are given in Table I.

Table I

Recovery of Plutonium After Ion Exchange Separation

Aliquot No.	Plutonium Recovery, %*
1	98.6
2	96.8
3	97.4
4	99.3
5	99.1
Average	98.2 (RSD = 1.1%)

*Corrected for [241]Am which is separated by ion exchange.

In order to simulate the salts in the acid-digested
scrubber solution, a salt solution was added to the plutonium
standard. The plutonium was separated by ion exchange and
alpha counted. The results are shown in Table II.

Table II

Recovery of Plutonium from a Synthetic Scrubber Solution

Aliquot No.	Plutonium Recovery, %*
1	98.0
2	100.9
3	101.7
4	98.3
Average	99.7 (RSD = 1.9%)

*Relative to the plutonium recovery for standards, Table I.

Thus, at the low levels of plutonium studied, satisfactory recovery was achieved by the ion exchange procedure. In standards spiked with salts to simulate actual scrubber solutions, the plutonium was again satisfactorily separated.

Several samples taken from the scrubber holding tank were diluted to yield a concentration of about 2,000 d/m/ml sample and aliquots were taken. To half of the aliquots, plutonium standard solution was added, and the resulting solutions along with the remaining unspiked aliquots were separated by ion exchange and alpha counted as described earlier. The standard addition technique was used to determine the recovery of the plutonium calculated as follows:

$$\text{Pu Recovery, } \% = \frac{G-N}{S} \times 100 \qquad (1)$$

where: G = gross alpha count (d/m/ml) of scrubber aliquot + standard.

N = net alpha count (d/m/ml) of scrubber aliquot alone.

S = average alpha count (d/m/ml) of standard alone.

Results from actual scrubber solution analyses are presented in Table III.

Table III

Recovery of Plutonium From Actual Scrubber Solutions
After Spiking

Aliquot No.	Net Plutonium Recovery, % Sample A*	Net Plutonium Recovery, % Sample B**
1	94.4	103.9
2	95.6	102.7
3	97.0	100.1
4	92.2	102.2
5	87.8	103.1
Average	93.4	102.4
RSD	3.44	1.25

*Sample A taken 11/22/77
**Sample B taken 8/23/79

Comparing the data in Tables II and III, it is apparent that the synthetic salt solution resembles the actual scrubber solution in regard to the effect of salt content on plutonium recovery. Slightly lower and less precise recoveries seen for Sample A above are probably a result of sample storage for two years prior to its analysis. The correlation between elapsed holding time and low results is unknown, but is currently under study. However, the method appears to be satisfactory for the determination of sub-microgram levels of plutonium.

REFERENCES

1. NBL Mini Anion Exchange, U.S. Department of Energy Procedure, NBL-PC-IE-3.

2. J. R. Weiss, C. E. Pietri, A. W. Wenzel, and L. C. Nelson, Jr., U.S. Atomic Energy Commission Report, NBL-265 (June 1972), p. 94.

3. G. E. Peoples, J. R. Weiss, and C. E. Pietri, U.S. Department of Energy Report, NBL-289 (January 1979), p. 82.

A RADIOCHEMICAL PROCEDURE SPECIFIC FOR THE
DETERMINATION OF PROTACTINIUM-231

C. R. Walker and B. W. Short, Technical Division,
Goodyear Atomic Corporation, Piketon, Ohio, USA.

ABSTRACT

Protactinium-231, a daughter product of uranium-235, can
be present at the Portsmouth Gaseous Diffusion Plant in detectable amounts in such sources as uranium oxide fluorination filter ash and UF_6 cylinder residues. Since its detection is important for health physics considerations, a method
for its analysis was required. To improve on lengthy existing procedures, a simple procedure specific for protactinium-
231 was developed: Protactinium is first extracted from a
sulfate-chloride medium into amyl acetate. It is then back-
extracted into a chloride-fluoride medium, complexed with
cupferron, and re-extracted into amyl acetate. A portion of
the amyl acetate is evaporated on a planchet, and protactin-
ium-231, the only alpha-emitting nuclide carried through the
process, is determined by alpha activity measurement on a
proportional counter.

INTRODUCTION

Because it is produced naturally by the radioactive decay of uranium-235, protactinium-231 (^{231}Pa) is present at
the Portsmouth Gaseous Diffusion Plant, a uranium enrichment
facility. The radionuclide is an alpha emitter (5.05 Mev)
with a half-life of 3.25×10^4 years and a specific activity
of 1.058×10^5 d/m/µg. Because of the undesirable biological
consequences that may result form exposure to this long-lived
bone-seeker, federal regulations set the maximum permissible
concentration (MPC) of soluble ^{231}Pa in an uncontrolled environment at 4×10^{-14} µCi/ml (0.089 d/m/m^3) for air and
9×10^{-7} µCi/ml (2 d/m/ml) for water. To ensure the safety
of the employees and compliance with state and federal regulations, many samples are analyzed for ^{231}Pa at the Portsmouth plant. Samples that can be high in ^{231}Pa include

residues from cylinders of UF_6 highly enriched in ^{235}U and
non-volatile residues left behind during the conversion of
uranium oxides enriched in ^{235}U to UF_6, i.e., fluorination
filter ash.

PROCEDURE

Overview

Soils, fluorination filter ash, and water samples are
solubilized in HCl and HF. Samples are then treated with
9 $\underline{N}$ H_2SO_4 to volatilize both the fluoride and silica in the
samples. It is important that no fluorides remain in the
samples, because their presence hinders the extraction of
protactinium into amyl acetate. The final residues dissolve
completely in 8 $\underline{N}$ HCl except when elements forming insoluble
chlorides or sulfates are present. From 8 $\underline{N}$ HCl, protactinium, along with any iron that is present, is easily extracted
into amyl acetate. The separation from iron is completed by
selectively back-extracting the protactinium with HCl-HF
solution. The HCl-HF solution is then treated with boric
acid to complex fluorides. Any remaining traces of iron are
complexed with _ortho_-phenanthroline, and other extractable
elements are complexed with ammonium citrate, which also acts
as a pH buffer. Protactinium is then complexed with cupferron and extracted into amyl acetate, which is evaporated on a
stainless steel planchet and alpha counted. Alpha spectrometry has shown that ^{231}Pa is the only alpha emitting radionuclide present in the final extract and that the protactinium recovery is essentially quantiative.

Sample Preparation

Soils and Fluorination Filter Ashes

Weigh about a one-gram portion of sample to the nearest
mg into a 75-ml platinum dish. Add 10 ml of concentrated HF
and 3 ml of concentrated HCl and fume to dryness. Repeat the
previous step. Fume the residue to dryness twice with 10 ml
of 9 $\underline{N}$ H_2SO_4 to expel fluorides.

Air Sample Taken on Filter Papers

Place a filter containing the sample in a 75 ml platinum
dish, wet it with 8 $\underline{N}$ HNO_3, and evaporate it to dryness. Add
2 g potassium pyrosulfate ($K_2S_2O_7$) and fuse at 700°C for 30
minutes. Let cool, add 10-15 ml 9 $\underline{N}$ H_2SO_4, and fume to dryness.

Aliquot 100 ml of sample into a 250-ml Teflon beaker and evaporate to dryness. Add 10 ml of 8 $\underline{N}$ HCl and 2 ml of concentrated HF; heat to dissolve the residue. Transfer the solution to a 75-ml platinum dish, evaporate it to dryness, and then fume to dryness twice with 10-15 ml of 9 $\underline{N}$ H_2SO_4.

Determination of Protactinium-231

Dissolve the residue from the appropriate preparation above in 50 ml of 8 $\underline{N}$ HCl with boiling. Transfer the solution to a 250 ml separatory funnel and add 30 ml of concentrated HCl. Dilute to 100 ml with 8 $\underline{N}$ HCl. Add 100 ml of amyl acetate and equilibrate for two minutes. Discard the aqueous phase. Wash the extract one minute with 25 ml of 8 $\underline{N}$ HCl and discard the wash solution. Add 5 ml of 8 $\underline{N}$ HCl-0.1 $\underline{N}$ HF to the extract and equilibrate for two minutes to back-extract the protactinium. Transfer the aqueous layer to a 100-ml beaker. Add 5 ml of 8 $\underline{N}$ HCl to the extract and equilibrate for one minute. Combine the aqueous portions. Add 5 ml of 20 percent ammonium citrate, 10 ml of 5 percent boric acid, and 5 ml of 50 percent hydroxylamine hydrochloride to the aqueous solution. Using bromocresol green as the indicator, adjust the pH to 3.5 with concentrated NH OH. Add 1 ml of 2 percent <u>ortho</u>-phenanthroline and cool to room temperature. Transfer the solution to a 125-ml separatory funnel and add 5 ml of 5 percent aqueous solution of cupferron and 5.0 ml of amyl acetate. Extract for two minutes and discard the aqueous layer. Wash the extract with 25 ml of 0.1 $\underline{N}$ H_2SO_4. If an emulsion is present, centrifuge in a 15-ml centrifuge tube. Transfer 4.0 ml of extract in 0.5 ml increments to a stainless steel planchet maintained at 250°C, evaporating to dryness after each increment. Bake the planchet at 550°C and alpha count in a proportional counter. The amount of protactinium-231 is calculated from the alpha activity.

RESULTS AND DISCUSSION

During previous work at the Portsmouth plant on the determination of neptunium and plutonium by extraction of their cupferron complexes into amyl acetate from 2 $\underline{N}$ HCl, it was found that protactinium and iron also complex with cupferron and extract into amyl acetate from a wide range of acidities. The major problems in determining protactinium are its incomplete dissolution from matrices such as soils and its separation from iron and from other radionuclides that are also complexed with cupferron and extracted. Iron-bearing counting residues reduce counting efficiency and give poor resolution of alpha energies in alpha spectrometry.

Other alpha-emitting radionuclides such as ^{234}U (4.77 Mev), ^{230}Th (4.68 Mev), ^{237}Np (4.78 Mev) and $^{239-240}$Pu (5.15 Mev) also extract and interfere with attempts to measure the ^{231}Pa alpha activity (4.73-5.05 Mev) by alpha sepctrometry.

Therefore, studies were undertaken to develop a procedure which would eliminate these difficulties and selectively separate protactinium from iron and from all other radionuclides.

Effect of Iron on Extraction and Determination of Protactinium

To study the effect of iron on the extraction and determination of ^{231}Pa and to develop a method to eliminate its interference, three experiments were performed.

First, five aliquots of a standard solution, each containing 136 pCi Pa, were spiked with 0.06-500 mg of iron and diluted to 100 ml with 8 $\underline{N}$ HCl. Protactinium, along with iron, was extracted into 100 ml of amyl acetate. Following back-extraction into 8 $\underline{N}$ HCl-0.1 $\underline{N}$ HF, protactinium was determined by cupferron extraction and alpha counting. Results given in Table I show that protactinium is quantitatively and perferentially extracted, even in the presence of 500 mg of iron.

Table I. Effect of Iron on the Extraction
of 136 pCi ^{231}Pa Into Amyl Acetate
from 8 $\underline{N}$ HCl

Fe (mg)	^{231}Pa % Recovery	Comments
0.06	95-100 (10)	Iron Extracted
50	91.2	Partial Extraction of Iron
100	91.2	Partial Extraction of Iron
250	97.8	Partial Extraction of Iron
500	102.0	Partial Extraction of Iron

( ) Number of Determinations

Second, a study was made to determine whether the ortho-phenanthroline complex of iron (II) could be used at certain pH ranges to prevent the extraction of iron in the presence of cupferron. Several standards, each containing 136 pCi

^{231}Pa and 60 µg iron were treated with hydroxylamine to reduce
the iron (III), boric acid and ammonium citrate. The pH of
the solutions was adjusted to 0.3-6.0, and 1 ml of a 2 percent aqueous solution of ortho-phenanthroline was added.
Protactinium cupferrate was extracted into amyl acetate,
evaporated to dryness, and alpha counted. Results of this
study, presented in Table II, show that protactinium is
quantitatively extracted over the whole range of 0.3-6.0 and
that 60 µg of iron is complexed in all cases except at pH 0.3.
A pH range of 3-4 was selected for further studies since at
higher pH (5-6), precipitates occasionally form and interfere
with protactinium recovery.

Table II. Effect of pH on Cupferron Extraction
of 136 pCi ^{231}Pa in Presence of 60 µg Iron

pH	% Recovery Of ^{231}Pa	Comments
0.3	106.6	Iron not complexed
1.0	91.9	Iron complexed
2.0	99.2	Iron complexed
3.0	98.5	Iron complexed
4.0	98.5	Iron complexed
5.0	74.3	Iron complexed
5.5	105.9	Iron complexed
6.0	98.5	Iron complexed

Third, the second experiment was repeated with larger
amounts of iron. Standards containing 200 pCi ^{231}Pa were extracted into amyl acetate and back-extracted with HCl-HF;
varying amounts of iron (0.06-10 mg) were then added. The
iron was reduced with hydroxylamine and complexed; then ^{231}Pa
was determined. The results in Table III show that even when
5 mg of iron were back-extracted from the amyl acetate, provided that sufficient quantity of ortho-phenanthroline was
added, the recovery of ^{231}Pa was 93 percent.

Extraction of Protactinium with Cupferron and Amyl Acetate

Cupferron is an excellent, though not very selective,
chelating reagent for a large number of cations, including
several radionuclides. Cupferron also chelates protactinium,
but in order to effect a separation of ^{231}Pa from other
nuclides such as uranium, thorium, and thorium daughters, a
preliminary separation is required. To demonstrate the
effectiveness of the preliminary separation, two aliquots of

Table III. Cupferron Extraction of 200 pCi
^{231}Pa in Presence of Iron at pH 3.5

Fe (mg)	ortho-phenanthroline (mg)	^{231}Pa % Recovery
0.06	20	95–100 (10)
0.5	40	94.5
2.5	200	91.0
5.0	300	93.5
10.0	400	88.5

( ) Number of Determinations

---

a ^{231}Pa standard were treated separately as follows. One was adjusted to a pH of 3.5 and extracted with cupferron into amyl acetate. The second was extracted into amyl acetate-8 $\underline{N}$ HCl, back-extracted with 8 $\underline{N}$ HCl-0.1 $\underline{N}$ HF, and, after adjustment of pH to 3.5, extracted with cupferron into amyl acetate. Both extracts were evaporated on stainless steel planchets and alpha-counted by proportional counting. Alpha spectrometry was then used to determine which radionuclides were recovered by the two procedures. The results, shown in Figure 1, demonstrate the effectiveness of the dual separation procedure. The initial extraction of ^{231}Pa into amyl acetate from 8 $\underline{N}$ HCl gives complete separation from the protactinium daughters.

Other experiments supported these findings. To demonstrate the extraction of other radionuclides with cupferron, a 300-pCi thorium 228-230 standard was extracted directly with cupferron at pH 3.5; the resulting complete recovery of the thorium isotopes indicated the non-selectivity of the complexing agent. However, another experiment demonstrated the selectivity for ^{231}Pa of the preliminary extraction step. An attempt was made to extract 5 g ^{232}Th(NO$_3$)$_4$ and its daughters into amyl acetate from 8 $\underline{N}$ HCl. Subsequent gamma counting of the extract gave no evidence that thorium or any other radionuclide had been extracted.

Analysis of Samples

Several samples analyzed by the previously described procedure are presented in Table IV to show the concentration levels of protactinium that may be encountered during recovery of uranium. Protactinium is not extracted with uranium during uranium recovery operations and it remains with the

84

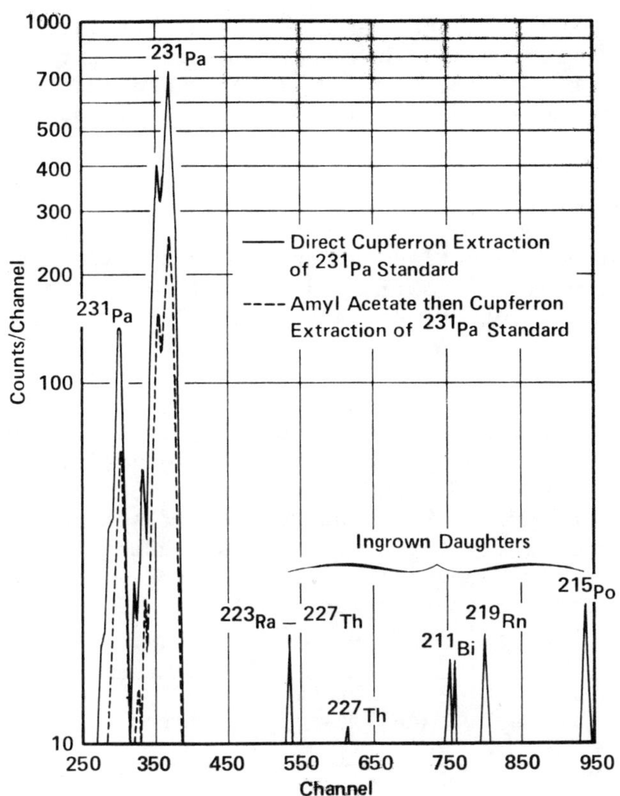

Figure 1. Alpha Spectrometry of ^{231}Pa and Daughters.

Table IV. Analysis of Samples for Protactinium-231

Sample	Type	Total Alpha, pCi	^{231}Pa Found, pCi
1	Filter Ash	2.5×10^6/g	3.5×10^4/g
2	Cylinder Wash Solution	2.5×10^4/ml	0.23×10^2/ml
3	Cylinder Wash Solution	2.5×10^4/ml	0.21×10^2/ml
4	Cylinder Wash Solution	1.2×10^4/ml	0.21×10^2/ml
5	Cylinder Wash Solution	3.5×10^5/ml	4.38×10^2/ml
6	Air Filter	1.5×10^4	0.24×10^2

raffinates that are discharged into the holding ponds. Oxide
fluorination filter ashes show higher concentration levels of
protactinium; therefore, care should be taken during their
handling or transfer so that protactinium does not become airborne.

Detection Limits

The limit of detection for environmental water samples
is 1×10^{-3} d/m/ml; for soils and oxide fluorination filter
ash, 1 d/m/g; and for UF_6 cylinder was solutions, 0.04 d/m/ml.

Precision Testing

To determine the reliability of the previously described
procedure for determining low levels of protactinium, four
standards containing 4-10 pCi ^{231}Pa were run through the complete analytical procedure. The results presented in Table V
show complete recovery for as little as 4 pCi ^{231}Pa. The
limit of error for determining 4-10 pCi ^{231}Pa is about 15
percent at the 95 percent confidence level. Somewhat better
precision can be expected at higher levels of ^{231}Pa.

Table V. Analysis of Standards

Sample	pCi ^{231}Pa Added	pCi ^{231}Pa Found	% Recovery
1	4.00	4.25	106
2	6.00	5.89	98
3	8.00	8.83	110
4	10.00	8.69	87

CONCLUSIONS

The simple procedure developed for determining ^{231}Pa by
separating it from iron and other radionuclides, extracting
its cupferron complex into amyl acetate and then counting
alpha activity is both selective and quantitative. Its applications include analysis of ^{231}Pa in aqueous wastes, soils,
and filter ash.

CARBON-14 IN REACTOR PLANT WATER

G. K. Knowles, Exxon Nuclear Idaho Company, Inc.,
Idaho Falls, Idaho 83401

ABSTRACT

The method for the analysis of ^{14}C in reactor plant
water and various waste streams previously used at the Idaho
National Engineering Laboratory has been shown to be ineffective for samples which contain organic compounds. The previous method consisted of acidification and refluxing of the
sample, precipitation of the liberated CO_2, and subsequent
analysis by the liquid scintillation method. The method was
simple but it did not convert all compounds containing ^{14}C in
the sample to CO_2. The new method, while it is based on the
previous method, has been improved by employing a strong oxidant, potassium persulfate and silver nitrate, for more complete oxidation of the organics to CO_2. The new method yields
^{14}C values that have typically been one to two orders of magnitude higher than the values obtained using the former method.
This indicates that most of the ^{14}C present in the current
reactor water samples being analyzed is associated with trace
amounts of organics.

INTRODUCTION

Carbon-14 is produced in the reactor coolant system of an
operating nuclear power plant by an (n,p) reaction with ^{14}N.
Accurate measurement of the ^{14}C content of the reactor coolant
and associated waters is important since the discharge of
these waters have a potential impact on the environment. Activity levels of ^{14}C in reactor waters vary from power plant
to power plant and within an individual plant depending upon
the type of water; i.e., reactor coolant, floor drain waste,
demineralizer waste, etc. Since the activity levels of the
samples vary widely, the measurement range must be large and
include a low detection limit. Liquid scintillation counting

of a ^{14}C sample offers an effective counting method with a wide measurement range. The detection limit can be increased by the use of larger sample volumes.

The method previously used at the Idaho National Engineering Laboratory (INEL) utilized the apparatus shown in Fig. 1. In the procedure, a large sample volume of 500 mL was used. The sample and Na_2CO_3 were placed in a reflux flask, acidified with nitric acid, and allowed to reflux overnight. The Na_2CO_3 spike acted as a carrier for any liberated CO_2. A vacuum was used to pull air through the system which carried the CO_2 into the $Ca(OH)_2$ bubbler where it precipitates as $CaCO_3$. A dioxane-base scintillation cocktail was added to the $CaCO_3$ and a geling agent, Cab-O-Sil,^{R} was used to suspend the $CaCO_3$ throughout the cocktail. The sample was counted by the liquid scintillation technique.

Four major problems were noted in the original procedure. Each problem and the corrective action taken is discussed. Afterwards, the new procedure is presented, along with comarative results and conclusions.

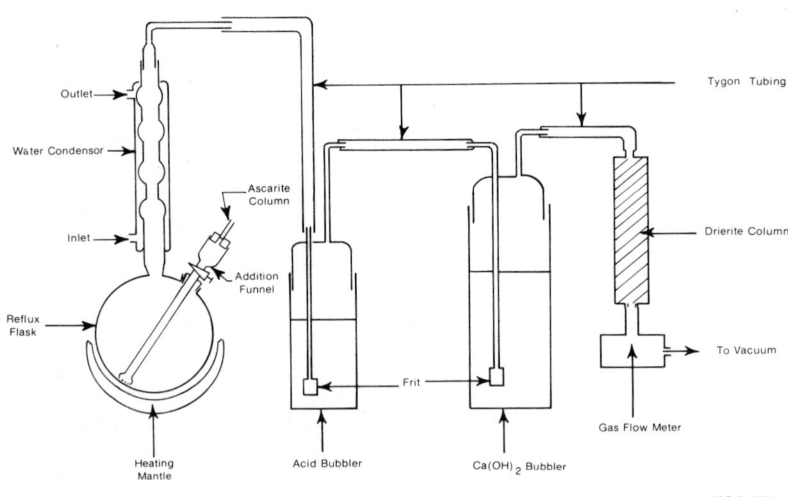

Figure 1. Apparatus Set-up

DISCUSSION OF PROBLEMS

1.   In the original procedure, a yield determination was not
made. To address this problem a known amount of Na_2CO_3
(106 mg) was added to the reflux flask, converted to CO_2,
and precipitated as $CaCO_3$. For the amount of Na_2CO_3 added,
the theoretical weight of $CaCO_3$ is 100 mG for 100% recovery;
however, in actual experiments 110 mG is obtained for a 90%
recovery of a ^{14}C tracer.

     Tests were performed to determine the source of the extra
weight. As a saturated solution of $Ca(OH)_2$ is used for the
precipitation of the CO_2, a possible reason for the extra
weight was the entrapment of $Ca(OH)_2$ with the $CaCO_3$ precipitate. Washing of the $CaCO_3$ precipitate, to remove any en-
trapped $Ca(OH)_2$, resulted in a loss of sample. Based on this
test the extra weight may or may not be $Ca(OH)_2$.

     Another test was to introduce known amounts of Na_2CO_3
and a ^{14}C spike directly into a saturated $Ca(OH)_2$ solution,
which had been filtered just prior to use. The dried $CaCO_3$
precipitate was weighed and compared to the theoretical weight.
For a 100% recovery of the ^{14}C spike, the theoretical and
actual recoveries were:

Theoretical (mG)	50	100	150	200
Actual (mG)	58	116	172	228

As shown, as more $CaCO_3$ is precipitated the extra weight also
increases indicating that some contaminant is being trapped in
the precipitate.

     The only crystal structure in the precipitate seen by
X-ray diffraction was $CaCO_3$. Semi-quantitative tests using
X-ray diffraction showed that not all of the weight was due to
$CaCO_3$ but the identity of the extra material could not be
determined.

     From the tests performed to date, the reason for the extra
material has not yet been determined. However, the percent
recovery for spikes, blanks, and many samples is consistent
and therefore, usable as long as it is consistant, regardless
of why it is higher than theoretically calculated. Additional
work is being done in this area.

2.   The refluxing period formerly used was excessively long,
typically sixteen-eighteen hours. The refluxing time was
shortened to two hours by varying the time, temperature, and
air flow to give a consistent recovery of a ^{14}C tracer.

3.   The dioxane-base scintillation cocktail is hazardous because it is a peroxide former. The cocktail is made by mixing
a primary and secondary scintillator, so each new batch has to
be tested for counting efficiency. Plastic counting vials are
used which allows evaporation of the cocktail. Breathing of
Cab-O-Sil[R] is hazardous to the lungs. The light, fluffy
nature of Cab-O-Sil[R] makes it difficult to mix a consistent
amount into the cocktail. It also separates from the cocktail
with time, leaving a liquid portion and a gel portion. As the
cocktail evaporates or the Cab-O-Sil[R] settles the counting
efficiency changes. These factors make it almost impossible
to make up reproducible samples or to recount an old sample.

These problems were solved with the use of Insta-Gel,[R] a
commercial ready-to-use liquid scintillation cocktail, and by
replacing the plastic vials with glass vials. Insta-Gel[R] has
as high a counting efficiency as the dioxane-base cocktail and
is consistent from lot to lot. The addition of water to
Insta-Gel[R] forms a gel which suspends the $CaCO_3$, thus eliminating the need for Cab-O-Sil.[R] However, changing the
Insta-Gel[R] to water ratio creates a quenching effect which
changes the counting efficiency. Since the samples are made
up the same way each time, the quench range used is small and
the change in counting efficiency is less than 1%. The weight
of solids added to the cocktail was checked as a quench factor.
Over the range of 50-300 mG of $CaCO_3$ the weight of the solids
does not effect the efficiency. For more information on
solgel type scintillation cocktails, the article by Benson[1]
is a good reference.

4.   In the original procedure the recovery of all of the [14]C
was unlikely with the chemicals used. Trace amounts of organics present would not be converted to CO_2 and therefore,
not collected as $CaCO_3$. A wet chemical oxidation procedure
was needed to convert the organics to CO_2. Although many
different reagents have been used for wet chemical oxidation,
Weinhouse[2] described an effective method for radiocarbon
studies of organic compounds using a persulfate oxidant. The
method with some modification has given efficiencies approaching 100% for milligram quantities of a wide variety of organics.[3,4,5] The potassium salt of the oxidant is inexpensive
and easy to use. The procedure described in this paper is a
modification of that developed by Weinhouse.[2] The method does
not insure that all of ^{14}C will be collected and counted but
the result should be very close to the actual value.

IMPROVED PROCEDURE

To the reflux flask add a known amount of sample (refer
to Fig. 1); to the acid bubbler add 60 mL of 0.6N HCl; and to

the $Ca(OH)_2$ bubbler add 25 mL methanol and 150 mL of a saturated solution of $Ca(OH)_2$, which has been filtered just prior
to use. The methanol aids the $CaCO_3$ precipitation and the
acid bubbler purifies the CO_2 gas stream. Assemble the apparatus as shown in Fig. 1 but without the addition funnel.
Add 1.0 mL of 1.0M Na_2CO_3, 5 G of $K_2S_2O_8$, 5 mL of 4% $AgNO_3$,
and a boiling chip into the reflux flask. Insert the addition funnel with the ascarite column. Start the condensor
water flow and an air flow of 65-75 cm^3/min. If the flow is
too rapid, CO_2 from the Na_2CO_3 carrier will evolve before the
persulfate can oxidize the organics in the sample.

Heat for two hours at 100-110°C, then turn off the heating mantles. After cooling ten minutes, remove the ascarite
column and disconnect the $Ca(OH)_2$ bubbler by first disconnect-
ing it from the acid bubbler, then from the drierite column.
This precaution is taken to prevent any $Ca(OH)_2$ from being
drawn back into the reflux flask. Shut off the air flow and
and the condensor water flow. Discard the sample in the reflux flask and add alkaline $KMnO_4$ to the flask. Allowing
$KMnO_4$ to set in the flask overnight and then rinsing with
oxalic acid decontaminates the flask.

Place the $Ca(OH)_2$ bubbler in an ultrasonic bath to break
up the $CaCO_3$. Centrifuge the $Ca(OH)_2$ into a conical centrifuge tube and retain the solid $CaCO_3$. Use methanol to
transfer the last of the $CaCO_3$ from the bubbler. Transfer
the $CaCO_3$ into a tare glass scintillation counting vial, centrifuge, and decant the liquid. Dry the $CaCO_3$ in an oven set
at 110°C for fifteen minutes. Allow the vial to cool to room
temperature and weigh. This weight is used to determine the
yield. Add 5.0 mL of water very low in activity, especially
^{3}H and ^{14}C, to the vial. Place the vial in an ultrasonic bath
to break up the $CaCO_3$ cake. Add 15 mL of Insta-Gel^R and
shake. A cloudy gel is formed which suspends the particles
of $CaCO_3$.

The vial is allowed to adapt to an equilibrium condition
in a liquid scintillation counter set at 12°C. This takes a
short time for the ^{14}C channel but the ^{3}H channel, which is
used as a check, may take up to eight hours to stabilize. For
a Packard 3385 instrument the following settings are used:

Channel	Window	Amplification	Nuclide
Red	30-450	50%	^{3}H
Green	50-550	7.5%	^{14}C
Blue	P-32(C-14)	Preset at factory	^{32}P

Spillover of the ^{14}C into the ^{3}H channel is about 22%. The counting efficiency is determined by processing a ^{14}C spike through the procedure. In this laboratory a ^{14}C spike with a $CaCO_3$ weight of 110 mG counts with an efficiency of 66±2%. This weight and efficiency should be determined by each laboratory. A sample weighing less than 110 mG is corrected to 110 mG by the following ratio: (dps) (110mG)/ weight of the sample. A sample weighing more than 110 mG due to organics converted to CO_2 is calculated as if it weighed 110 mG because no effective method for a yield determination has yet been developed for these samples. A minimum uncertainty of 10% is applied to a sample greater than 1.2 dps above background. The uncertainty increases to 20% at the detection limit of .08 dps. The major portion of the uncertainties is due to the inaccuracy in measuring the yield. Precision of the method is within the uncertainty as determined by replicate analyses.

DISCUSSION OF RESULTS

The new procedure was used for samples from three reactor power plants with different operating conditions. The samples were either reactor coolant water or water from various waste streams. The samples were also analyzed by refluxing an acidified sample per the previous method used at the INEL. The analysis differs from the previous method in that a yield was determined and Insta-Gel^R was used in place of the dioxane-based cocktail. Table I shows the results obtained from the analysis as well as a ratio of the two methods. In the Table, "new" indicates the persulfate method and "old" denotes the nitric acid method. The smallest difference between the persulfate and the nitric acid method is a factor of four and most of the samples are at least one order of magnitude higher. This shows conclusively that the acid method was inadequate and that most of the ^{14}C is in the trace organics of a sample. The less-than values used in the Table represent the minimum detection limit in dps; since this value is divided by the sample volume, the larger the sample used the lower the detection limit. Blanks for the two methods were identical.

CONCLUSION

This method can readily be used at nuclear power plants. It would involve buying some apparatus. The major item, the liquid scintillation counter, is already in use at plants to count ^{3}H samples. The major problems are that the persulfate oxidation efficiency is not known but should approach 100% and the inaccuracy of the yield determination. In spite of these problems, the procedure is simple, fast, and will

Table 1. Comparison of Methods

Plant 1

Sample No.	New	Old	Ratio New/Old
1	2.8E-2	< 2.6E-4	> 107
2	2.4	2.8E-3	857
3	1.OE-3	< 2.6E-4	> 4
4	1.7	3.1E-1	5
5	3.1E-1	2.3E-2	14
6	8.5E-3	< 2.6E-4	> 33
7	6.4E-1	2.2E-2	29
8	2.0E-2	< 1.0E-3	> 20
9	1.6E-1	1.5E-3	107
10	2.2E-2	< 7.7E-4	> 29
11	9.7E-3	< 5.1E-4	> 19
12	8.2E-3	< 5.1E-4	> 16

Plant 2

Sample No.	New	Old	Ratio New/Old
1	1.5E-1	< 5.1E-4	> 294
2	1.1E-1	5.1E-4	216
3	4.9E-2	< 5.1E-4	> 96
4	1.7E-1	1.5E-3	113
5	6.0	2.5E-2	240
6	3.9E-1	7.7E-3	51
7	4.6E-1	2.1E-2	22
8	2.3E-1	2.3E-2	10
9	4.4E-1	1.1E-1	4
10	4.4E-1	7.7E-2	6
11	5.1E-1	1.3E-2	39
12	5.9E-1	3.3E-3	179

Plant 3

Sample No.	New	Old	Ratio New/Old
1	2.4E-2	< 1.7E-4	> 141
2	2.1E-2	5.5E-4	38
3	2.6E-2	2.8E-4	93
4	2.1E-2	< 1.7E-4	> 123
5	8.2E-2	< 1.7E-4	> 482
6	3.0E-3	< 1.7E-4	> 18
7	9.6E-3	< 1.7E-4	> 56
8	1.1E-2	< 1.7E-4	> 65
9	1.5E-1	7.7E-4	195

Sample Results are in d/s/mL

ICPP-A-4332

give results more closely approaching the actual ^{14}C content
and certainly give a more accurate result than the previous
method.

REFERENCES

1. R. H. Benson, Int. J. App. Radiat. Isotopes, _27_, 667
    (1976)

2. S. Weinhouse, _Isotopic Carbon_, M. Calvin, Editor,
    John Wiley and Sons, New York, _94_ (1949)

3. J. M. Baldwin and R. E. McAtee, Michrochem J., _19_, 179
    (1974)

4. J. Katz, S. Abraham, and N. Baker, Anal. Chem., _26_,
    1503 (1954)

5. P. D. Goulden and P. Brooksbank, Anal. Chem., _47_, 1943
    (1975)

RADIOANALYSIS OF TECHNETIUM-99 AT THE OAK RIDGE
GASEOUS DIFFUSION PLANT

T. L. Rucker and W. T. Mullins. Analytical Services
Department, Technical Services Division, Oak Ridge
Gaseous Diffusion Plant, Oak Ridge, Tennessee, USA.

ABSTRACT

Technetium-99 (Tc-99) is determined in various media by
measuring its beta activity using liquid scintillation spectrometry. The use of liquid scintillation spectrometry reduces the time and simplifies preparation of the samples for
counting. Some types of samples are analyzed by direct
counting procedures which involve no chemical separation.
For samples which require a separation of the technetium from
uranium and its daughters, a simple hydroxide precipitation
removes the interfering elements. This separation technique
is faster and more reliable than previously used solvent extraction methods and provides adequate specificity for use at
the Oak Ridge Gaseous Diffusion Plant (ORGDP).

INTRODUCTION

The ORGDP provides efficient enrichment of uranium for
use as a nuclear fuel. Tc-99, a fission product, was introduced into the plant in small amounts when uranium hexafluoride prepared from spent reactor fuel was fed into the enrichment cascade. The plant is monitored for Tc-99 to protect the health and safety of employees and to preserve the
environment. Tc-99 is measured in the various gaseous,
liquid, and solid materials shown in Tables 1, 2, and 3.

LIQUID SCINTILLATION SPECTROMETRY

Because of the generally low levels of technetium, most
of the analyses are radiochemical. Current procedures use
liquid scintillation spectrometry to measure the Tc-99
beta activity. The liquid scintillation spectrometers

Table 1

GASES IN WHICH Tc-99 IS MEASURED

Uranium Hexafluoride (UF_6)
Air (Work Area and Environment)
Vent Gas (from Purge Cascade Scrubber)

Table 2

LIQUIDS IN WHICH Tc-99 IS MEASURED

Water (Surface and Effluent)
Scrubber Solutions (from Purge Cascade)
Raffinate (Waste Acid from Decontamination)
Urine

Table 3

SOLIDS IN WHICH Tc-99 IS MEASURED

Uranium Oxide (U_3O_8, from Scrap Recovery)
Trapping Materials (Al_2O_3, NaF, MgF_2)
Soil and Sediment
Solid Scrubber Waste (from Purge Cascade)
Metals

---

presently being used are three-channel analyzers which allow
different energy regions to be monitored. This aids in determining whether interfering nuclides may be present in the
sample.

An advantage of liquid scintillation counting is the
simplicity of sample preparation. A 1- to 2-ml sample aliquot is mixed with 15 ml of liquid scintillator for counting.

The samples are typically counted for 20 min, after
which they are spiked with a standard Tc-99 solution and

counted for an additional minute. Counting the spiked sample
provides a measure of the counting efficiency (typically 80%)
and the amount of signal quenching by the sample. The
counter backgrounds range from 20 to 25 counts per minute.
Equations 1, 2, and 3 show that under these conditions the
minimum detectable activity (MDA) of Tc-99 is 4.2 dpm.

(1)   MDA = 3σ above the background activity

(2) $\sigma \approx \dfrac{\sqrt{\text{background count}}}{\text{time counted}} = \dfrac{\sqrt{25 \text{ cpm} \times 20 \text{ min}}}{20 \text{ min}} = 1.1 \text{ cpm}$

(3) $\text{MDA} = \dfrac{3 \times 1.1 \text{ cpm}}{.80 \text{ cpm/dpm}} = 4.2 \text{ dpm}$

SAMPLE PRETREATMENT

The samples must be converted to an aqueous form prior
to analysis. The sample types which require pretreatment are
shown in Table 4.

Table 4

PRETREATMENT OF SAMPLES

Sample Type	Pretreatment
Uranium Oxide (U_3O_8)	Dissolve in HNO_3
Air (sampled through paper filters)	Leach with HNO_3
Vent Gas (purge cascade scrubber)	Pass through 1% KOH impingers
Soil	Leach with HNO_3
Nickel Metal	Dissolve in HNO_3
Water	Evaporate for concentration
Uranium Hexafluoride (UF_6)	Hydrolyze to UO_2F_2

The sample types which originate in aqueous form and require no pretreatment are shown in Table 5.

ANALYSIS INVOLVING HYDROXIDE PRECIPITATION

Uranium and its daughters are the primary radionuclides
which give possible interference in the beta counting of

Table 5

SAMPLES WHICH REQUIRE NO PRETREATMENT

Raffinate (waste HNO_3 from decontamination facility)
Scrubber Solution (KOH solution from purge cascade)
Urine

Tc-99 at ORGDP. A separation of technetium from uranium and
its daughters is achieved by making a hydroxide precipitation. After the sample has been prepared in aqueous form, a
sample aliquot is acidified with nitric acid and potassium
persulfate is added to ensure dissolution of the technetium
and its oxidation to Tc^{7+}. If fluorides are suspected to be
present, lanthanum nitrate is added as a complexing agent.
Ferric nitrate is added and the uranium and its daughters are
coprecipitated with the iron by making the solution basic
with sodium hydroxide. The solution is centrifuged and a
portion of the supernate is measured by liquid scintillation
spectrometry.

It is recognized that some beta emitting radionuclides
will not be separated from technetium by this method. These
include ^{24}Na, ^{90}Sr, ^{137}Cs, ^{36}Cl, ^{131}I, ^{35}S, and ^{32}P. However, these radionuclides are not present in sufficient quan-
tities at ORGDP to cause an interference problem. This separation technique is faster and more reliable than solvent extraction methods used in the past and provides adequate
specificity to justify its use at ORGDP.

ANALYSES INVOLVING NO CHEMICAL SEPARATION

Some types of samples which have a low uranium content
can be analyzed by direct counting procedures which involve
no chemical separation. These include urine and some water
and air samples. A 1- to 2-ml portion of the sample or leach
solution is pipetted directly into the scintillation vial,
mixed with the liquid scintillator and the gross beta count
is measured on the liquid scintillation spectrometer.

When considerable uranium is present in air filter
samples, the gross count can be corrected for the uranium
contribution by measuring a second portion of the leach solution by alpha spectrometry. The uranium alpha activity in
the air filter is multiplied by a beta/alpha factor (2.5),
previously determined by measuring the beta activity of
uranium by liquid scintillation spectrometry. The product is

subtracted from the beta activity in the air filter, and the
net activity is calculated as Tc-99, as shown in equation 4.

(4) ^{99}Tc dpm = gross beta dpm $-$ 2.5 $\times$ U alpha dpm

DETECTION LIMITS

The detection limits for the various types of samples
are shown in Table 6.

Table 6

DETECTION LIMITS FOR VARIOUS SAMPLES

Type Sample	Detection Limit
UF_6 and U_3O_8	165 dpm/g U
Raffinate and Scrubber Solutions	22 dpm/ml
Water and Urine	6 dpm/ml
Air	330 dpm/filter
Vent Gas	4 dpm/ft^3
Trapping Material	22 dpm/g
Soil	33 dpm/g
Metals	132 dpm/g

The detection limit is determined by multiplying the minimum
detectable activity of 4.2 dpm by the various aliquot and
dilution factors.

PRECISION AND ACCURACY

Quality control measures have been established for many
types of samples by routine analysis of control samples. The
control data are shown in Table 7.

CONCLUSION

The methods which have been described are specifically
designed for the types of samples which are encountered at
ORGDP. They provide a method of obtaining reliable results
while keeping the time and cost of analyses at a minimum.
They provide adequate specificity for use at ORGDP but may
not be universally applicable.

Table 7

TECHNETIUM-99 CONTROL SAMPLES

Sample	Activity, dpm Added	Found	Number of Determinations	Recovery, %	Rel. Std. Dev., %
Vent Gas	4940	5244	4	106	4
U_3O_8	1477	1637	30	111	10
Soil	4569	4617	4	101	4
Water	37	38.7	49	105	15
Urine	13	13	40	100	24
Air	13775	14132	16	103	9
Nickel	55100	54984	5	100	3

THE DETERMINATION OF TECHNETIUM-99 BY LIQUID
SCINTILLATION COUNTING

C. R. Walker, B. W. Short, H. S. Spring, Technical
Division, Goodyear Atomic Corp., Piketon, Ohio, USA.

ABSTRACT

A rapid and reliable method for analyzing technetium-99
in a wide variety of environmental samples, including waters,
soils, stream sediments, and vegetations, has been developed.
The procedure entails oxidizing the technetium to the heptavalent state and dissolving it in 6 $\underline{N}$ sulfuric acid. From
that medium the technetium is quantitatively and selectively
extracted into tributyl phosphate. A portion of the extract
is then added to a scintillation cocktail, and technetium-99
activity is measured by liquid scintillation counting. Since
a relatively large sample can be handled, the method can detect as little as 0.016 pCi ^{99}Tc/ml water, 0.1 pCi ^{99}Tc/g
soil or sediment, and 0.2 pCi ^{99}Tc/g vegetation. The procedure has also been adapted to analyzing urine samples, in
which technetium activity as low as 0.12 pCi/ml can be detected.

INTRODUCTION

To help protect the environment and to assure the health
and safety of employees, many samples must be analyzed for
technetium-99 at the Portsmouth Gaseous Diffusion Plant.
Technetium-99, a weak beta emitter (0.292 Mev) with a long
half-life (2.1 x 10^5 years), is a major product of nuclear
fission. For uranium-235 fission, the yield of technetium-99
is about six percent. Technetium produced in this manner is
not completely separated from irradiated uranium that is
reprocessed to recover usable uranium-235, and since it forms
volatile fluorides that accompany uranium as UF_6 in the gaseous diffusion process, technetium-99 contamination of gaseous
diffusion plant equipment occurs. Thus, during the handling
and decontamination of process equipment and the recovery of
deposited uranium, its presence must be monitored for both

environmental and industrial hygiene considerations. Samples
analyzed to verify conformance to state and federal regulations for releases of and exposure to technetium-99 include
aqueous effluents, stream sediments, soils, airs, and urines.

The method previously used at the Portsmouth Gaseous
Diffusion Plant for analysis of technetium-99 involved the
solvent extraction of technetium as its crystal violet complex and subsequent beta counting of an evaporated portion of
the extract with a proportional counter. This procedure
proved adequate for a large variety of samples, but interferences from nitrate limited sample sizes and resulted, in some
instances, in unsatisfactory limits of detection.

When a liquid scintillation counter became available,
other solvent extraction systems were considered. For this
purpose, a solvent was required that would quantitatively extract the technetium from other radionuclides, produce a
colorless extract, and be chemically compatible with scintillation counting. Based on the extensive work of Boyd and
Larson[1] on the solvent extraction of technetium, tributyl
phosphate (TBP) was selected as the solvent. Preliminary
work verified that TBP met the above criteria and that it
gave an extraction efficiency of 98-99 percent. The effects
of other radionuclides and chemical interferences and the
elimination of emulsion problems and chemiluminescence were
also studied. For these studies a reliable procedure for
analyzing technetium-99 by liquid scintillation counting was
developed.

PROCEDURES

Sample Preparation

Soils and Sediments

Weigh a 40-g portion of sample into a 300 ml porcelain
dish and ignite at 500°C for 30 minutes to destroy organic
matter. Add 75 ml of 6 $\underline{N}$ H_2SO_4 and 10 g potassium persulfate
($K_2S_2O_8$). Boil for 20 minutes to solubilize the technetium.
Filter the cooled leachate through two Whatman #41 filter
papers into a 250 ml separatory funnel. Wash the residue
with sufficient 6 $\underline{N}$ H_2SO_4 to give a final volume of 225-250
ml.

Vegetation

Weigh a 15-g portion of dried and ground vegetation into
a 300-ml porcelain dish and wet with 75 ml concentrated NH_4OH.
Evaporate the sample to dryness and then bake it at 550°C on

a hot plate until charred. Place the sample in a muffle furnace at 515°C and ignite for one hour. Add 75 ml of 6 $\underline{N}$
H_2SO_4 and 10 g $K_2S_2O_8$. Boil for 20 minutes to solubilize the
technetium. Treat the leachate as described for soils samples.

Waters

Place 200 ml of homogenized water into a 250-ml separatory funnel. Add 40 ml concentrated H_2SO_4 and sufficient 1
percent $KMnO_4$ solution to maintain a pink color.

Urines

Pipette a 25-ml aliquot of urine into a 250-ml beaker.
Add 50 ml 9 $\underline{N}$ H_2SO_4 and 10 g of $K_2S_2O_8$. Heat to boiling for
20 minutes and transfer to a 100-ml separatory funnel.

Extraction of Technetium

Allow the solution produced in any of the preparation
procedures given above to cool to room temperature in the
separatory funnel. (For soil or vegetation samples only, add
2 ml concentrated HF to complex silica.) Pipette 5.0 ml concentrated TBP that has been previously equilibrated with 9 $\underline{N}$
H_2SO_4 into the separatory funnel. Stopper and shake for two
minutes to extract technetium. Discard the aqueous phase and
wash the extract for one minute first with 25 ml 6 $\underline{N}$ H_2SO_4-2%
HF and then with 25 ml 4 $\underline{N}$ HCl. Allow the phases to separate
for ten minutes so that sufficient extract will be available
for counting. Discard the aqueous phase and transfer the extracts to a 15-ml centrifuge tube. Centrifuge for 2-3 minutes to obtain a clear extract.

Scintillation Counting

With a Finnpipette, pipette 4 ml of extract into a scintillation vial, and add 12 ml of Insta-Gel scintillation
cocktail. Shake vigorously to mix and wipe off exterior of
vial. Place in counter and begin counting sample after one
hour.

RESULTS AND DISCUSSION

Tributyl phosphate is an excellent solvent for extracting several radionuclides, including technetium, uranium,
thorium, and some transuranics. Boyd and Larson[1] showed that
technetium could also be extracted into TBP from 1 $\underline{N}$ sulfuric
acid solution. Since strong anionic complexes that are not
extracted into TBP are formed between sulfate and any of the

above-mentioned radionuclides except technetium, selective extraction of technetium into TBP from sulfuric acid is possible. Consequently, the sulfuric acid solvent medium was chosen for further investigation.

Effects of Sulfuric Acid Concentration

Preliminary extraction of 2×10^4 pCi of technetium-99 into 5 ml TBP from 1 $\underline{N}$ sulfuric acid solution gave excellent extraction efficiencies, but because the phase separation was poor, the decontamination from other nuclides entrained in the extract was not complete. When extractions from higher concentrations of sulfuric acid were tried, the phase separations improved. To evaluate the technetium extraction efficiency, raffinates from the extractions, along with solutions from the wash steps, were analyzed for technetium by a crystal violet extraction-beta counting procedure. Results of the evaluation are found in Table 1.

Table I. Effects of Sulfuric Acid Concentration
on the Extraction of 2×10^4 pCi Tc-99

Normality H_2SO_4	Technetium Found (pCi)			% Tc Extracted
	Raffinates	H_2SO_4 Wash	4 N HCl Wash	
3	a	a	a	99
4.5	a	a	a	99
6	a	a	66	99
7	128	30	80	98.8
8	176	26	68	98.7
9	258	46	64	98.2

aThe raffinate and wash solutions were too turbid to perform technetium analysis.

A clear extract was obtained from acid concentrations of 6 – 9 $\underline{N}$ H_2SO_4; 6 $\underline{N}$ was selected for further studies since that concentration gave an extraction efficiency of about 99 percent.

Counting Efficiency

Aliquots of a standard solution containing 2.22×10^3 pCi of technetium-99 were extracted into 10 ml TBP from 6 – 9 $\underline{N}$ H_2SO_4 solutions. The extracts were washed with the same concentration of H_2SO_4 and then with 4 $\underline{N}$ HCl. Aliquots of 1,

2, and 3 ml of the TBP extract were added to Insta-Gel scintillation cocktail and counted in a scintillation counter. Several aliquots of the above standard were also extracted into 5 ml TBP from 6 $\underline{N}$ H_2SO_4 and, for these, 4 ml of each extract were counted. The results listed in Table II show that varying sample volume from 2 to 4 ml extract produced no significant difference in counting efficiency. Taking 4 ml of a 5 ml TBP extract was chosen as the standard procedure since the counting of 80 percent of samples taken for analysis would improve the sensitivity of the method.

Table II. Effect of Amount of TBP Extract
on Counting Efficiency

Normality H_2SO_4	Counting Efficiency – Percent			
	1 ml Extract Counted	2 ml Extract Counted	3 ml Extract Counted	4 ml Extract Counted
6	94.2	89.7	89.3	90.0[a]
7	92.8	90.4	88.8	----
8	94.1	91.4	90.8	----
9	90.1	90.9	90.0	----

[a]This % efficiency represents an average of more than 20 determinations.

Elimination of Chemiluminescence

In many types of samples, technetium is found in a less soluble, reduced state; therefore, before it can be completely dissolved, it must first be oxidized to the more soluble heptavalent state. Potassium persulfate ($K_2S_2O_8$) has proven to be an excellent oxidizing agent for this purpose. When $K_2S_2O_8$ is used as oxidant, however, peroxides are formed. These undesirable components dissolve in TBP and are not readily removed by washing with 6 $\underline{N}$ H_2SO_4; and, if left in the extract to be counted, they cause severe chemiluminescence, which gives false counting events. Since complete decay of chemiluminescence requires several hours, this phenomenon is a problem when rapid determination of low levels of technetium (<1 pCi) is necessary. Therefore, procedures for its elimination were investigated. The addition of stannous chloride in 0.1 $\underline{N}$ HCl, as reported by Newman,[2] was investigated, and found to reduce the chemiluminescence considerably, but has the disadvantage of adding a step to the procedure. However, since technetium is also extracted into TBP from HCl solutions, 4 $\underline{N}$ HCl was tried as a wash solution for

the extract, and chemiluminescence was found to be reduced to
a point that even blank solutions could be counted within an
hour. The effects of the 4 $\underline{N}$ HCl wash on the chemiluminescence of TBP extracts from solutions containing 10 g $K_2S_2O_8$
with and without technetium are shown in Figures 1 and 2.
HCl reacts with peroxides to liberate chlorine, thereby reducing peroxide to free oxygen and eliminating chemiluminescence. Also, the apparent enhancement of the chemiluminescence in the presence of 2×10^3 pCi Tc suggests that the
radioactivity of technetium-99 is causing secondary reactions
between peroxides and the scintillator.

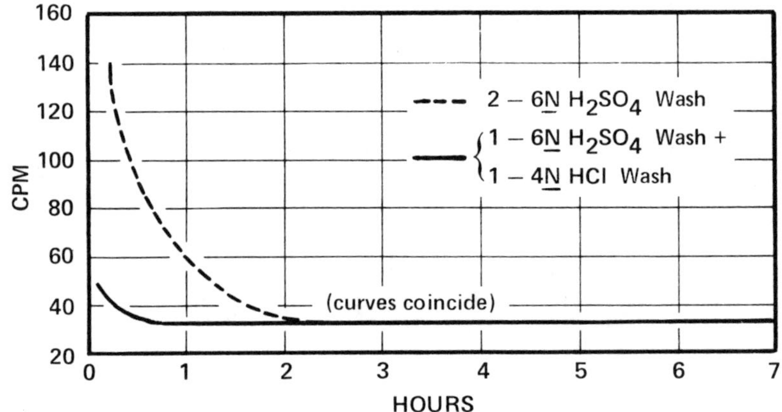

Figure 1. Effect of H_2SO_4 and HCl Washes of TBP
Extracts of Chemiluminescence – Background

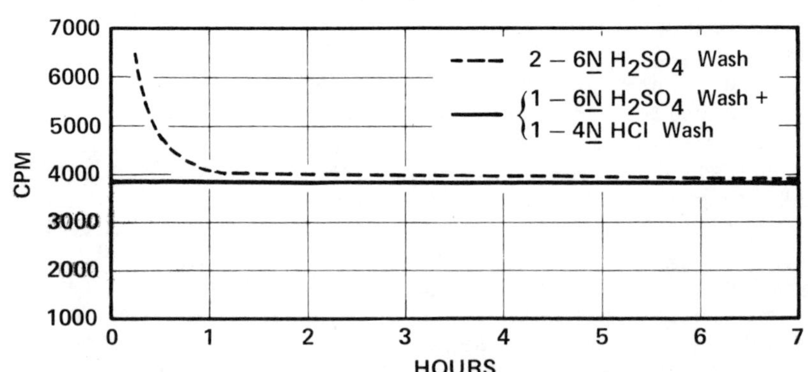

Figure 2. Effect of H_2SO_4 and HCl Washes of TBP
Extracts on Chemiluminescence – 2×10^3 pCi ^{99}Tc

Analysis of Soil Samples

A major consideration in the analysis of soils or sediments is the valence state of technetium. Heptavalent technetium is readily soluble in dilute acids, whereas the lower valence states produce insoluble compounds within the matrix of the soil or sediment. Another consideration is the presence of organic matter which, unless destroyed, will extract into TBP and cause both color and chemical quenching in liquid scintillation counting. Thus, three soils were analyzed by several approaches to evaluate the effects of ignition at 500°C and of oxidation with $K_2S_2O_8$ on the results. The results, as shown in Table III, definitely show that large quantities of soils should first be ignited to oxidize most of the organic matter and then leached with 6 $\underline{N}$ H_2SO_4-10 g $K_2S_2O_8$ to solubilize technetium prior to extraction. The H Numbers (H[#]) tabulated are an indication of scintillation counting efficiency. High H Numbers (i.e., more than 125) indicate color or chemical quenching in the scintillation medium.

Table III. Study of Sample Treatment Variables in Analysis of Soil for ^{99}Tc

Sample No.	Wt. Taken (grams)	Sample[a] Treatment	H[#]	pCi Tc/g
1	5	No oxidant in leach	155	37
	20	No oxidant in leach	246	10
	5	10 g $K_2S_2O_8$ added	124	105
	20	10 g $K_2S_2O_8$ added	253	53
	2	10 g $K_2S_2O_8$ added	100	112
	20	Ignited + 10 g $K_2S_2O_8$	96	117
2	20	NH_4OH added, ignited, oxidized	92	99
	20	Ignited, oxidized	92	100
3	20	NH_4OH added, ignited, oxidized	88	156
	20	Ignited, oxidized	88	160

[a]All samples were leached with 6 N H_2SO_4

Analysis of Environmental Samples

Numerous samples of soil, sediment, vegetation and effluent water are analyzed routinely at the Portsmouth Gaseous Diffusion Plant for technetium. Previously, these analyses were accomplished by complexing the technetium with crystal violet, extracting it into chloroform, and then beta counting

an evaporated portion of the extract. Several samples that
had been previously analyzed by this method and found to contain low concentrations of technetium were chosen for analysis by TBP extraction-liquid scintillation counting. The
results obtained by both methods and tabulated in Table IV
show the acceptability of this new procedure for low-level
technetium determinations.

Table IV. Analysis of Environmental
Samples for Technetium-99

| Samp. No. | Type | Method of Analysis | |
		Crystal Violet	Liquid Scintillation
1	Soil	<2.2 pCi/g (2)	0.38-0.45 pCi/g (5)
2	Soil	<2.2 pCi/g (2)	0.45-0.52 pCi/g (5)
3	Soil	3.8-4.4 pCi/g (2)	4.2-4.9 pCi/g (5)
4	Vegetation	4.7-6.3 pCi/g (2)	5.4-6.2 pCi/g (4)
5	Water	0.31-0.36 pCi/ml (5)	0.31-0.32 pCi/ml (8)

( ) Number of determinations.

With the liquid scintillation method, technetium determination could be made on two of the soil samples that were
below the crystal violet detection limit of 2.2 pCi/g Tc.
The results obtained with the new procedures on the other
soil, vegetation, and water samples are in excellent agreement with those obtained by the crystal violet method. They
prove that TBP extraction of technetium and subsequent determination by liquid scintillation counting is at least as good
as the crystal violet technique and that it can accommodate
much larger sample sizes.

Detection Limits

Low detection limits on the order of 0.1 pCi Tc/g require large amounts of sample. In the previously used crystal violet procedure, sample sizes were limited to 5 g for
soil, 2 g for vegetation, and 50-75 ml for water. In the new
liquid scintillation procedure, sample sizes are increased by
a factor of eight for soil and vegetation and by a factor of
four for waters. The improved detection limits, along with
sample size for routine analysis, are found in Table V.
These limits of detection are not necessarily the lowest
obtainable by this procedure. For water, a sample size of
200 ml was chosen because the technetium can be extracted
easily from 250 ml solution in 250-ml separatory funnels.

Further increases can be made by evaporation of the sample.
Larger samples of soil and vegetation can be analyzed if an
evaporation step is incorporated after the technetium is
leached.

Table V.   Detection Limits

Type of Sample	pCi/gram	pCi/ml
Soil (40 g)	0.1	--
Vegetation (15 g)	0.2	--
Water (200 ml)	---	0.016
Urine (25 ml)	---	0.12

Urine Analysis

Our standard procedure for analyzing technetium in urine
starts by screening the samples by direct liquid scintillation counting. A 2-ml aliquot of urine is added to 12 ml
scintillation cocktail and the counts are measured in two
channels. The first channel is set to detect low energy
emissions from technetium-99 and any other weak alpha or beta
emitters and the second to determine if any high energy alpha
or beta is present. For this preliminary measurement, the
limit of detection is about 2 pCi ^{99}Tc/ml urine. If a positive result of more than 7 pCi/ml above background is found,
an additional aliquot of the sample is evaporated to dryness
on a stainless steel planchet with NH_4OH present and then
beta counted. If a positive result is again obtained, the
residue on the planchet is fumed with perchloric acid to
volatilize Tc and then recounted to confirm the presence of
technetium-99.

To achieve a lower detection limit, a method involving
the TBP extraction of technetium and subsequent liquid scintillation counting has been developed. The method involves
the preliminary oxidation of the organic matter in the urine
without loss of technetium. The urines are made 6 $\underline{N}$ in
H_2SO_4, and 10 g $K_2S_2O_8$ is added to each aliquot. The samples
are then boiled for 20 minutes, cooled, and extracted into 5
ml TBP. Four milliliters of extract are added to a scintil-
altor and counted. To evaluate the method, aliquots of five
urine samples were taken in duplicate and one aliquot from
each sample was spiked with 12.8 pCi Tc. The results of the
analyses of these aliquots are found in Table VI. The

recovery of technetium ranges from 93–115% at 0.5 pCi/ml.
Analysis of 50 ml urine is more difficult since two or more
oxidations with $K_2S_2O_8$ are required.

Table VI. Technetium-99 in Urine

Sample	Original cpm[a]	Spike cpm[a]	Spike pCi	pCi Found
1	31.2	53.0	12.8	14.8
2	34.2	55.2	12.8	12.4
3	34.5	57.9	12.8	13.7
4	35.1	59.6	12.8	14.0
5	33.1	52.3	12.8	11.9

[a]Count includes 33 cpm background.

CONCLUSIONS

The TBP extraction-scintillation counting method developed for the analysis of technetium-99 has proven to be successful for many types of environmental samples as well as
for urines. The limits of detection are, for many samples,
lower than those obtained by the crystal violet method by a
factor of eight with no loss in precision or accuracy.

REFERENCES

1. G. E. Boyd and Q. V. Larson, J. Phys. Chem., __64__, 988
    (1960).

2. F. Newman, Comments: International Symposium on Current
    Status of Liquid Scintillation Counting, Cambridge, MA,
    1969.

ANALYSIS OF TWO RADIOACTIVE NEPHELINE SYENITE GLASS
HEMISPHERES RETRIEVED FROM THE CHALK RIVER NUCLEAR
LABORATORIES' WASTE MANAGEMENT SITE

J. D. Chen. Analytical Science Branch, Atomic Energy
of Canada Limited, Whiteshell Nuclear Research Establishment, Pinawa, Manitoba, Canada.

ABSTRACT

Only slight changes occurred to two nepheline syenite
glass hemispheres containing high-level fission product
wastes which were buried for twenty years in fine sand below
the water table at the Chalk River Nuclear Laboratories.
After retrieval and shipment to the Whiteshell Nuclear
Research Establishment, the hemispheres were photographed,
weighed and sectioned into samples for analysis by a variety
of techniques. Surface changes due to weathering were
observed to a depth of one to two nanometres by X-ray photoelectron spectroscopy.

INTRODUCTION

In support of high-level radioactive waste disposal
studies, two glass hemispheres were retrieved from Chalk
River Nuclear Laboratories (CRNL) Perch Lake waste management
site for examination at Whiteshell Nuclear Research Establishment (WNRE). The purpose of their retrieval was to permit their examination for the degree of "weathering" and for
any loss of integrity or release of radioactivity.

The glass was based on nepheline syenite, a naturally-
occurring aluminosilicate mineral composed of 60% by weight
SiO_2, 24% Al_2O_3, 10% Na_2O, 5% K_2O and 1% calcium, iron,
magnesium, and titanium oxides. The 12.7-cm diameter hemispheres were prepared by combining a mixture of 85% nepheline syenite and 15% lime with fission-product solution in a
ceramic crucible. The mixture was heated to 100°C to dryness, denitrated at 900°C and melted at 1350°C to form a
glass. Fission product loading in the first hemisphere was

111

7.7 Ci   and in the second hemisphere was 27.7 Ci.[1,2,3]   The
first hemisphere was buried in August, 1958, and the second,
more highly-active, hemisphere in May, 1960.   The first
hemisphere was received at WNRE on October 18, 1978, and the
second hemisphere on November 29, 1978.

The glass hemispheres were unloaded from the shipping
containers into a shielded cell, which had been thoroughly
decontaminated and lined with plywood, cardboard and polyethylene sheet to ensure against radioactive contamination
of the glass hemispheres by residues from earlier work.   To
determine if sampling could be done without contamination,
an inactive replicate glass hemisphere was sampled in the
cell by chipping with a hammer and chisel.   The resulting
pieces were analyzed for α, β and γ activity.   No activity
could be detected above background, indicating that no
radioactive contamination occurred in the sampling process.

In the shielded cell the specimens were examined visu-
ally and photographed, basic physical measurements were
taken and samples chipped from the glass hemispheres for a
variety of analytical measurements.   The hammer-and-chisel
chipping technique was chosen over a cutting method because
it produced suitable specimens which were free from foreign
materials or lubricants.

Visual Examination and Physical Property Measurements

The first glass hemisphere in its carrier is shown in
Figure 1.   Water found at the bottom of the container was

Figure 1.   Glass Hemisphere #1

introduced at CRNL to keep the glass surface moist during
shipment. Severe chipping on the top outside edge of the
hemisphere probably occurred during removal of the glass
from the original crucible.[2] No damage could be detected
due to handling or shipment from CRNL. The curved surface
had a rough appearance with some of the original crucible
material strongly adhering to the glass. A layer of sand
covered the top surface and was easily removed with a brush.
Figure 1 shows the smooth shiny top surface with surface
bubbles. The second glass hemisphere had a thick layer of
sand on the top surface and a thin coating on the rest of
the surface. Figure 2 shows that the second glass hemisphere
was considerably more chipped around the top edge than the
first, this chipping extending two thirds of the way to the
bottom. The sand was removed from the glass by light brush-
ing, revealing the characteristic crucible material on the

Figure 2. Glass Hemisphere #2

unchipped curved surface. The top surface, Figure 2, had a
dull pitted appearance and was not shiny like the first
hemisphere.

The masses and densities of the two spheres were, as
shown below, very similar and the radiation field of the
second hemisphere was three times that of the first.

	Mass (g)	Density (g/cm^3)	Radiation Field, γ (60 cm, Gy/h)*
Hemisphere #1	1990	2.593	0.02
Hemisphere #2	1904	2.652	0.06

*1 Gy/h = 100 rad/h

Radiochemical Measurements of the Bulk Glass

To determine the fission product loading of the bulk glass, several samples from various parts of each glass hemisphere were analyzed for γ- and α-emitting isotopes and for ^{90}Sr. The γ-emitting nuclides were determined by direct analysis of small specimens using a Ge(Li) γ-spectrometer. A number of glass chips were dissolved in a nitric/hydrofluoric acid solution to determine α-emitting actinide isotopes and ^{90}Sr. Aliquots of the acid solution were deposited on stainless steel counting trays and analyzed using a silicon surface barrier α-spectrometer and a zinc sulphide detector to determine the concentration of the α-emitting nuclides. For ^{90}Sr analysis, chemical separations were required followed by β-counting with a 2π end window gas proportional counter. The radiochemical results are summarized in Table I.

Table I

Radioisotope Assays – WNRE 1979

CRNL GLASS HEMISPHERE #1

SAMPLE	ISOTOPE ACTIVITY (Curies/hemisphere)					
	^{137}Cs	^{90}Sr	^{154}Eu	^{241}Am	^{239}Pu	^{244}Cm
1	4.3	1.6				
3	4.1					
8	4.3	1.5		1.2×10^{-3}	ND	5.2×10^{-5}
9	3.9	1.2				
25A	4.2	0.52		2.0×10^{-3}	ND	4.0×10^{-5}
28B	3.9	0.35	5×10^{-4}	1.9×10^{-3}	ND	1.4×10^{-4}
AVERAGE	4.1	1.03		1.7×10^{-3}		7.7×10^{-5}

CRNL GLASS HEMISPHERE #2

SAMPLE	^{137}Cs	^{90}Sr	^{154}Eu	^{241}Am	^{239}Pu	^{244}Cm
51A1	10.4	2.9		2.0×10^{-2}	6.1×10^{-4}	2.9×10^{-5}
53A1	11.3	3.1				
53E1	10.4	3.0	7×10^{-3}	9.3×10^{-3}	4.5×10^{-4}	1.3×10^{-4}
53F1	11.9	2.5		2.3×10^{-2}	2.0×10^{-3}	ND
AVERAGE	11.0	2.9		1.7×10	1.0×10^{-3}	

The major γ-emitting isotope in the glass hemispheres was ^{137}Cs. Its distribution in both glass hemispheres was quite uniform. The total activities in the first and second glass hemispheres were 4.1 and 11.0 Ci, respectively. Concentrations of ^{154}Eu obtained from a single sample from each hemisphere were about four orders of magnitude lower than ^{137}Cs. The nuclide ^{134}Cs was just detectable in the two specimens. The ^{90}Sr data showed more variation in distribution than ^{137}Cs. The total ^{90}Sr activities in the first and second hemispheres were 1.0 and 2.9 Ci, respectively. Alpha analysis of solutions of the glass specimens showed that approximately 95% of the α activity was due to ^{241}Am and the remaining 5% was due to ^{244}Cm in the first hemisphere and to ^{239}Pu and ^{244}Cm in the second hemisphere.

The original radiochemical measurements were carried
out in 1958 and 1959 at CRNL on samples obtained from fission
product waste tanks.[1,2,3] Approximately two litres of solution from the waste tanks were incorporated into each glass
hemisphere. The CRNL results were compared with WNRE data
obtained by direct analysis of specimens from the hemispheres. Good agreement was obtained only for ^{90}Sr in the
first hemisphere and for ^{137}Cs in the second hemisphere.
The difference in the two sets of data may be attributed to
evaporation losses incurred during glass fabrication and
variability in sampling from the waste tanks which contained
significant concentrations of solid material.

Alpha-Track Analysis at the Surface Interface

The α-track method was used in an attempt to determine
any gradient in α-emitting nuclide concentration between the
surface and the bulk of the first hemisphere which may have
arisen because of surface leaching. A plastic film was used
as a detector. The film, placed in contact with a sectioned
and polished glass sample, captured α particles emitted by
radionuclides from the surface of the glass. The α particles
caused tracks of damaged molecules which were then prefer-
entially dissolved by a chemical etchant, leaving well-defined holes. The relative concentration of α-emitting
nuclides was determined by counting the holes.

Two samples from the first hemisphere were analyzed
using the α-track method. Cellulose nitrate film, LR-115,
was placed on the specimens and exposed for about five
hours. The developed films were examined under a microscope
to observe the α-track density as a function of distance
across the surface of the samples. The results showed that,
within the resolution of the technique (100 μm), no
α-particle track gradient could be observed.

X-Ray Photoelectron Spectroscopy

X-ray photoelectron spectroscopy (XPS) is a surface
analysis technique which analyzes a depth of one to two
nanometres of the outer surface of the glass over an area of
one to three square millimetres.

Each glass sample was mounted in the spectrometer and
measured at two or three different locations on the surface
to compare the homogeneity of composition. This was followed
by argon-ion bombardment to remove approximately two nanometres of the surface followed by a second XPS analysis.
Four samples from the first hemisphere were examined but
only one sample from the second was examined because of its
high radioactivity.

Three exposed and one bulk sample were analyzed from
the first hemisphere and the elemental intensity ratios are
given in Table II with a 2σ precision of less than ± 10%.

Table II

XPS Elemental Intensity Ratios for First Glass Hemisphere

SAMPLE DESIGNATION	ORIGIN	INTENSITY RATIOS		
		Na/Si	Ca/Si	K/Si
18-SPOT 1	BULK	0.20	0.62	0.09
SPOT 2	"	0.36	0.59	0.05
SPOT 3	"	0.37	0.64	—
SPOT 3-SPUTTERED	"	0.15	0.61	0.02
14-SPOT 1	CHIPPED SURFACE	0.14	0.42	<0.01
SPOT 2	"	0.07	0.17	<0.01
SPOT 3	"	0.18	0.16	<0.01
SPOT 3-SPUTTER 1	"	0.10	0.16	0.03
SPOT 3-SPUTTER 2	"	0.07	0.24	0.02
10-SPOT 1	TOP SURFACE	0.04	0.17	0.05
SPOT 2	"	0.12	0.21	0.05
SPOT 2-SPUTTERED	"	0.06	0.24	0.03
22-SPOT 1	TOP SURFACE	0.05	0.46	—
SPOT 2	"	0.11	0.22	—

The Ca/Si ratios were quite constant over the three
spots analyzed for the bulk sample, #18, and did not change
following ion sputtering. The Na/Si and K/Si were more variable and the intensity ratios decreased after ion sputtering.

Two different types of exposed surface were examined.
The unaltered top surface, #10 and #22, and the chipped surface, #14, formed during removal from the original crucible.
The unaltered surface may have been depleted or enriched in
certain elements during cooling from the melt and the chipped
surface should have been representative of the original bulk
composition before exposure to burial conditions. Both
sample types showed Ca/Si ratios which were substantially
lower than the bulk specimen, #18. This suggests that surface segregation in the melt was not a factor in the observed
changes in elemental concentration. The Na/Si and K/Si ratios were noticeably lower on surface samples but appeared to
vary with position.

Minor concentrations of strontium, cesium, chromium,
zinc, iron and lead were detected in a number of outer surface specimens. The concentration of these minor elements at
the surface of the glass may have been due to segregation in
the melt. In the bulk only zinc and iron were detected.

The sample analyzed from the second hemisphere was from
the unaltered top surface, #52. The unexposed side was assumed to be representative of the bulk glass phase. The ele-

mental intensity ratios from the spectral analysis are given
in Table III. In the bulk phase the K/Si ratio was similar
to those in the first hemisphere, but the Ca/Si ratio was
significantly lower. In addition, minor concentrations of
uranium, strontium and lead were identified in the bulk
phase that were not detected in the bulk phase of the first
hemisphere.

Table III

XPS Elemental Intensity Ratios for Second Glass Hemisphere

SAMPLE DESIGNATION	ORIGIN	INTENSITY RATIOS				
		Na/Si	Ca/Si	K/Si	Pb/Si	Zn/Si
52C-SPOT 1	SURFACE	0.08	—	—	0.16	0.09
52C-SPOT 2	SURFACE	0.08	≤0.27	≤0.27	0.08	0.08
52C-SPOT 3	BULK	0.40	0.10	0.08	0.04	≤0.03
52C-SPOT 4	BULK	0.50	0.07	0.07	—	0.02

The outer one to two nanometres of the surface of the
sample were strongly depleted in sodium and the relative
concentrations of lead and zinc were two to four times as
great as their concentrations in the bulk. This increase in
concentration could be explained by either (a) segregation
during fabrication or (b) dissolution and reprecipitation on
the glass surface. The latter mechanism for elemental surface enrichment has been observed in our laboratory studies
on inactive glasses.[4]

In summary, X-ray photoelectron studies have shown surface changes in both glass hemispheres. The second hemisphere showed larger changes in surface composition. This
effect was expected, since the second hemisphere was a
poorer quality glass because of the higher loading of fission
products, iron and uranium.

Scanning Electron Microscopy and Energy Dispersive X-Ray
Spectrometry

Three specimens from the outer surface of the first
hemisphere and one from the second hemisphere were examined
by scanning electron microscopy (SEM) and energy dispersive
X-ray spectrometry (EDXS). For quantitative studies, three

specimens were metallographically polished, and cross-sectional X-ray line scans were taken for calcium, silicon, aluminum and sodium.

Chip #1 from the top corner of the first hemisphere was examined for topographical variation between the top, the outer curved and the fresh-fractured surfaces for characteristic leaching features previously observed in laboratory experiments with borosilicate glass.[5] The specimen showed no evidence of a leach zone between the top, the fresh-fractured or outer curved surfaces, as shown in Figure 3. X-ray line scans of specimen #4D, Figure 4, showed depletion of calcium from the surface to a depth of about 0.3 mm. Figure 5 shows an X-ray scan of sample #10C where the calcium

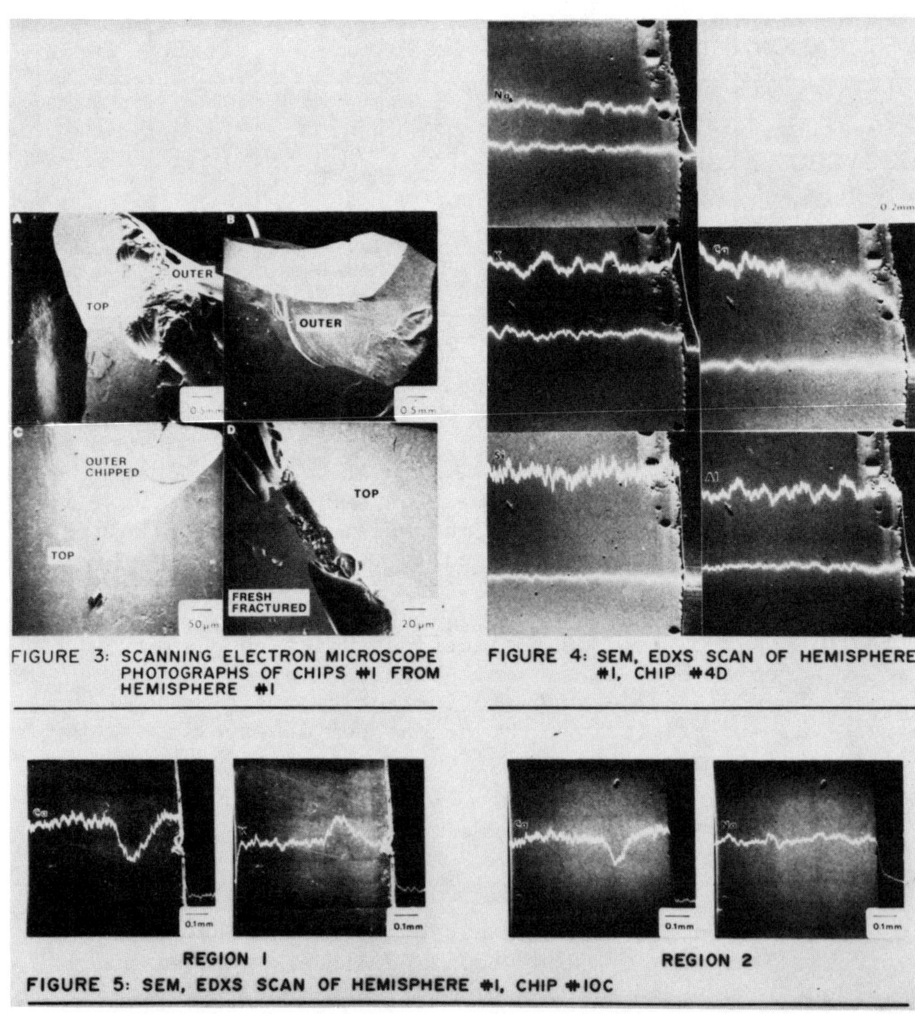

FIGURE 3: SCANNING ELECTRON MICROSCOPE PHOTOGRAPHS OF CHIPS #1 FROM HEMISPHERE #1

FIGURE 4: SEM, EDXS SCAN OF HEMISPHERE #1, CHIP #4D

FIGURE 5: SEM, EDXS SCAN OF HEMISPHERE #1, CHIP #10C

concentration had decreased about 0.2 mm from the surface
with a subsequent rise to the bulk concentration at the
surface. In the calcium deficient region, potassium showed
an increase in concentration, while sodium showed no change.

A chip from the second glass hemisphere was similarly
analyzed and no apparent elemental variations could be
observed.

SEM and EDXS studies showed distinct variations in calcium and potassium at the surface or 0.2 mm from the surface.
These effects may be due to segregation during glass manu-
facturing.

Other Analyses

A specimen from the first glass hemisphere was analyzed
by X-ray fluorescence to determine the major elements in the
glass. The analysis confirmed the presence of silicon, calcium, aluminum and potassium.

Analysis of crushed bulk glass from the first glass
hemisphere using a Guinier-Simon X-ray diffraction camera
showed no evidence of devitrification of the glass.

CONCLUSIONS

Two glass hemispheres were examined by a variety of
analytical techniques. Elemental inhomogeneities were observed in the outermost glass surface by the SEM/EDXS and
XPS methods.

A calcium depletion from the surface to a depth of
0.3 mm was observed in a specimen from the first glass
hemisphere by SEM/EDXS analysis. In another specimen,
calcium depletion and potassium enrichment were observed
together 0.2 mm below the top surface. The observed depletion of calcium at the surface is considered to be too
severe to be explained by leaching. It is more probable
that the elemental variations observed by SEM/EDXS occurred
during fabrication. No changes in concentration of elements
could be observed in the second glass hemisphere.

XPS methods detected depletion of some elements from
the surface to a depth of about two nanometres in all surface specimens in both glass hemispheres. In the second
hemisphere, the higher fission-product waste loading appeared
to increase the depletion of elements from the surface.
Trace elements were, however, enriched at the surface, and

this has been attributed to segregation in the melt and/or
dissolution by the leach water and reprecipitation on the
glass surface.

Long-term leaching tests on a number of glass samples
from these hemispheres are now underway by the Storage and
Disposal Branch at WNRE. Future studies may include the
determination, by secondary ion mass spectrometry, of elemental concentrations as a function of surface depth.

This work has shown that the CRNL glass hemispheres
have undergone little overall change during a burial period
of twenty years in sand in continuous contact with flowing
water. Nepheline syenite glass under the burial conditions
appears to be an excellent host material for the immobiliza-
tion of high-level fission-product waste.

ACKNOWLEDGEMENTS

The author thanks his associates M.A. Ryz, W.A. Boivin,
N.S. McIntyre, F.W. Stanchell, F.E. Doern, A.G. Wikjord,
N.J. Pearson, P. Taylor and O.E. Acres for analytical data
contributions.

REFERENCES

1.   A.R. Bancroft and J.D. Gamble, Atomic Energy of Canada
      Limited Report, AECL-718 (1958).
2.   A.R. Bancroft, Can. J. Chem. Eng., 38, 19 (1960).
3.   W.A. Merritt, Nuclear Technology, 32, 88 (1977).
4.   N.S. McIntyre, unpublished data.
5.   F.E. Doern, unpublished data.

USE OF RADIOACTIVE TRACERS FOR SELECTION OF RARE
EARTH PRECIPITANTS AND IGNITION TEMPERATURES*

N. L. Smith, E. S. Delucchi, and D. T. Mecozzi.
Nuclear Chemistry Division, Lawrence Livermore
Laboratory, Livermore, California, USA.

ABSTRACT

We have found variations in the specific activity of ignited radioactive-labeled rare earth oxide samples. The variations appear to depend on the precipitating agents and
temperatures. Using various precipitating agents and different ignition temperatures, samples of ^{88}Y-, ^{168}Tm-, and
^{173,174}Lu-labeled oxides were produced from stock solutions.
Observed specific activities were compared to the known specific activities of the starting solutions. At 800°C ignition temperatures, errors of 6% to 7% can be obtained for yttrium precipitated with 8-hydroxyquinoline or ammonia. Thulium and lutetium quinolates and cupferrates ignited at 800°C
are in error by 3% to 5%. Our results show that temperatures
in excess of 1000°C are required for complete ignition of
rare earth quinolates, cupferrates, hydroxides, and chlorides.

INTRODUCTION

Experiments are being done at Livermore to determine
optimum conditions for precipitating and igniting some rare
earth elements and yttrium to their oxides.

The possibility of conventional gravimetric rare earth
assays[1,2,3,4] giving incorrect values was first noticed with
yttrium. Lawrence Livermore Laboratory and Los Alamos Scientific Laboratory radiochemists independently became aware
of the problem and came to the same conclusions as to the
causes.

---

*Work performed under the auspices of the U.S. Department of
Energy by the Lawrence Livermore Laboratory under contract
number W-7405-ENG-48.

These two laboratories use ^{88}Y in their nuclear device diagnostic work. The radioisotope is mixed with a known amount of stable yttrium carrier, then various purification steps are performed to remove interfering radioactivities and mass contaminants, and finally the ^{88}Y-labeled Y_2O_3 is radioassayed. This type of procedure has been used for many years. Recently, however, the development of Ge(Li) pulse-height-analyzers has permitted ^{88}Y counting of solutions containing many other radioactivities, without the need to purify the yttrium fraction. When the ^{88}Y content of puri-fied, ignited ^{88}Y$_2$O$_3$ samples was compared to ^{88}Y counted directly in solution with no purification, the oxide samples gave 5% to 10% lower values than direct counting.

After exploring several possible causes of the dis-agreement, the major source of error was found to be pre-cipitation-ignition variations.

EXPERIMENTAL AND RESULTS

High-purity yttrium metal, and thulium and lutetium oxides, were used along with added radioactive tracers to prepare three stock solutions. One contained ^{88}Y, another ^{168}Tm, and a third solution was ^{173,174}Lu, all with known amounts of tracer and carrier. The specific activity (atoms of tracer per milligram of carrier) was therefore known.

We used various precipitants and different ignition temperatures to produce ^{88}Y$_2$O$_3$, ^{168}Tm$_2$O$_3$, and ^{173,174}Lu$_2$O$_3$ samples for counting. Table I shows the sample prepara-tion flow chart.

TABLE I

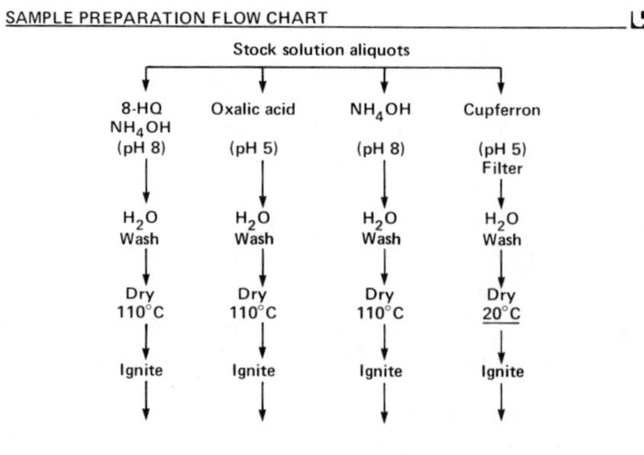

Table II shows experimental conditions and results for yttrium samples.

TABLE II

YTTRIUM PRECIPITANT-IGNITION TEMPERATURE RESULTS
(PERCENT OF THEORETICAL SPECIFIC ACTIVITY)

	800°C	900°C	1000°C	1100°C	1176°C	1250°C
8-hydroxy-quinoline	93.9%	98.1%	98.0%	99.6%	97.6%	—
Oxalic acid	98.7%	100.3%	99.8%	100.2%	99.8%	100.0%
NH$_4$OH	93.2%	97.7%	98.2%	98.3%	98.2%	—
Cupferron	—	—	99.6%	—	—	99.9%

30 mins at temperature

During the time of the ^{88}Y disagreement, we were precipitating yttrium with 8-hydroxyquinoline and igniting at 800°C for 20 to 30 minutes. The results shown in Table II explain the 5% to 10% discrepancies we had noticed.

Our conclusions for yttrium are that precipitation with oxalic acid and ignition at 1000°C for 30 minutes is satisfactory.

Since other rare earths may be affected similarly, we repeated the work, using thulium and then lutetium radioactive-labeled stock solutions. Table III shows our results for thulium. No marked low results are apparent for thulium as there were with yttrium.

TABLE III

THULIUM PRECIPITANT-IGNITION TEMPERATURE
RESULTS (PERCENT OF THEORETICAL
SPECIFIC ACTIVITY)

	800°C	1000°C	1100°C	1250°C	Fisher burner
8-hydroxy-quinoline	97.3%	98.4%	99.6%	—	97.0%
Oxalic acid	98.9%	99.0%	99.3%	99.8%	—
Cupferron	96.9%	98.6%	98.7%	99.4%	97.8%

30 mins at temperature

Table IV shows lutetium results. These data show a
definite temperature dependence, requiring ignition temperatures of more than 1000°C to approach a constant specific
activity. Where yttrium and thulium oxalates produce analyt-
ic quality oxides at the lower temperatures, lutetium oxalate
ignitions gave slightly low results from 800°C to 1250°C.
This is unexpected, and will be verified as soon as possible.

TABLE IV

LUTETIUM PRECIPITANT-IGNITION TEMPERATURE RESULTS (PERCENT OF THEORETICAL SPECIFIC ACTIVITY)

	800°C	1000°C	1100°C	1250°C
8-hydroxy-quinoline	94.9%	96.6%	98.6%	98.3%
Oxalic acid	95.7%	96.0%	96.8%	96.8%
Cupferron	95.6%	96.6%	99.0%	100.3%

30 mins at temperature

REFERENCES

1. C. Duval, Inorganic Thermogravimetric Analysis, El-
    sevier, New York, 1953, 400-416.

2. W. W. Wendlandt, Anal. Chem., 30, 58 (1958).

3. W. W. Wendlandt, Anal. Chem., 31, 408 (1959).

4. I. M. Kolthoff and P.J. Elving, Treatise on Analytical
    Chemistry, Part II, 8, 52-53 (1963).

GAMMA SPECTROMETRY
AND ACTIVATION ANALYSIS

A COMPARISON OF Ge(Li) WELL AND N-TYPE COAXIAL DETECTORS
FOR LOW ENERGY GAMMA-RAY ANALYSIS OF ENVIRONMENTAL
SAMPLES

Colin G. Sanderson. Environmental Measurements
Laboratory, U. S. Department of Energy, New York,
New York, U. S. A.

ABSTRACT

Ge(Li) well and large n-type germanium coaxial detectors
have made possible environmental gamma-ray analysis below 30
keV. Large surface area n-type detectors are ideally suited
for the measurement of large low-level samples in Marinelli
(reentrant) beakers. However, when sample size is limited or
simple pre-concentration procedures can be performed, Ge(Li)
well detectors are preferred.

INTRODUCTION

Current inventories of germanium gamma-ray detectors
list over a half-dozen different types. Commercially available are: Planar; Coaxial, both true and closed end; lithium
drifted germanium [Ge(Li)] and hyperpure germanium (HPGe)
p-type and n-type; Ge(Li) and HPGe well configurations. If
one is primarily interested in low energy environmental
gamma-ray spectrometry, the choice of type of detector is
narrowed down to either a Planar, a well or an n-type coaxial.
P-type Ge(Li) and HPGe, both true and closed end, have
significant dead layers which limit the penetration of low
energy gamma-rays.

For example, more than 50% of the 60 keV gamma-rays from ^{241}Am would be attenuated by the 500 μm Ge dead layer present in both Ge(Li) and HPGe closed end coaxial configurations. While the face of the true coaxial Ge(Li), with a 200 μm dead layer would attenuate only about 20% of these gamma rays, its thick side walls would limit the use of Marinelli beakers for low energy environmental analysis.

The n-type HPGe coaxial detector, shown in Figure 1, has a very insignificant dead layer of less than 1 μm. Low-energy gamma-ray detection for this type of detector is therefore only limited by internal fabrication materials, the end cap, and sample geometry.

Detector surface area which is directly related to low-energy efficiency can be best exploited by putting the detector inside the sample with Marinelli beakers.

If one does not have sufficient sample so that the detector can be placed inside the sample, then the sample can be placed inside the detector - as is done with a well (Figure 1). The nearly 4π geometry yields very high counting efficiencies for low-energy gamma rays since only a few millimeters of Ge are required for total energy transfer. Both HPGe and Ge(Li) wells have been made with essentially no internal dead layer in the well. Low energy gamma-ray spectrometry therefore depends upon the material of which the end cap (well) is made and the less than 2 cm^3 sample volume that can be accommodated.

EXPERIMENTAL

Figure 2 presents efficiency versus energy curves for detectors with aluminum and beryllium wells. At energies below 80 keV the curves reflect the attenuation effect of the 0.5 mm Al and Be side walls of the well end cap. A Be well detector fabricated for the Los Alamos Scientific Laboratory provides an 85% counting efficiency at 88 keV and 65% at 12 keV. For comparison the efficiency curve for an n-type coaxial with Marinelli beaker geometry is also shown.

The 10 to 12 times lower efficiency indicated for the n-type coaxial is more than offset by the much larger sample

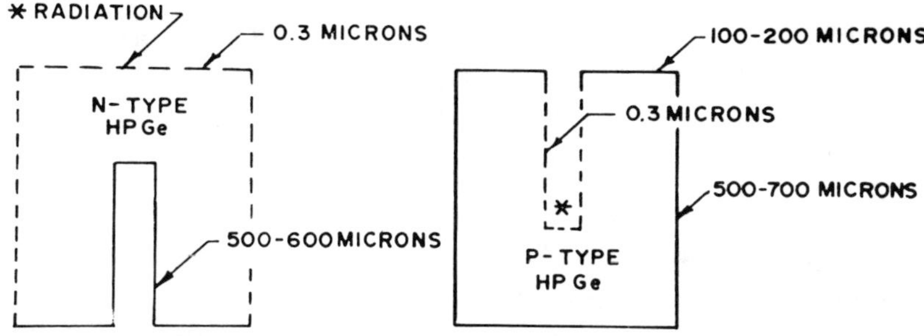

Figure 1.   N-type germanium coaxial and P-type germanium well gamma-ray detectors.

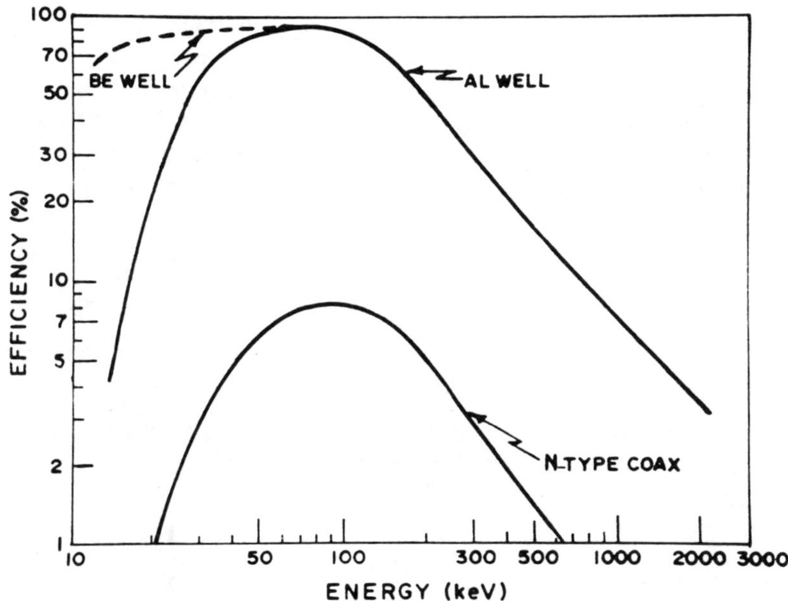

Figure 2.   Gamma-ray detection efficiency vs. energy for coaxial and well detectors.

that can be counted when a Marinelli beaker is used. However,
large samples usually mean higher backgrounds, so that even
though 300 to 400 times more sample can be counted on a
coaxial detector with $\frac{1}{10}$ the efficiency of a well, the
resulting lower limits of detection may only vary by factors
of 2 to 10.

A lower limit of detection (LLD) of 0.02 pCi/g at the
95% confidence level was measured for ^{241}Am in soil using a
15% n-type coaxial with 96 cm^2 total surface area and a 1000
minute count time. See Table I. By comparison, when the
maximum sample size of 2 grams was counted in a 14% well for
1000 minutes, the LLD was 0.086. However, if one is willing
to apply some simple chemistry in order to concentrate 20 to
1000 grams of soil prior to gamma-ray counting the LLD can be
lowered by an order of magnitude or more. For instance, at
the Environmental Measurements Laboratory (EML), ^{241}Am,
separated from 1 kg of acid leached soil by oxalate precipitation, ion exchange and $Fe(OH)_3$ scavenging provides limits
of detection below 0.002 pCi/g. The absolute counting
efficiency at 60 keV at the face of the n-type coaxial with a
18 mm diameter source is about 25% while with a well detector
the efficiency is about 80%, therefore, the well detector will
have a lower limit of detection advantage by almost a factor
of 10.

TABLE I

COMPARISON OF LOW ENERGY LIMITS OF DETECTION
FOR ^{241}Am

860 g Soil in	2.0 g Soil in
Marinelli Beaker 15% N-Type HPGe	14% Ge(Li) Well
0.020 pCi/g	0.086 pCi/g

The characterization of lake sediments for polychlo-
rinated biphenyls, polyaromatic hydrocarbons, and trace
metals requires some means of accurately dating the sediment
core. Cesium-137 from atmospheric fallout can provide very

good age estimates back to about 1950, while ^{226}Ra-^{210}Pb
analysis can extend this time scale back to about 1880. The
simplest way to determine ^{210}Pb is to measure its 46 keV
gamma ray, however, most core increments are too small for
Marinelli beaker counting, therefore, 6 cm diameter Petri
dishes holding 30 to 40 g of sample are counted at the face
of the n-type coaxial. The resulting LLD, as shown in Table
II, is significantly lower than the usual .5 to 1 pCi/g
^{226}Ra-^{210}Pb equilibrium activity encountered. In this
instance, the n-type coaxial is superior to the well solely
because of sample size. Two grams of sediment in the well
yields a LLD of only 1.8 pCi/g.

TABLE II

COMPARISON OF LOW ENERGY LIMITS OF DETECTION
FOR ^{210}Pb

37 g Lake Sediment on Top of Detector 15% N-Type HPGe	2.0 g of Sediment in 14% Ge(Li) Well
0.28 pCi/g	1.8 pCi/g

Characterizing the degree of equilibrium of uranium and
decay products in aerosols at ore processing mills continues
to be of interest at our Laboratory. Dust particles collected on filter papers have been directly analyzed by low
energy gamma-ray spectrometry. The 9 cm diameter filter
paper samples were counted by wrapping them over the end cap
of the coaxial detector and holding them in place with a
rubber band. Well counting was accomplished by carefully
folding and rolling half of the 9 cm filter into a small
cylinder and inserting the filter encased in a thin rubber
bag into the well.

The resulting LLD's shown in Table III indicate that
greater sensitivity is obtained with the well. In this case,

the fact that only half the sample is being counted is more
than offset by the well detector's greater counting efficiency.

TABLE III

COMPARISON OF LOW ENERGY LIMITS OF DETECTION -
NATURAL SERIES

Nuclide	15% N-Type HPGe[a] (pCi/filter)	14% Ge(Li) Well[b] (pCi/filter)
^{234}U	25	8.0
^{235}U	25	2.2
^{238}U	4.0	1.9
^{230}Th	40	15
^{226}Ra	3.8	4.5
^{210}Pb	3.9	1.9

---

[a] 9 cm diameter filter paper over detector

[b] 1/2 of 9 cm filter paper in well

CONCLUSION

The choice of which detector to use for low level, low
energy gamma-ray spectrometry can be predicated solely on
sample size. If large amounts of sample are available,
Marinelli beakers and n-type coaxial detectors will provide a
better LLD (even when sample attenuation is considered) and
will minimize the uncertainty due to small sample inhomoge-
neity. However, if sample size is limited or chemical precon-
concentration of radionuclides is carried out the well
detector will provide better LLD than the n-type coaxial.

GERMANIUM DETECTOR EFFICIENCY CALIBRATION FOR
RADIOLOGICAL MONITORING OF NUCLEAR PLANT EFFLUENTS

D. W. Nix and N. E. Scott, Sequoyah Nuclear Plant,
Tennessee Valley Authority, Daisy, Tennessee

ABSTRACT

A brief comparison is given of various efficiency calibration methods as they apply to intermediate to small size
laboratories. The choice of radionuclides for calibration
standards, sources of errors in the efficiency determination,
and fitting functions for efficiency curves are discussed.
Much of the paper reviews previously published work; however,
original work on fitting functions applied to a wide variety
of sample geometries using various manufacturers' detectors
is presented.

INTRODUCTION

Gamma-ray spectrometry utilizing germanium detectors is
the primary method used to quantitate the release of radionuclides from nuclear plants. The efficiency calibration of
the germanium detector directly affects the accuracy of all
quantitative release data. Therefore, the method of calibration, the choice of standards, sources of errors in the
efficiency determination, and the choice of the fitting
function should be considered carefully.

CALIBRATION METHODS

Three primary methods of efficiency calibration are in
use: (1) Theoretical Monte-Carlo methods, (2) theoretical
efficiency models with experimentally determined parameters,
and (3) experimental determinations.

Theoretical calculations of efficiency detection by
Monte-Carlo methods [1,2] have been limited to point or disc
sources because calculations for more complex sample config-

urations are difficult, time-consuming, and relatively
inaccurate. This method has not been used except in research
laboratories and will not be discussed further.

The Lawrence Livermore Laboratory has developed a model
that combines theoretical calculations with experimentally
defined parameters to describe the overall efficiency of a
germanium detector as a function of parameters such as gamma-
ray energy, source-to-detector distance, dimensions of the
detector, etc.[3] The semiempirical model is accurate to one
to two percent for small-volume sources and accurate to ten
percent for large-volume sources placed close to the detector.
This model's usefulness is limited because it is only
applicable to point, disc, and cylindrical geometries; and
many laboratories use Marinelli beakers. Effective use
of the semiempirical model also requires a sizable computer
for data analysis and a staff with considerable knowledge
of germanium detectors to make the necessary measurements.*

The most common approach to calibration of germanium
detectors is the experimental determination of full-energy
peak (FEP) efficiency as a function of energy for each
counting configuration. Practical calibration standards for
each counting configuration must be prepared for appropriate
calibrated radionuclides. The composition of these standards
should approximate as closely as possible, with respect to
density and attenuation factors, the actual samples that are
to be analyzed after calibration.

Experimental determination of the FEP efficiency as a
function of energy for each counting configuration does not
require the sophisticated computer codes or the measurement
of detector characteristics (dead layers, dimensions of the
active volume of the detector, position of the detector
within the cryostat endcap, etc.) that are required for the
first two methods. Also, the experimental method is not
subject to the uncertainties associated with the theoretical
or semiempirical models chosen to calculate the detection
efficiency. However, the experimental method is time
consuming because of the number of samples that must be
prepared and counted; but the time required to prepare and
count samples may be less than that required to set up the
necessary computer programs and perform and verify the
calculations suggested for the first two methods.

---

*Science Applications, Inc., Rockville, MD, offers a
 germanium detector calibration service using a semiempirical model.[4]

CALIBRATION SOURCES

Appropriate gamma-emitting radionuclides must be selected for use as standards in calibration. Characteristically, calibrated solutions of these radionuclides should be readily available in a pure form and have an accurately measured, relatively long half-life (greater than 30 days).

Radionuclides that are used to determine efficiency can be classified into two groups: (1) Those radionuclides with only a few prominent gamma-rays and (2) those radionuclides with many prominent gamma-rays. Debertin et.al.[5] and Hirshfeld et.al.[6] list many radionuclides with only a few prominent gamma-rays that have been extensively used for efficiency calibrations.

Several radionuclides can be used either singly or in combination to determine an efficiency curve for a germanium detector over an energy range of 10 to 2800 keV. The energy range of interest must be adequately covered by calibration points so that interpolations between calibration points are accurate.

Many radionuclides of the second category have been suggested; but the complexity, magnitude, and uncertainty of the summing corrections lead most laboratories to use several radionuclides with few prominent gamma-rays. Even some commonly used radionuclides (^{134}Cs, ^{94}Nb, ^{46}Sc, ^{88}Y, ^{60}Co, ^{24}Na, and ^{140}La) require summing corrections.

The National Bureau of Standards (NBS) issues mixed-radionuclide emission-rate standards in point-source form and in solutions of different concentrations.*[7,8] These standards provide 11 principal gamma-rays covering the energy range of 88 to 1836 keV in reasonably well-spaced interfals. The standards were originally developed for calibrating germanium detectors for environmental analyses.[7] The solutions are easy to use and offer the added advantage of traceability to NBS. Only two radionuclides, ^{60}Co and ^{88}Y, must be corrected for summing. An excellent description of these standard solutions and potential problems associated with their use have been given by Coursey.[7]

____

*Amersham Co., 2636 S. Clearbrook Drive, Arlington Heights, Illinois 60005, sells a mixed-radionuclide emission-rate standard that is traceable to NBS and is available in February, June, and October. This standard has the identical composition of the NBS standard.

EFFICIENCY DETERMINATION AND SOURCES OF ERRORS

After an experimental method for determining efficiency
has been selected, count-rate data must be acquired by using
the appropriate calibration sources. Before actual standardization is begun, the spectroscopy system should be thoroughly
tested to ensure electronic stability. The calibration sources
then must be carefully prepared with a total activity low
enough that random summing is negligible.

The standard source should be carefully positioned in
the detector shield, and spectra should be acquired which
contain a statistically sufficient number of counts in each
FEP of interest. Once the calibration spectra have been
obtained, the net count rate must be determined for each FEP
of interest. Any method that yields consistent results may
be used to determine net count rate, but the method selected
should yield the highest efficiency while minimizing the
background contribution to the FEP region.

When source-to-detector distance is 10 cm or less,
correlated photon-summing corrections should be considered
for all gamma-rays being used in the calculations that are
emitted in cascade with other gamma-rays. The summing
corrections for a 3.5-L Marinelli beaker and a 15-percent
germanium detector is two to three percent for the γ-rays
emitted by ^{60}Co and ^{88}Y. The summing for a filter paper
placed on the endcap of a detector is considerably larger.

Log-log plots of efficiency data, with energy as the
ordinate and efficiency as the abscissa, can be used to
display a wide range of energies. These plots show a fairly
linear curve above about 150 to 300 keV for most germanium
detectors.

ANALYTICAL EFFICIENCY EXPRESSIONS

When efficiency data are available for a sufficient
number of energies in the energy region of interest, a method
of representing the efficiency as a function of energy must
be chosen. Graphical methods or spline fits can be used;
however, it is often desirable to express the efficiency in
some analytical form as a function of gamma-ray energy.
Such expressions can be readily programmed and are adaptable
to automatic data analyses.

Least-squares fitting procedures are used to fit the
efficiency data to an analytical expression. However, the
analytical expressions must be carefully selected to avoid
introducing systematic divergences from the observed data.

Several semiempirical formulas involving the physical characteristics of the detector have been proposed[9-11] to fit germanium detector efficiency curves. Due to incomplete charge collection, local variations in efficiency,[12] and uncertainties in the actual geometry of coaxial and closed-end coaxial detectors, empirical analytical functions are often preferred.[13,14] Table I lists some proposed empirical analytical functions with appropriate literature references for each expression. The energy ranges over which these functions give adequate fits can be found in the listed references or in the review papers of McNelles and Campbell[14] or Singh.[15] When the available degrees of freedom are sufficiently large (i.e., the number of data points should exceed the number of fit parameters by at least one, preferably more), the data can be broken into two separate regions for fitting if fitting the data for one extended region leads to systematic divergences.

TABLE I. ANALYTICAL FUNCTIONS USED TO FIT
EFFICIENCY DATA FOR GERMANIUM DETECTORS

Function/formula	References
$\ln \varepsilon = a_1 + a_2 \ln E$	5
$\ln \varepsilon = a_1 \ln (1.022/E) + a_2 \ln (1.022/E)$	16
$\ln \varepsilon = a_1 + a_2 \ln E + a_3 (\ln E)^2$	17
$\ln \varepsilon = a_1 \ln E + a_2 (\ln E)^2 - a_3/E^3$	18
$\ln \varepsilon = \sum_{j=1}^{n} a_j (\ln E)^{j-1}$	3
$\varepsilon = a_1 \exp(a_2 E) + a_3 \exp(a_4 E)$	19
$\varepsilon = (a_1/E)^{a_2} + a_3 \exp(-a_4 E) + a_5 \exp(-a_6 E) + a_7 \exp(-a_8 E)$	14
$\ln \varepsilon = a_1 E^{-3} + a_2 E^{-2} + a_3 E^{-1} + a_4 + a_5 E$	6
$\ln \varepsilon = a_1 + a_2 E + a_3 E^2 + a_4 E^3 + a_5 E^4 + a_6 E^5$	20

where: $\ln$ = natural log,

exp = exponential function,

ε = absolute FEP efficiency,

$a_1, a_2, a_3, \ldots$ = fit parameters.

All the expressions shown in Table I are similar, and several are special cases of the more general equations. The simple log-log fit (entry number 1 in Table I) can produce reasonable results in the 300 to 1800 keV region for many detectors. The reduction in efficiency observed at lower energies is often difficult to represent with any analytical function.

Most of the empirical analytical functions have been tested using point sources; however, very few functions have been extensively tested using source configurations which are used in environmental and nuclear plant laboratories (e.g., liter bottle, 14-cc serum vial, 3.5-L Marinelli beaker, 2-in. filter paper, 2-in. charcoal filter, etc.) Therefore, a test of the fifth entry in Table I was made. About 100 efficiency curves determined using the NBS mixed-radionuclide gamma-ray emission-rate standard were fit using first through sixth order equations. The three data points less than 279 keV were omitted from the first order fit. An analysis of the deviation of the fitted results from the experimentally determined efficiency points showed that the fifth- and sixth-degree polynomials fit most efficiency data to within ± 3 percent. Only a slight improvement was observed for the sixth order fit; therefore, the lower order fit was selected. The data analysis also indicated that the first order fit could be useful in -etecting erroneous efficiency points above 166 keV.

A further test of the first- and fifth-degree fit was made. Three detectors from different manufacturers were calibrated for five different source configurations and three source-to-detector distances (1, 3, and 10 cm). The October 1978 and February 1979 issues of the Amersham mixed-radionuclide gamma-ray emission-rate standards were used to determine the efficiency curves. Corrections were not made for correlated photon summing. The 514 keV data points determined using the October 1978 standard were omitted from the fit. The deviation of the fitted values from the experimentally determined values showed that: (1) About one half of the efficiency curves contained no points with a deviation greater than ± 2 percent, and most of the other efficiency curves contained no points which deviated greater than 3.5 percent; (2) the first degree fit can be effectively used to detect erroneous efficiency points above 166 keV; and (3) for most efficiency curves, the 661.6-keV data point showed the worst deviation. Considering the uncertainty in the gamma emission rate, typically two to four percent, and the errors introduced by correlated photon-summing, the efficiency data is adequately represented by the fifth order equation.

ACKNOWLEDGEMENTS

We would like to thank the Tennessee Valley Authority (TVA) Radioanalytical Laboratory at Muscle Shoals and the TVA Browns Ferry Nuclear Plant Radioanalytical Laboratory for providing us with efficiency data. We would also like to thank Marvin Adams and James Gerstle for their programming assistance.

Much of the review portion of this paper has been extracted from Nix, Powers, and Kanipe.[21] Reference 21 discusses many other topics relevant to gamma-ray spectrometry using germanium detectors.

REFERENCES

1. G. Aubin, J. Barrett, G. Lamoureux, and S. Monaro, "Calculated Relative Efficiency for Coaxial and Planar Ge(Li) Detectors", Nucl. Instrum. Meth., <u>76</u>:85-92, (1969).

2. B. F. Peterman, S. Hontzeas, and R. G. Rystephanick, "Monte Carlo Calculations of Relative Efficiencies of Ge(Li) Detectors", Nucl. Instrum. Meth., <u>104</u>:461-468, (1972).

3. R. Gunnink, and J. B. Niday, <u>Computerized Quantitative Analysis by Gamma-Ray Spectrometry</u>, Vol. 1, Description of the GAMANAL Program. UCRL-51061, Lawrence Livermore Laboratory, Livermore, California, (1972) 75 pp.

4. J. E. Cline, "A Technique of Gamma-Ray Detector Absolute Efficiency Calibration for Extended Sources", in <u>Proceedings of Computers in Activation Analysis and Gamma-Ray Spectroscopy</u>, Mayaquez, Puerto Rico, (1978).

5. K. Debertin, U. Schotzig, K. F. Walz, and H. M. Weib, "Efficiency Calibration of Semiconductor Spectrometers-- Techniques and Accuracies", in <u>Proceedings of the ERDA X- and Gamma-ray Symposium</u>, Ann Arbor, Michigan, (1976) pp. 59-62.

6. A. T. Hirshfield, D. D. Hoppes, and F. J. Schima, "Germanium Detector Efficiency Calibration with NBS Standards", in Proceedings of the ERDA X- and Gamma-ray Symposium, Ann Arbor, Michigan, (1976) pp. 90-93.

REFERENCE (contd)

7.   B. M. Coursey, "Use of NBS Mixed-Radionuclide Gamma-Ray Standards for Calibration of Ge(Li) Detectors Used in the Assay of Environmental Radioactivity", in _Proceedings of Measurements for the Safe Use of Radiation_ (NBS Special Publication 456), National Bureau of Standards, Gaithersburg, Maryland, (1976) pp. 173-179.

8.   L. M. Cavallo, B. M. Coursey, S. B. Garfinke, J. M. R. Hutchinson, and W. B. Mann, "Needs for Radioactivity Standards and Measurements in Different Fields", Nucl. Instrum. Meth., 112:5-18, (1973).

9.   T. Paradellis and S. Hontzeas, "A Semiempirical Efficiency Equation for Ge(Li) Detectors", Nucl. Instrum. Meth., 73:210-214, (1969).

10.   R. S. Moxatt, "A Semiempirical Efficiency Curve for a Ge(Li) Detector in the Energy Range 50-1400 keV", Nucl. Instrum. Meth., 70:237-244, (1969).

11.   F. Hajnal and C. Klusek, "Semiemprical Efficiency Equation for Ge(Li) Detectors", Nucl. Instrum. Meth., 103:113-115, (1972).

12.   I. S. Sherman and M. G. Strauss, "Local Efficiency Variations in Coaxial Ge(Li) Detectors", Nucl. Instrum. Meth., 103:113-115, (1972).

13.   V. Zobel, J. Eberth, U. Eberth, and E. Eube, "^{226}Ra as a Calibration Standard for Ge(Li) Spectrometers", Nucl. Instrum. Meth., 141329-336, (1977).

14.   L. A. McNelles and J. L. Campbell, "Absolute Efficiency Calibration of Coaxial Ge(Li) Detectors for the Energy Range 160-1300 keV", Nucl. Instrum. Meth., 109:241-251, (1973).

15.   R. Sigh, "Validity of Various Semiempirical Formulae and Analytical Functions for the efficiency of Ge(Li) Detectors", Nucl. Instrum. Meth., 136:543-549, (1976).

16.   W. R. Kane and M. A. Mariscotti, "An Emprical Method for Determining the Relative Efficiency of a Ge(Li) Gamma-Ray Detector", Nucl. Instrum. Meth., 56:189-196, (1967).

REFERENCES (contd)

17. B. Numann, "Techniques for Fast Evaluation of Ge(Li)
     Detector Gamma-Ray Spectra with Small Computers", Nucl.
     Instrum. Meth., 108:237-241, (1973).

18. J. B. Willett, "High Precision Relative Efficiency
     Curves for Measurements of Gamma-Ray Relative Intensities
     and Internal Conversion Coefficients", in Proceedings
     of the International Conference on Radioactivity in
     Nuclear Spectroscopy, Nashville, Tennessee. Gordon and
     Breach Publishing Co., New York, (1972), pp. 1317-1321.

19. L. V. East, "Precision Measurement of Gamma-Rays from
     ^{94}Nb Decay", Nucl. Instrum. Meth., 93:193-195, (1971).

20. W. H. Zimmer, "Automatic Radioactive Inventory System",
     IEEE Trans Nucl. Sci., 20(1):352-360, (1973).

21. D. W. Nix, R. P. Powers, and L. G. Kanipe, "Application
     of Germanium Detectors to Environmental Monitoring",
     in print, Division of Environmental Planning,
     Tennessee Valley Authority, Muscle Shoals, Alabama.

APPLICATIONS OF GAMMA SPECTROMETRY IN THE LABORATORIES
DEPARTMENT AT THE SAVANNAH RIVER PLANT

C. D. Denard, Savannah River Plant, E. I. du Pont de
Nemours and Company, Aiken, South Carolina, USA.

ABSTRACT

Gamma spectrometry provides important data for environmental control, reactor operation, and the chemical separation processes at the Savannah River Plant (SRP). Much of
these data are provided by the Laboratories Department from
samples submitted by production departments. Routine samples
are analyzed using Ge(Li) detectors with automatic data reduction by computers. Samples requiring special handling are
analyzed using a variety of detectors, including low energy
germanium detectors for the isotopic analysis of plutonium
and a Ge(Li) detector for monitoring plutonium content in
waste before it leaves the laboratory. A wide range Gamma-X^{TM}
detector is under evaluation for use with many of the special
samples.

This paper describes some of the gamma spectrometry systems used in the laboratory. A variety of high resolution
gamma detectors and how they are selected for specific applications will be discussed. Calibration techniques using
National Bureau of Standards (NBS) mixed radionuclide standards will also be discussed.

INTRODUCTION

Gamma spectrometry has been used at SRP throughout its
years of operation. NaI(Tl) detectors were first used with
single channel analyzers for the determination of specific
radionuclides. When multichannel analyzers (MCA) became
available they usually had 50 to 400 channels and were used
with NaI(Tl) detectors. The spectra obtained on these early
MCA's were usually analyzed by photopeak summation as had been

done with the single channel analyzers. This technique required a fairly high degree of purity for the nuclide being determined. More complex spectra were analyzed by summing several photopeak areas and using simultaneous equations for determining the individual radionuclides. These simultaneous equations were solved by hand calculations which effectively limited this technique to determination of three radionuclides such as ^{131}I, ^{103,106}Ru, and ^{95}Zr-^{95}Nb. More complex spectra were analyzed using a spectrum stripping technique employing standard spectra stored on magnetic tape. This technique required a great deal of skillful judgment and careful alignment of instruments. Certain radionuclides, such as ^{103}Ru and ^{106}Ru and ^{95}Zr and ^{95}Nb, could not be easily separated using this technique, even with the greatest skill, due to the resolution of the NaI(Tl) detector.

During this period (1954 - 1966), most of the calibrations were related to a standard gross gamma counter. A purified radionuclide source would be counted on the standard gamma counter and the results of this count would be used as the value of the source. This source would then be used for spectrum stripping. Results of sample analyses were reported in the relative terms of counts per minute.

When high resolution gamma detectors became commercially available, a small Ge(Li) detector was purchased for evaluation. The advantages of the improved resolution were soon realized and this led to the purchase of a more efficient co-axial Ge(Li) detector and a larger MCA (1,600 channels). Soon there was a demand from the production departments for more routine samples to be analyzed by high resolution gamma spectrometry. To meet this demand, a computerized system to provide automatic data reduction of the gamma spectra was installed in 1971. This system has been in routine use since early 1972. The efficiency of the coaxial Ge(Li) detectors that have been used with this system has ranged from 6 to 10%. All detectors have been calibrated using mixed radionuclide gamma standards from NBS. Results are routinely reported in absolute terms of disintegrations per minute.

Three different computerized gamma analysis systems are now in use. A portable high resolution gamma spectrometry system has also been assembled for use in field problems. The combination of these systems gives the versatility to analyze large numbers of routine samples as well as special handling of nonroutine samples.

EQUIPMENT

PDP-15 System

This is a dual processor system, manufactured by Digital Equipment Co., which consists of a PDP-15/35 and a PDP-15/10. There is 16K words of shared memory and a total 80K words of

memory in both systems. The PDP-15/10 is used primarily for
controlling the nuclear hardware and the display oscilloscope.
This computer has its own terminal and paper tape reader which
allows it to be operated independently of the PDP-15/35. However, there is no data reduction capability in the PDP-15/10.

The nuclear hardware systems for the two 10% coaxial
Ge(Li) detectors are conventional anticoincidence counting
systems[1] which are connected to 4,096 channel analog-to-
digital converters (ADC's). The ADC's are interfaced directly
into the PDP-15/10. Two nuclear hardware systems for alpha
spectra analyses are also included in the PDP-15/10.

The PDP-15/35 provides automatic data reduction using the
GASPAN[2] programs with additional window routine programming
for certain applications. The analytical results are printed
on a line printer. Data acquisition and analysis are initiated
through a terminal on the PDP-15/35. Teletypes[TM] were originally used as terminals but have been replaced with
DECwriter[TM] terminals to reduce the cost of repairs.

Other peripherals include dual DECtape[TM] drives, a
punched card reader, a magnetic tape drive, three fixed head
disks, and two disk packs. Storage space for over 20 million
words is provided by the five disk systems. This storage space
contains programs and data that are used by the many other
tasks that are performed by this system on a real-time sharing
basis.

D-112 System

This computerized system was built by Lawrence Livermore
Laboratory (LLL) under the Safeguards program. It was designed
for plutonium isotopic analysis using gamma spectrometry[3] and
for the determination of plutonium and uranium concentrations
by x-ray fluorescence and gamma spectrometry. This system has
been programmed at SRP for the gamma analysis of fission product solution and basin and moderator samples.

This system was built around a Digital Systems Co. D-112
computer with 12K words of memory. It is equipped with a
paper tape reader and punch, two disk cartridges with storage
space for 1.2 million words on each, and a display oscilloscope with controls.

The nuclear hardware includes an ADC which is interfaced
to the computer. This ADC is capable of receiving signals from
several detector systems. A digital stabilizer is also included in the system for use in the plutonium isotopic
analysis. A low energy high purity germanium detector designed
at LLL is used for the isotopic analyses. Other low energy
detectors can also be used with this system. A 10% coaxial
Ge(Li) detector is used for the analyses of fission product,
basin, and moderator samples. Si(Li) detectors are used for
x-ray fluorescence analyses. These detectors are used on the
system one at a time as they are needed.

Complete data reduction is provided for the plutonium
isotopic and basin sample analyses. However, hands-on identification of radionuclides is needed for fission product and
moderator sample analyses with computer calculations for the
individual nuclides which were identified. This provides a
good hands-on system for nonroutine samples. The x-ray fluorescence analysis is completely hands-on. Data acquisition is
initiated from a Teletype^{TM} and the analytical results are
printed by this Teletype^{TM}.

PDP-11/34 System

This computer with 48K words of memory was purchased from
Digital Equipment Co. under the Safeguards program. Additional storage space for 64K words has recently been purchased
and installed. The system is equipped with two disk cartridges
with storage space for 7 million words on each. DECwriter^{TM}
and DECscope^{TM} computer terminals are used to initiate data
acquisition and analysis. The results are printed by a
DECprinter^{TM}.
The nuclear hardware systems are built around HIAS, Inc.
N 100 pulse height analyzers. The ADC's provide the signals
for the N 100's which are interfaced to the computer. Each
N 100 can accommodate two counting systems with 4,096 channels
each. A display oscilloscope can be controlled by the N 100
for spectra examination.
Programs used with the PDP-15 and D-112 have been adapted for this computer which gives it the analytical capability
of both. The Gamma-X^{TM} detector is being used on this system
with plans for an 8,192 channel N 100. Plans are to analyze
waste for transuranium elements with this system using a remotely located computer terminal, an N 100, and a display
oscilloscope.

Portable System

A portable system has been assembled using a hand-held
high purity coaxial germanium detector. The power supplies
and amplifiers are housed in a mini-bin. A portable 1,024
channel MCA is used for data acquisition and display. The data
can be stored on a tape cassette for later use or printed with
a small portable printer.

CALIBRATION

Coaxial Detector

The high energy coaxial detectors are calibrated with
mixed radionuclide gamma standards from NBS. The standards
contain 11 gamma energies which are shown in Table I. Typical
gamma spectra of these standards are shown in Figure 1.

A point source standard is used to calibrate for dry
mount geometry up to 10 cm from the detector. Counting times
must be varied for different positions to obtain desired
counting statistics. The point source must be placed in a
configuration comparable to the dry mounts used for samples.

Two liquid geometries are used routinely. Low activity
samples are often sampled in 250 ml plastic bottles and 8 oz
glass bottles. The detectors are calibrated for these bottles
so that the samples can be analyzed as received. A solution
source standard is used for the calibrations. The solution
source supplied by NBS does not conform to the size and shape
of the bottles, therefore, a weighed portion of the solution
source is transferred to fill the bottles which are then used
as standards for calibration purposes. Efficiency curves are
plotted for each geometry and this curve is fitted to the
equation:

$$(1) \qquad \chi = C_1 (y^{C_2} + C_3 e^{C_4 y})$$

by varying the coefficients C_1, C_2, C_3, and C_4. In this equation χ represents efficiency and y represents the gamma
energy. After the best possible fit is obtained, the coefficients are placed into a computer file which enables the computer to calculate the efficiency for the gamma energy, detector, and geometry that is needed.

TABLE I. NBS MIXED RADIONUCLIDE GAMMA STANDARDS

Radio-nuclide	Gamma Energy, keV	Half-life		Point Source, γ/m	Solution Source, $\gamma/m/ml$
^{109}Cd	88	463.9	d	4.24×10^4	404
^{57}Co	122	272.4	d	7.57×10^4	720
^{139}Ce	166	137.7	d	4.55×10^4	432
^{203}Hg	279	46.62	d	1.43×10^5	1,364
^{113}Sn	392	115.0	d	1.36×10^5	1,294
^{85}Sr	514	64.85	d	2.27×10^5	2,156
^{137}Cs	662	30.0	y	1.18×10^5	1,116
^{88}Y	898	106.66	d	8.44×10^5	8,016
^{60}Co	1,173	5.271	y	2.99×10^5	2,840
^{60}Co	1,332	5.271	y	2.99×10^5	2,844
^{88}Y	1,836	106.66	d	8.83×10^5	8,384

Low Energy Detectors for Pu Isotopic Analysis

Specially designed vials are used for the isotopic analysis of liquid plutonium samples. Solutions of plutonium with known isotopic composition are used in these vials for calibration purposes. Some of the plutonium is obtained from NBS and some of it is SRP material which has been analyzed by mass spectrometry.

Waste Monitor

The waste monitor is calibrated with standards that are made from weighed portions of plutonium or neptunium oxides. A slurry of the oxides is made with contact cement and acetone. The slurry is quantitatively transferred to a plastic bottle or a metal can. The receiving container is rotated so that the slurry will adhere to the sides in a thin layer to reduce self attentuation. After the cement has dried, the container is sealed and packaged for contamination control. If less than 2 grams of oxide are used in a 1 liter container, the self attentuation is negligible for plastic bottles and usually less than 10% for most metal cans.

Gamma-X^{TM} Detector

The Gamma-X^{TM} detector can be calibrated in the same way as any coaxial detector using the NBS mixed standards. However, for most routine high energy applications it is necessary to place an adsorber over the detector to reduce its low energy efficiency. For low energy applications, appropriate standards such as plutonium with known isotopic composition are needed.

SAMPLE ANALYSIS

Routine Samples

Approximately 1,200 samples are analyzed routinely each month with the PDP-15 system. The precision for most of these analyses is within ±10%. Most of the samples are from various points in the chemical separation process. The analysis of these samples on an around-the-clock basis provides important information for process and environmental control. The samples are analyzed for fission product content after proper dilutions are made and dry mounts are prepared. The spectrum of a typical sample containing fission products is shown in Figure 2.

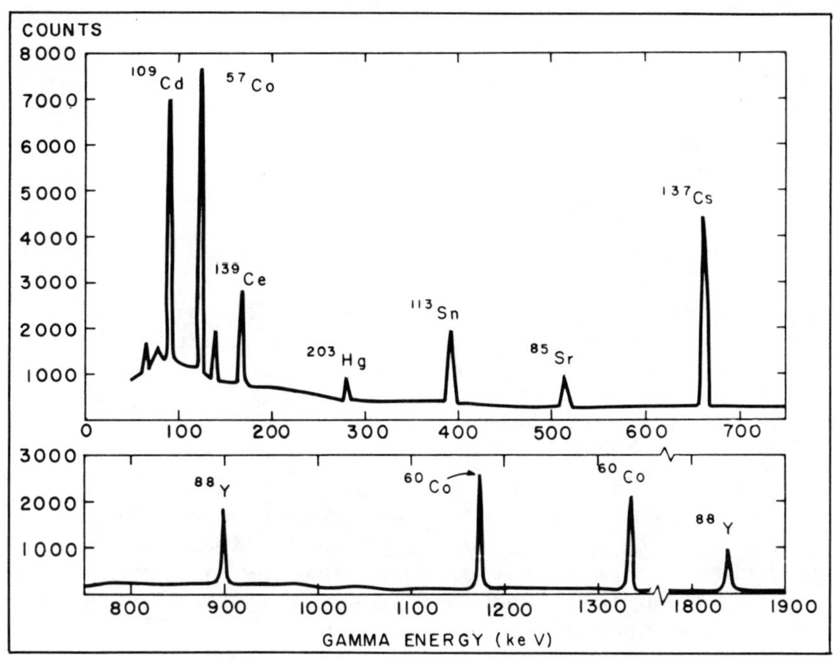

FIGURE 1.   A SPECTRUM OF A MIXED RADIONUCLIDE GAMMA STANDARD

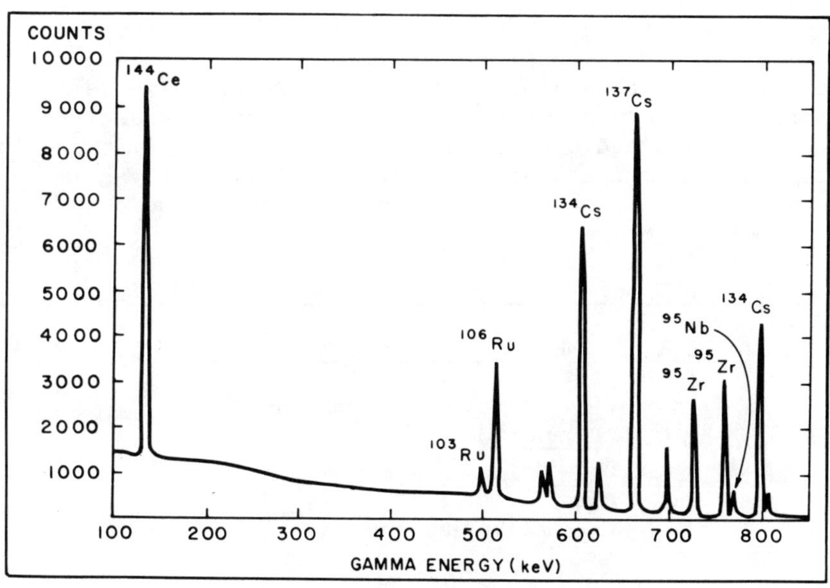

FIGURE 2.   THE SPECTRUM OF A TYPICAL FISSION PRODUCT SAMPLE

Samples from the reactor basins are also analyzed routinely. The analysis of these samples provides information for environmental control. The samples may contain activation products and fission products. The spectrum of a typical basin sample is shown in Figure 3. In addition to the ^{51}Cr, ^{134}Cs, and ^{137}Cs shown in Figure 3, each sample is also specifically analyzed for ^{144}Ce, ^{95}Zr, ^{95}Nb, and ^{131}I.

Reactor Moderator Samples

The gamma analysis of moderator samples provides data related to reactor component failure and deionizer efficiencies. Due to the complexity of the gamma activity in these samples, many of the radionuclides require chemical separation. Chemical separation is usually required for the determination of ^{91,92}Sr, ^{237}U, ^{238,239}Np, and ^{141,143}Ce. Determinations of ^{24}Na, ^{51}Cr, ^{140}Ba, ^{131}I, ^{95}Zr, and ^{95}Nb are usually made without chemical separation; however, several days aging may be required to allow for the decay of short-lived radionuclides. Around-the-clock analyses of these samples are not provided, but they are normally analyzed on day shift Monday through Friday and on special request.

Laboratory Waste

Waste that is generated in the laboratory must be monitored for plutonium and neptunium content to ensure proper handling and storage. Table II shows the gamma energy which is used for each transuranium element for which the waste is monitored. The transmission method for determining self attentuation is used with plutonium or neptunium sources. The precision obtained by this method is usually ±25%. ^{237}Np is determined from its daughter radiation. Low energy gamma analysis of waste has been avoided because of the high density of most of the waste.

TABLE II. GAMMA ENERGIES USED FOR WASTE MONITOR

Radionuclide	Gamma Energy, keV
^{239}Pu	129.3
^{238}Pu	152.7
^{237}Np (^{233}Pa)	311.9

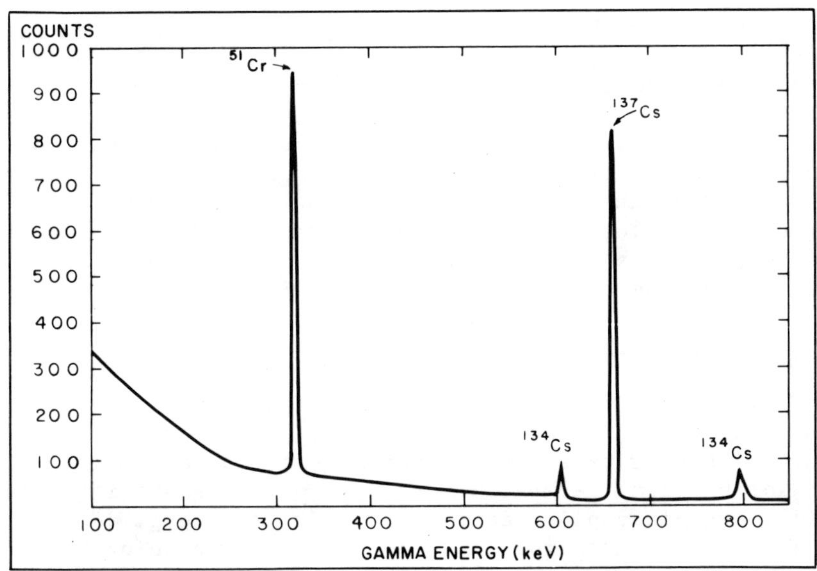

FIGURE 3.   THE SPECTRUM OF A TYPICAL REACTOR BASIN SAMPLE

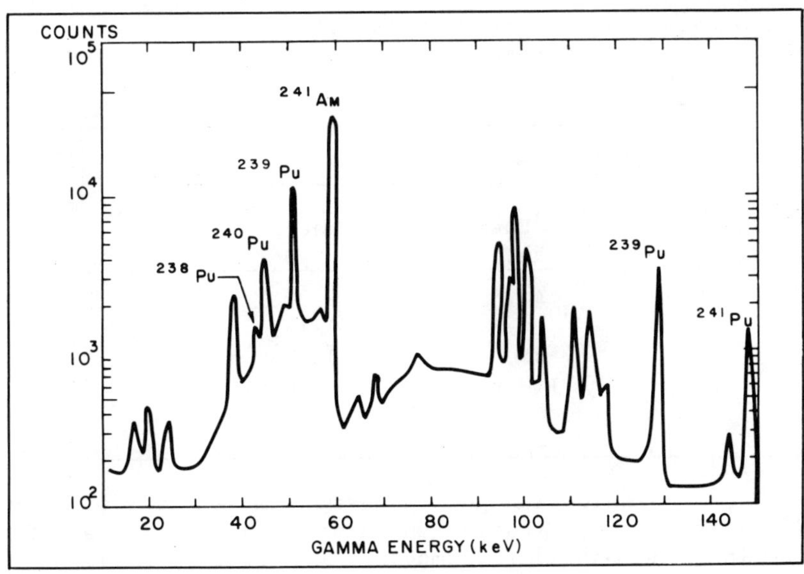

FIGURE 4.   LOW ENERGY GAMMA SPECTRUM OF PLUTONIUM

Plutonium Isotopic Distribution

Presently, the only routine application of low energy gamma spectrometry is the analysis of plutonium isotopic distribution. A typical spectrum is shown in Figure 4. Typical results, precision, and the gamma energies normally used are shown in Table III. Thin copper adsorbers have been used to improve the spectra. After plutonium ages, the ^{241}Am content increases and the gamma energies below 60 keV cannot be used. For these samples, cadmium adsorbers are used to attentuate the 60 keV gamma and the complex area around 100 keV is used for isotopic determination.

DETECTOR SELECTION

All of the detectors which are used routinely are 5 to 10% coaxial Ge(Li) detectors with the exception of the low energy detector used for plutonium isotopic determination. It is a high purity germanium detector which was specifically built for this application. Almost any low energy detector could be calibrated for this application, including the versatile Gamma-X^{TM} detector.

The Gamma-X^{TM} is an ideal detector for special samples that require radionuclide identification. With the proper use of adsorbers and standards, this detector can be used for both high and low energy gamma spectrometry.

TABLE III. PLUTONIUM ISOTOPIC DISTRIBUTION

Radio-nuclide	Gamma Energy, keV	Isotopic Distribution	Precision
^{238}Pu	43.5	0.0001	±10%
^{239}Pu	51.6	0.935	±0.1%
^{240}Pu	45.2	0.058	±1.2%
^{241}Pu	148.6	0.007	±2.5%

The hand-held high purity germanium detector is used for radionuclide identification and gamma energy ratio applications in locations where sampling is difficult or impossible. This system is quite useful in determining the origin of radiation sources.

High purity germanium detectors are a must in locations
where liquid nitrogen may not be readily available. They are
also ideal for applications where the detector may not be used
for long periods of time. Commercial shipping of these detectors is also much easier than it is for those requiring
liquid nitrogen.

Even with these high resolution detectors, an NaI(Tl) is
by far the best choice when good precision and high efficiency
is needed for the determination of relatively pure radionuclides such as in standard preparation. Two 3 × 3 in. NaI(Tl)
detectors are maintained for just such specialized applications.

REFERENCES

1. ORTEC Application Note AN 34, Experiments in Nuclear
   Science, July 1976.
2. V. Barnes, United Kingdom Atomic Energy Authority, Wind-
   seale Works, Seascale, Cumberland, England: <u>GASPAN - An
   Advanced Computer Code for the Analysis of High Resolution
   Gamma-Ray Spectra</u>.
3. R. Gunnink, J. B. Niday, and P. D. Siemens, Lawrence
   Livermore Laboratory, Rept. UCRL-51577 (1974).

This paper was prepared in connection with work under Contract AT(07-2)-1 with the U.S. Department of
Energy. By acceptance of this paper, the publisher and/or recipient acknowledges the U.S. Government's right to
retain a nonexclusive royalty-free license in and to any copyright covering this paper, along with the right to
reproduce and to authorize others to reproduce all or any part of the copyrighted paper.

GAMMA SPECTROMETRY AND CALIBRATION METHODS
USED IN NEUTRON DOSIMETRY

R. L. Malewicki, R. R. Heinrich, and R. J. Popek.
Chemical Engineering Division, Argonne National
Laboratory, Argonne, Illinois, 60439, USA.

ABSTRACT

Neutron activation rates are calculated from measured
gamma-ray spectral data obtained from calibrated lithium
drifted germanium [Ge(Li)] detectors. The calibration techniques, which include energy, efficiency, pulse height analyzer data reduction, and geometry factors, are discussed.
Problems encountered when analyzing highly radioactive samples, specifically random coincidence summing, sample size,
air absorption, and use of absorbers, are also discussed
briefly. To illustrate the accuracy achievable, the reduction of gamma-ray spectral data and the calculation of reaction rates by three independent laboratories are presented
for both fission and nonfission foils to a precision of
± 1-2% (1σ). The application of these neutron activation
rates to neutron dosimetry is discussed. An unfolded neutron
spectrum is illustrated, along with energy sensitivity limits
for many reactions of dosimetry interest.

INTRODUCTION

The Analytical Chemistry Laboratory (ACL) of Argonne
National Laboratory (ANL) is playing a significant role in
providing accurate neutron dosimetry measurements in support
of national fast reactor and fusion reactor programs. In
order to provide reliable foil reaction rates necessary for
neutron dosimetry, considerable care must be taken in calibration methods. Recognizing the need for the capability of
determining consistent and reliable experimental values for
reaction rates, the USAEC, in 1971, established the

Interlaboratory LMFBR Reaction Rate (ILRR) program.[1,2] The
initial goal of this program was an accuracy to better than
±5% at the 95% confidence level for the principal fission reactions, namely ^{235}U, ^{238}U, and ^{239}Pu. Accurate measurement
of other fission and nonfission reactions were required, but
to a lesser accuracy--between ±5 and 10% at the 95% confidence level. Argonne National Laboratory was one of eight
laboratories participating in this program. Others included
Oak Ridge National Laboratory, Hanford Engineering Development
Laboratory, National Bureau of Standards, Aerojet Nuclear
Company, Los Alamos Scientific Laboratory, Atlantic Richfield
Hanford Company, and Atomics International.

To achieve the accuracy required by ILRR, detailed evaluation of the radiometric method, which depends upon absolute
gamma counting by means of Ge(Li) detectors was required.
Presented are the detector calibration techniques, corrections used in adjusting the counting data, and the results of
an interlaboratory comparison of both fission and nonfission
reaction rates.

EQUIPMENT

The ACL utilizes three Ge(Li) detectors, each with a
relative efficiency of about 8% and a resolution of about
2.0 KeV for the 1.33-MeV line of ^{60}Co. The detectors are
connected to either a 4096-channel, computer-assisted pulse-
height analyzer or to computers that have been programmed to
operate as pulse-height analyzers. Besides collecting a
4096-channel spectrum for two Ge(Li) detectors simultaneously,
the computer system allows control of an automatic sample
changer and on-line data analysis. Both analyzer systems
are equipped with a read-write magnetic tape data storage
capability for off-line computer analysis of the Ge(Li)
gamma-ray spectra.

DETECTOR CALIBRATION

The largest uncertainty in gamma-ray counting generally
resides in the detector efficiency versus gamma-ray energy
calibration. Absolute gamma-ray standards are obtained for
detector calibration from the National Bureau of Standards
and Amersham-Searle Corporation.* Their disintegration rates
are typically determined by 4Π beta counting or coincidence
techniques. In general, the precision between duplicate
absolute standards has been ±1.2% and the quoted accuracy

---

*Amersham-Searle Corporation, Des Plaines, Ill., 60005.

ranges from 0.8 to 2.9%. The absolute efficiency calibration
of each detector is determined at a fixed position 10 or 20
centimeters from the cryostat face, and all detectors are
intercalibrated in an energy range from 0.1 to 2.0 MeV.

The values for efficiency versus gamma-ray energy determined at the absolutely calibrated position are fit to a
fifth order polynomial, the coefficients of which are used
in a computer analysis program for determining the efficiency
of any gamma-ray within the energy range of calibration. The
same computer code for analyzing Ge(Li) spectra is used both
for unknown samples and for detector calibration. This as-
sures that each unknown spectrum will be analyzed in the same
manner as the standards used for calibration.

The ACL counts samples whose activities vary by several
orders of magnitude; therefore, it is necessary to calibrate
counting positions (shelves) at distances ranging from 5 to
500 centimeters from the detector. Absolute calibration at
each position is impractical because of the large number of
varying standard source strengths that would be required.
Our approach is to relate all geometry positions to the absolutely calibrated position by shelf factors. From the relative count rates of 32 gamma-ray energies ranging from 0.122
to 1.8 MeV, shelf factors for 11 different positions have
been established. A weighted least squares fit of these data
indicates that for distances between 5 and 80 centimeters a
constant shelf factor can be used with an accuracy within
±1%. At distances greater than 80 centimeters, the deviation
of the shelf factor with energy is characteristic of air absorption. After applying air-absorption corrections, the
shelf factor is independent of energy to better than 1%.

The use of various geometry positions enables counting
at rates below 2000 counts per second, and thus avoids common
problems associated with random summing, gain shifts, resolution degradation, etc.

CORRECTIONS

Many of the samples that are counted require several
corrections to the Ge(Li) data. Gamma-ray self-absorption
corrections are necessary for thick samples or heavy-element
samples for which the linear absorption coefficient is large.
For most cases, the first order approximation may be used:

$$(1) \qquad I = I_o\left(\frac{1 - \exp(-\mu\rho x)}{\mu\rho x}\right)$$

where I is the observed intensity of the gamma-ray, I_o is the
desired intensity, x is the thickness of the sample in centimeters, ρ is the density of the sample in g/cc, and μ is the
mass absorption coefficient in cm^2/g. In the case of wire
dosimeters, self-absorption corrections are determined by the
first order approximation:

$$(2) \qquad I = I_o \, \exp\!\left(-\frac{8}{3\pi}\,\mu\rho R\right)$$

where R is the wire radius, in centimeters. This approximation is valid if $\mu\rho R < 0.1$ and R is very small compared with
the source-to-detector distance. Samples that contain certain fission products with significant beta activity or those
that contain many low-energy gamma-rays are usually counted
through external absorbers. External absorber corrections
are made by the equation:

$$(3) \qquad I = I_o \, \exp(-\mu\rho x)$$

This equation is also used for correction of absorption due
to encapsulation and for making air absorption corrections
for samples counted at a source-to-detector distance of >80
centimeters.

Since our detectors are calibrated with "point-source,
zero-thickness" standards, corrections must be made for samples that are large in diameter and/or relatively thick. The
closer such a sample is to the detector, the greater this
correction. Jaffey[3] has mathematically addressed the magni-
tudes of these sample size and thickness corrections. Thus,
by knowing the Ge(Li) detector size and the location of the
detector in the cryostat, one can use this mathematical approach to determine the necessary sample-related corrections.

DATA ANALYSIS

The computer code used in the analysis of Ge(Li) spectra
is a modified version of GAMANAL written by R. Gunnink
et al.[4,5] Basically, the code locates peaks within the spectrum by describing a tangent to the peak and noting where the
tangent changes signs. Peak boundaries are determined by
noting the forward and backward direction where the slope of
the peak either levels off or changes signs. If an overlap
of several peaks occurs, the boundaries are determined in the
same manner, but after the last peak of the cluster. Background is determined by subtracting the area under the defined
boundaries from the gross counts under the peak. A statistical uncertainty is then assigned to the net count under the

peak. Peak energies are determined by relating the peak
position or channel number to a unique polynomial equation
which describes the nonlinearity of the counting system.

An auxiliary library program provides information on
nuclide numbers, half-lives, parent-daughter relationships,
accurate gamma-ray energies and intensities, efficiency calibrations, and geometry factors. From this information,
counts per minute are converted to photons per minute, peaks
are identified, disintegrations per minute are calculated,
and confidence levels are assigned. The success of this code
depends primarily upon accurate energy versus efficiency calibration and accurate nuclear data, namely, gamma-ray energies and intensities.

RESULTS

Reaction rates were determined by three independent laboratories on both nonfission and fission dosimeters. Each
laboratory had its own set of dosimeters of common origin
which were irradiated in a fast-flux reactor under nearly
identical conditions. All participating laboratories agreed
to use the same gamma-ray intensities and half-lives in their
calculations. The resulting radioactive products had gamma-
ray energies ranging from 159 KeV to 1596 KeV with the majority of the gamma-rays possessing different energies than
those of the primary standards used in the detector calibrations. All of the samples counted by our laboratory were
counted to better than 1% statistics on three different detectors. These results were then averaged to obtain the best
possible reaction rate. A composite of results for the three
laboratories for both types of dosimeters is summarized in
Tables 1 and 2.

The quoted uncertainties (given in percent) are each
laboratory's estimates of the absolute uncertainties on its
measurements. In our uncertainty analysis, we have included
detector efficiency, gamma-ray intensity, half-life, precision, self-absorption corrections, external absorber corrections and counting statistics. The most significantly varying uncertainty was found to be associated with the absolute
uncertainty in the gamma-ray intensity, the second most significant uncertainty being the detector calibration. Even
with this type of uncertainty assessment, the data from all
laboratories are remarkably consistent. The average value
given in the tables represents the mean of the results from
the three reporting laboratories. No considerations have
been given to the individual laboratory's estimated absolute
uncertainties in determining this mean. Therefore, the

Table I. Interlaboratory Comparison of
Fission-Product Reaction Rates

Reaction	Reaction Rate (10^{-15} atoms·atom^{-1}·s^{-1})			
	Lab A(±%)[a]	Lab B(±%)[a]	Lab C(±%)[a]	Avg.±1σ(%)[b]
^{235}U(n,f)^{140}Ba-La	6.64±2.1	6.75±1.7	6.86±2.0	6.75±1.6
^{103}Ru	3.70±4.0	3.75±3.8	3.81±3.9	3.75±1.5
^{95}Zr	7.13±2.4	7.17±1.9	7.18±2.7	7.16±0.4
^{238}U(n,f)^{140}Ba-La	.250±2.1	.250±1.7	.253±1.9	.251±0.7
^{103}Ru	.270±4.0	.265±3.8	.263±4.2	.266±1.4
^{95}Zr	.213±2.4	.213±1.9	.224±2.0	.217±2.9
^{239}Pu(n,f)^{140}Ba-La	7.00±2.8	7.11±1.6	7.13±1.9	7.08±1.0
^{103}Ru	9.57±4.0	9.36±3.8	9.51±4.1	9.48±1.1
^{95}Zr	6.35±2.4	6.40±1.9	6.41±2.2	6.39±0.5
^{237}Np(n,f)^{140}Ba-La	1.96±3.3	2.07±1.9	--	2.02±3.9
^{103}Ru	2.04±4.1	2.18±4.8	--	2.11±4.7
^{95}Zr	2.12±3.8	2.09±2.5	--	2.11±1.0

[a]Estimated absolute uncertainty.

[b]Standard deviation.

Table II. Interlaboratory Comparison of
Nonfission Reaction Rates

Reaction Rate (10^{-15} atoms·atom^{-1}·s^{-1})			
Lab A(±%)[a]	Lab B(±%)[a]	Lab C(±%)[a]	Avg.±1σ(%)[b]
^{197}Au(n,γ)^{198}Au	19.0±2.6	18.7±1.8	19.0±1.3	18.9±0.9
^{58}Ni(n,p)^{58}Co	1.32±2.3	1.34±1.6	1.34±1.9	1.33±0.9
^{54}Fe(n,p)^{54}Mn	.979±2.7	.986±2.3	.975±2.5	.980±0.6
^{58}Fe(n,γ)^{59}Fe	.329±4.3	.334±4.0	.340±3.8	.334±1.6
^{45}Sc(n,γ)^{46}Sc	1.46±3.7	1.44±1.6	1.45±2.1	1.45±0.7
^{46}Ti(n,p)^{46}Sc	.137±3.0	.143±2.8	.143±3.1	.141±2.5
^{47}Ti(n,p)^{47}Sc	.239±4.8	.227±4.3	.231±4.0	.232±2.6
^{238}U(n,γ)^{239}Np	11.9±3.8	12.4±2.2	12.1±2.8	12.1±2.1

[a]Estimated absolute uncertainty.

[b]Standard deviation.

uncertainty represents solely the standard deviation of the
mean, or a measure of the precision of the results and not a
measure of accuracy.

DOSIMETRY APPLICATION

The term neutron dosimetry, as used in this work, is
defined as the technique of characterizing the magnitude and
the energy distribution of a neutron flux. Of the various
dosimetry methods available, the multiple foil activation
method is best suited to high intensity neutron environments,
such as mixed fission reactors (ORR), fast reactors (EBR-II),
T(d,n) sources (RTNS-I), and Be(d,n) sources. For convenience, consider dosimetry divided into two phases; the first
consisting of foil activation-rate measurements and the
second, methods for determining the flux, fluence and spectral
shape from the activation-rate data. As was presented, measurements of activation data can be made relatively precise
and accurate. However, the methods used to deduce flux and
spectral information require assumptions and data which can
result in uncertainties far greater than those associated
with the reaction-rate data.

The principle of the foil activation method can be expressed in the form:

$$(4) \qquad A_i = \int_0^\infty \phi(E)\sigma_i(E)dE \qquad i = 1,2,3\ldots n$$

where A_i is the saturated activity of a particular foil i,
$\phi(E)$ is the unknown neutron flux, and $\sigma_i(E)$ is the cross section for the reaction under consideration. The product of
$\sigma_i(E)\phi(E)$ defines the sensitivity region of the reaction for
the measured flux. If the differential (point-wise energy
dependent) cross sections are well known for each foil reaction over the energy range of their sensitivity and a good
approximation (physics calculation) of the neutron spectrum
is available, then the unknown neutron spectrum can be unfolded using appropriate computer algorithms and numerous
iterations. Computer codes which are commonly used to perform this analysis include SAND-II,[6] SPECTRA,[7] and CRYSTAL-
BALL.[8] Illustrated in Figure 1 is an unfolded Be(d,n) neutron
spectrum obtained from 27 foil reaction rates. The solid
lines represent the 90% sensitivity limits of the various
reactions. Neutrons were generated by stopping 16-MeV deu-
terons within a thick beryllium target. Foils were located
at 2.01 cm from the target face and at zero degrees to the
incident deuteron beam. Input neutron spectra for the SAND-II

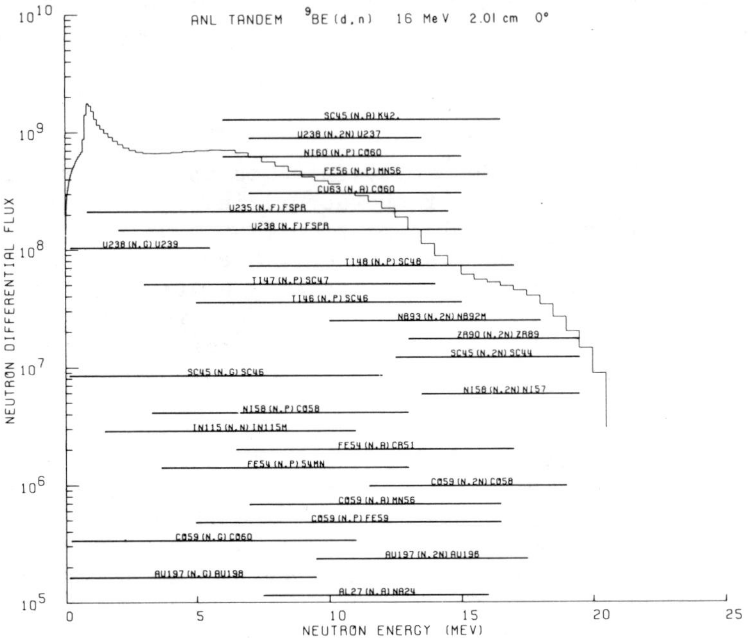

Figure 1.   Be(d,n) Neutron Spectrum Unfolded
from 27 Foil Reaction Rates.

code were obtained experimentally by the time-of-flight (TOF)
technique.

SAND-II is the analysis code used most frequently in our
laboratory. However, a relatively new code, STAY'SL,[9] which
obtains the neutron distribution, as well as uncertainty estimates, is currently being tested and appears to have some
advantages. Currently, in our laboratory, uncertainties on
the experimental neutron distribution are obtained from the
code SANDANL,[10] which consists of Monte Carlo analysis of
estimated uncertainties on cross sections, input neutron
spectra, and reaction rates. Until recently, reaction cross
sections were considered to be the greatest contributor to
neutron distribution uncertainties, especially at higher neutron energies (>20 MeV) where differential cross section data
are sparse. Recent work[11,12] using Be(d,n) sources for the
integral testing of reaction cross section data has indicated
that integral cross sections most common to dosimetry are accurate to about ±10%. It is our opinion that uncertainties
in the input neutron distribution used in the computer code
analysis contribute most significantly to the uncertainty of
the measured neutron distribution.

162

ACKNOWLEDGMENTS

The authors wish to express their appreciation to the participants of the ILRR counting laboratories for the use of their counting data in the interlaboratory comparison. These laboratories include HEDL, EGG-Idaho, and ANL. Work was performed under the auspices of the U.S. Department of Energy.

REFERENCES

1. "LMFBR Reaction Rate and Dosimetry 1st Quarterly Progress Report," W. N. McElroy, Ed., HEDL-TME 71-131, Hanford Engineering Development Laboratory, June 1971.

2. W. N. McElroy, Nucl. Tech., 25, 177 (1975).

3. A. H. Jaffey, Rev. Sci. Instrum., 25, 349 (1954).

4. R. Gunnink, H. B. Levy, and J. B. Niday, "Identification and Determination of Gamma Emitters by Computer Analysis of Ge(Li) Spectra," UCID-15140, Lawrence Radiation Laboratory, May 1967.

5. R. Gunnink and J. B. Niday, "Computerized Quantitative Analysis of Gamma-Ray Spectrometry," UCRL-51061, Vol. I-V, Lawrence Livermore Laboratory, March 1972.

6. W. N. McElroy et al., "SAND-II, Neutron Flux Spectra Determinations by Multiple Foil Activation Iterative Method," TSIC Computer Code Collection, CCC-112, Oak Ridge National Laboratory, Radiation Shielding Information Center, May 1969.

7. C. R. Greer et al., "A Technique for Unfolding Neutron Spectra from Activation Measurements," TSIC Computer Code Collection, CCC-108, Oak Ridge National Laboratory, Radiation Shielding Information Center, December 1967.

8. F. B. K. Kam and F. W. Stallman, "Crystal Ball," ORNL-TM-460, Oak Ridge National Laboratory, June 1976.

9. F. G. Perey, "Least Squares Unfolding: The Program STAY'SL," ORNL-TM-6062, ENDF-254, Oak Ridge National Laboratory, October 1977.

10. G. R. Odette, Trans. Am. Nucl. Soc., 22, 803 (1975); revised by L. R. Greenwood.

11. L. R. Greenwood, R. R. Heinrich, R. J. Kennerly, and
     R. Medrzychowski, Nucl. Tech., $\underline{41}$, 109 (1978).

12. L. R. Greenwood, R. R. Heinrich, M. J. Saltmarch, and
     C. B. Fuller, Nucl. Sci. Eng., to be published.

TELLURIUM INTERFERENCE IN ^{129}I ACTIVATION ANALYSIS

J. H. Kaye, F. P. Brauer, and R. S. Strebin, Jr.
Physical Sciences Department, Analytical and Nuclear
Research Section, Pacific Northwest Laboratory,
Richland, Washington, USA.

ABSTRACT

The major sources of environmental ^{129}I are artifically-induced fission processes. Environmental levels of ^{129}I are routinely measured around nuclear facilities, especially fuel reprocessing facilities. The measurement of ^{129}I at low levels is required for effluent control, environmental assessment and tracer studies in the natural environment. Low level ^{129}I analysis has also been used for measurement of natural-fission-produced ^{129}I in minerals and ores as well as meteorites. The most sensitive method for ^{129}I measurements is neutron activation analysis of iodine separated from a suitable sample. A potential interference in the activation analysis method is the production of ^{129}I (^{130}I) from Te impurities in the irradiated ampoule. Procedures have been developed and applied to measurement of the magnitude of the potential Te interference. This paper discusses the procedures and results of Te interference studies in ^{129}I activation analysis.

INTRODUCTION

Iodine-129 measurements are currently being made in a number of laboratories in order to gain a better understanding of the behavior of this isotope in the environment. The most sensitive method for its measurement currently available is thermal neutron activation analysis, a procedure first applied to the analyses of ore samples[1] and later developed for the analysis of various biological materials by Studier, et al.[2] The procedure first involves isolation of the iodine from the sample, followed by irradiation with thermal neutrons to yield the 12.4-hour activation product ^{130}I, radiochemical separation and purification (of the iodine) and finally measurement of the resulting ^{130}I radioactivity.

There are four reactions with thermal neutrons which can potentially interfere:

1. $^{127}I(n,\gamma)^{128}I(n,\gamma)\ ^{129}I(n,\gamma)^{130}I$

2. $U, Pu(n,f)^{129}I(n,\gamma)^{130}I$

3. $^{133}Cs(n,\alpha)^{130}I$

4. $^{128}Te(n,\gamma)^{129}Te \xrightarrow[T_{\frac{1}{2}}=70\ min]{\beta-}\ ^{129}I(n,\gamma)^{130}I$

The first reaction involves successive capture of three neutrons by natural iodine in the irradiated sample. The interference from ^{127}I is usually not serious because the half-life of ^{128}I is so short (25 minutes) that it largely decays to ^{128}Xe before capturing a second neutron. However, this mode of ^{130}I production may be important for cases in which a large amount of natural iodine is present in the sample, such as in analysis of thyroid tissues or sea water. Interference from this reaction can be minimized by irradiating at a reduced flux, since the production of ^{129}I from ^{127}I varies as the square of the neutron flux.

The second interfering reaction can occur if the irradiated sample contains uranium or any other fissionable nuclide which produces ^{129}I. This can then capture a neutron to give ^{130}I, so that a high estimate for ^{129}I is obtained. Iodine-130 is not produced directly by fission in any significant yield. If uranium or other fissionable nuclides are present then ^{133}I and ^{135}I will also be produced. These nuclides can be measured by counting methods in order to estimate the quantity of fission-produced ^{129}I.

A third possible interference is due to the n,α reaction on natural cesium in the irradiated sample. The amount of cesium in most samples is low, however, and cesium is separated during the pre-irradiation chemical purification process so this interference may generally be ignored.

The fourth potential source of interference is due to capture of a neutron by the stable isotope ^{128}Te, followed by beta decay to ^{129}I. The purpose of this study was to determine the importance of this interfering reaction.

The method employed at this laboratory for pre-irradiation separation of iodine involves oxidation of the sample and collection of the liberated iodine on charcoal.[3] The charcoal trap is then oxidized and the volatilized iodine

is collected on a second, smaller charcoal trap. This
second trap is evacuated, heated, and the iodine driven to
the other end which is cooled in liquid nitrogen. This
section of tubing is then sealed off giving an ampoule ready
for irradiation.

It seemed conceivable that under these circumstances
oxides of tellurium could end up in the final ampoule. The
volatility of tellurium and some of its semi-volatile
compounds are listed in Table 1.[4] A preliminary tracer study
was done to see how serious this problem might be. The results are shown in Table 2. The amount of tellurium which
could get into the ampoule was found to be largest for
samples which were almost entirely consumed during the combustion process, such as vegetation. Much less was observed
for samples such as soil where the majority of the sample remains in the combustion tube after oxidation. In general,
the tracer seemed to settle on any available surface downstream from the high temperature region of the furnace.
Results of the tracer study alone were not sufficient because
the amount of tellurium present in the original samples was
not known. A search of the literature revealed that the concentration of tellurium in most materials of interest had
never been measured. A calculation shows that as little as
0.1 μg of natural tellurium under normal irradiation conditions would be sufficient to account for the observed
activity level in laboratory blanks. This is equivalent to
about 10^8 atoms of ^{129}I. In this study typical samples of
vegetation and one meteorite (Allende) were analyzed for ^{129}I
content and for tellurium content by two different methods.
The two methods for measuring tellurium were compared, and
the amount of ^{129}I produced from tellurium was estimated.

EXPERIMENTAL

<u>Post-Irradiation Procedure for Separation of Tellurium and Iodine</u>

The post-irradiation procedure for separation of iodine
has been well established, and it was desired to adapt it for
recovery of tellurium with minimum modification. This turned
out to be straightforward. After irradiation, the ampoule
was crushed and carriers for iodine, tellurium and bromine
were added. The bromine and iodine were then distilled off,
as described elsewhere.[3] Tellurium remains in the distillation pot. The solution was filtered to remove quartz frag-
ments, and concentrated NH_4OH was added until brown MnO_2
precipitated (the bromine distillation requires addition of
permanganate). Four drops of 30% H_2O_2 were added and the
solution was heated. Four drops of 2% aerosol were added,

then SO_2 gas was passed through the solution to precipitate tellurium. The precipitate was washed with 40 ml hot deionized water saturated with SO_2. Then it was dissolved in 1 ml of hot concentrated HCl containing two drops of concentrated HNO_3. It was diluted to 10 ml volume with 3$\underline{M}$ HCl. The solution was passed through a column 4 cm high and 7 mm in diameter containing 1.5 ml of strong anion exchange resin (Bio-Rad AG-1x8(100-200 mesh) which had been pre-washed with 10 ml of deionized water and 10 ml of 3 $\underline{M}$ HCl.

The column was rinsed with 15 ml of 3 $\underline{M}$ HCl, then the tellurium was eluted with 30 ml of 0.3 $\underline{M}$ HCl. The solution was neutralized with NH_4OH, 4 drops of 2% aerosol were added, it was heated on a hot plate and tellurium was precipitated with SO_2. The precipitate was filtered, washed with 30 ml of hot deionized water saturated with SO_2, dried under a heat lamp and weighed. The mean chemical yield was 70 percent. The major loss was due to incomplete precipitation of tellurium during the first SO_2 addition step.

<u>Counting Equipment and Procedure for Measurement of Tellurium</u>

Decay of tellurium isotopes was followed by counting with a small GM counter interfaced to a printing scaler. The counter was surrounded on three sides by an anticoincidence guard detector. Q-gas (98.7% helium, 1.3% iso-butane) was passed through the main counter and guard counter at a flow rate of 0.03 ℓ-min^{-1}. The counter background was about 0.15 c-min^{-1}. The scaler contents were printed out every 20 minutes onto adding machine paper tape.

From the decay curves plotted from the counting data, the isotopes ^{127}Te, ^{131}Te and ^{131m}Te were identified and the ^{127}Te activity level was estimated. The amount of tellurium in the irradiated ampoule was determined by comparing the ^{127}Te counting rate 600 minutes after the end of the irradiation with the counting rate of samples of telluric acid irradiated and processed in the same manner. Corrections for chemical yield were made by weighing each sample.

The ^{130}I and ^{131}I levels in the same samples were also measured, using methods described elsewhere.[3] Samples containing 0.1, 1.0 and 10.0 µg of Te as telluric acid were irradiated and the ^{130}I and ^{131}I counting rates were measured for calibration purposes. The irradiations were performed for approximately 16 hours at the Hanford N Reactor, at a thermal neutron flux of about 10^{14} n-cm^{-2}-sec^{-1}.

RESULTS AND DISCUSSION

The results of this study are presented in Tables 3-5.
In Table 3, the amount of Te in each sample was estimated by
comparison of the ^{127}Te activity with that of the 0.1 µg Te
standard after chemical yield adjustment. The average of the
^{129}I results for the 1 and 10 µg Te standards was then used
to estimate the atoms of ^{129}I in a given irradiated sample
from the Te level for that sample. The ^{129}I value for the
0.1 µg Te standard is obviously in disagreement with the
values for the other two standards and this value was disregarded.

In Table 4 the amount of ^{129}I due to tellurium is
estimated from the ^{131}I activity. The average number of atoms
of ^{129}I per microgram of tellurium was calculated and divided
by the average ^{131}I counting rate per microgram of tellurium,
again only using data for the 1 and 10 µg tellurium standards.
This gave 2.2×10^5 atoms of ^{129}I per c-min^{-1} of ^{131}I. In this
way the amount of ^{129}I in each sample due to tellurium can be
estimated from the observed ^{131}I activity extrapolated back
to the end of irradiation and corrected for chemical yield.

In Table 5 the total observed amount of ^{129}I and the
fraction of ^{129}I due to tellurium are given for each sample.
Estimates of the amount of ^{129}I produced based on measurements of both ^{127}Te and ^{131}I activities are included. The
most significant observation from Table 5 is that the contribution to the amount of ^{129}I in the sample due to tellurium
is generally negligible. The only exception is for the case
of the Allende meteorite, for which up to 23 percent of the
^{129}I may come from activation of tellurium in the irradiated
sample.

The ^{129}I levels in blanks generally are on the order of
3×10^8 atoms. It would appear that an insignificant part of
this is due to tellurium in the irradiated sample. Again the
exception is for the case of the meteorite samples. After
being oxidized, the powdered meteorite residue was re-spiked
with ^{125}I tracer and re-run as a second sample. For this
case both the ^{127}Te and ^{131}I measurement methods indicate
that a significant part of the observed ^{129}I may have been
due to activation of tellurium.

Agreement between the two measurement methods is rather
poor, as can be seen from Table 5. The method based on
counting of ^{131}I is both more sensitive and more selective,
since beta-gamma coincidence techniques can be used to discriminate against interfering radionuclides. The method
based on following the decay of ^{127}Te requires that the

sample to be counted be free of other beta emitters with a similar half-life.

CONCLUSIONS

Two methods have been successfully developed and tested for determining the amount of interference in ^{129}I analysis caused by natural tellurium in the irradiated sample. Both methods indicate that for the vegetation samples analyzed Te interference does not pose a significant problem. Activation of Te likewise does not appear to account for the amounts of ^{129}I found for laboratory blanks. However, up to about one fourth of the ^{129}I measured in the Allende meteorite sample may be due to interference from activation of ^{128}Te. The method based on counting ^{131}I is the more sensitive and selective of the two methods, because in addition to obtaining half-life information one can discriminate against other radionuclides based on gamma spectroscopic measurement of the 364 KeV ^{131}I gamma-ray.

TABLES

TABLE 1. VOLATILITY OF TELLURIUM COMPOUNDS[*]

Compound	Melting Point	Boiling Point
Te metal	452° C	1390° C
$TeCl_4$	224	380
TeO_2	733	1245
TeO_3	d395	----

[*]from Handbook of Chemistry and Physics, 56th Ed (1975-1976).

TABLE 2. RESULTS OF PRELIMINARY STUDY ON TELLURIUM VOLATILITY WITH ^{129m}Te TRACER

Sample Type	Fraction of initial tracer in 1st charcoal absorber	Fraction initial tracer in 2nd charcoal absorber	Fraction of initial tracer in final ampoule
Vegetation	0.16	0.062	1.3×10^{-3}
Soil	8.4×10^{-3}	2.9×10^{-3}	2.0×10^{-4}

TABLE 3. ATOMS OF ^{129}I ESTIMATED FROM AMOUNT OF
TELLURIUM IN IRRADIATED SAMPLES

Sample I.D.	Sample Type	^{127}Te Activity 600 min. after end of irradiation (c-min^{-1})	Chemical Yield (percent)	Te in irradiated ampoule (μg)+	^{129}I from Te+++ (atoms)
68993	10 μg Te		--	--	1.06×10^{10} #
68994	1 μg Te	1.14×10^{5}	85	1.12	1.64×10^{9} #
68997	0.1 μg Te	1.15×10^{4}	96	0.10	(5.31×10^{9})* #
56755	quartz wool blank	3.13×10^{2}	94	2.8×10^{-3}	3.8×10^{6}
28079	charcoal blank	2.18×10^{2}	58	3.1×10^{-3}	4.2×10^{6}
34114	vegetation, leaves	1.01×10^{3}	45	$\leq 9.4 \times 10^{-3}$ ++	$\leq 1.3 \times 10^{7}$
56689	vegetation pine needles	6.50×10^{2}	85	6.4×10^{-3}	8.6×10^{6}
56776	vegetation, forage	2.70×10^{3}	27	8.4×10^{-2}	1.1×10^{8}
46231	meteorite, Allende	6.79×10^{4}	91	6.2×10^{-1}	8.4×10^{8}
46233	meteorite, Allende, re spiked	9.20×10^{3}	43	1.8×10^{-1}	2.4×10^{8}

* This sample was contaminated somehow and result was disregarded.

+ All samples compared with 0.1 μg Te standard

Measured values

++ This sample was contaminated with activities other than Te radioisotopes

+++ Average of values obtained for 1 and 10 μg standards, 1.35×10^{9} atoms ^{129}I per μg Te, was used to calculate atoms of ^{129}I from quantity of Te in each sample.

TABLE 4. ATOMS OF ^{129}I ESTIMATED FROM ^{131}I COUNTING RATE

Sample I.D.	Sample Type	^{131}I Activity at end of irradiation (c-min^{-1})	^{129}I from Te + (atoms)
68993	10 μg Te	6.9×10^4	1.06×10^{10} [+][#]
68994	1 μg Te	5.4×10^3	1.64×10^9 [+][#]
68997	0.1 μg Te	7.0×10^2	(5.31×10^9) [*][#]
56755	quartz wool blank	8.0	1.8×10^6
28079	charcoal blank	2.4×10^1	5.3×10^6
34114	vegetation, leaves	1.3	2.9×10^5
56689	vegetation, pine needles	6.6	1.5×10^6
56776	vegetation, forage	1.3×10^2	2.9×10^7
46231	meteorite, Allende	2.3×10^3	5.1×10^8
46233	meteorite, Allende re-spiked	2.8×10^2	6.2×10^7
46232	blank before running meteorite 46231	3.9	8.9×10^5

[#] Measured value

[+] Average of measured ^{131}I and ^{129}I values for 1 and 10 μg Te standards used to estimate atoms of ^{129}I from ^{131}I activity (2.2×10^5 atoms ^{129}I/c-min^{-1} ^{131}I).

[*] This sample was contaminated somehow, and result was disregarded.

TABLE 5. CONTRIBUTION OF TELLURIUM TO TOTAL
OBSERVED ^{129}I IN A NUMBER OF SAMPLES

Sample I.D.	Sample Type	Total ^{129}I, blank corrected (atoms)	Fraction Due to Te		Ratio
			Based on ^{127}Te activity	Based on ^{131}I activity	$\frac{^{127}\text{Te Method}}{^{131}\text{I Method}}$
56755	Quartz wool blank	2.9×10^8	0.013	0.0062	2.1
28079	Charcoal blank	1.5×10^9	0.0028	0.0035	0.80
34114	Vegetation leaves	2.8×10^8	≤ 0.046	0.0010	≤ 46
56689	Vegetation, pine needles	6.8×10^9	0.0013	0.00022	5.8
56776	Vegetation, forage	7.6×10^{10}	0.0014	0.00038	3.7
46231	Meteorite, Allende	3.6×10^9	0.23	0.14	1.6
46233	Meteorite, Allende, re-spiked	2.0×10^8	1.20	0.31	3.9
46233	blank before running 46231	3.3×10^8	-----	0.0027	-----

This paper is based on work performed under Department of Energy Contract No. EY-76-C-06-1830. The authors would like to thank Dr. Louis Rancitelli for suggesting this study and giving valuable suggestions. The authors also wish to thank NASA (Contract NAS-9-12589) for financial support for the measurement of Te interference levels for the Allende meteorite.

REFERENCES

1. B. C. Purkayastha and G. R. Martin, "The Yields of ^{129}I in Natural and in Neutron-Induced Fission of Uranium," Can. J. Chem., pp. 293-299, 1956.

2. M. H. Studier, C. Postmus, Jr., J. Mech, R. R. Walters and F. N. Sloth, "The Use of ^{129}I as an Isotopic Tracer and its Determination Along with Normal ^{127}I by Neutron Activation--The Isolation of Iodine from a Variety of Materials," J. Inorg. Nucl. Chem., Vol. 24, pp. 755-761, 1962.

3. F. P. Brauer and H. Tenny, "^{129}I Analysis Methodology," Battelle, Pacific Northwest Laboratories Report No. BNWL-SA-5287, 1975.

4. R. C. Weast, Editor, "Handbook of Chemistry and
    Physics," 56th Ed., pp. B-148-149. 1975-1976.

DETERMINATION OF TECHNETIUM-99 IN MIXED FISSION
PRODUCTS BY NEUTRON ACTIVATION ANALYSIS

L. C. Bate, Analytical Chemistry Division, Oak Ridge
National Laboratory, Oak Ridge, Tennessee 37830

ABSTRACT

A method has been developed for analysis of ^{99}Tc in
fission product mixtures. The analysis consists of a chemical separation of ^{99}Tc, neutron irradiation of the isolated
^{99}Tc, and gamma-ray spectrometric determination of the
induced ^{100}Tc radioactivity.

Technetium-99 is chemically separated from most fission
products by a cyclohexanone extraction from basic carbonate
solution. Technetium-99 is stripped into water by addition
of carbon tetrachloride to the cyclohexanone phase. A final
step in the separation procedure is adsorption of ^{99}Tc on
an anion exchange column which provides additional decontamination and places the ^{99}Tc in a concentrated form for neutron
activation analysis.

Neutron irradiations of the isolated ^{99}Tc were made in
the pneumatic tube facility at the High Flux Isotope Reactor
at a flux of 5×10^{14} n/cm^2/sec for 11 seconds. Induced ^{100}Tc
radioactivity was determined immediately after irradiation
using gamma-ray spectrometry to measure the 540 and 591 keV
lines. Sensitivity of the analysis under these conditions
is approximately 5 ng, and samples of up to about 100 ml
volume can be easily processed. The method has been
successfully applied to reactor fuel solutions and off-gas
traps containing 6.5×10^{-4} to 240 μg ^{99}Tc/ml.

INTRODUCTION

The fate of ^{99}Tc in the chemical reprocessing of reactor
fuels is important in relation to fission product inventory
and eventual disposal or recovery of the isotope. Technetium-
99 is a major constituent in fission product mixtures from

high burn-up fuels because of its high fission yield (6%) and long half-life (2.1×10^5 years). The behavior of ^{99}Tc in chemical reprocessing operations is complicated by the presence of technetium in multiple oxidation states and the volatility of technetium under some conditions. Chemical analyses are occasionally required to establish or verify the behavior and/or fate of ^{99}Tc in reprocessing operations.

Analysis of ^{99}Tc in fission product solutions by direct counting techniques is not possible as ^{99}Tc emits only a 292 keV beta in its decay; chemical isolation is required before beta counting measurements can be made (1).

Neutron activation analysis (NAA) is another technique which has been used for ^{99}Tc analysis (2-5). Submicrogram quantities of ^{99}Tc can be determined by the measurement of the ^{100}Tc produced by n,γ reaction. The thermal and resonance neutron cross sections for the neutron capture reaction are 19 and 340 barns, respectively, and the half-life of the ^{100}Tc product is 15.8 seconds (6). Technetium-100 has two prominant gamma-rays (540 and 591 keV) with 6.6 and 5.4% abundances which permits analysis by gamma-ray spectrometry.

This paper describes a method for NAA of ^{99}Tc which is applicable to highly radioactive fission product mixtures. In the method, ^{99}Tc is separated from most fission products by solvent extraction with cyclohexanone. Either ^{95m}Tc or ^{99m}Tc is used as a tracer to establish chemical recovery in the separation. Neutron activation analyses are performed by concentration of the separated ^{99}Tc on a small ion-exchange resin column and irradiation of the resin in a pneumatic tube irradiation facility. Samples are counted immediately after irradiation because of the short half-life of ^{100}Tc. Computer programs are used to process the gamma-ray spectral data and to obtain NAA results by absolute techniques.

EXPERIMENTAL

Equipment and Reagents

A modified polyethylene irradiation insert is used as an ion exchange column for preconcentration of the ^{99}Tc prior to NAA. Figure 1 shows the modified insert and a polyethylene rabbit that contains the insert during irradiation. Approximately 20 holes of 0.4 mm diameter are drilled through the bottom of the insert to allow passage of the solution through the resin bed. The inserts are washed in 3 $\underline{M}$ HNO$_3$ after the holes are drilled to remove aluminum and other contaminants from the insert surfaces.

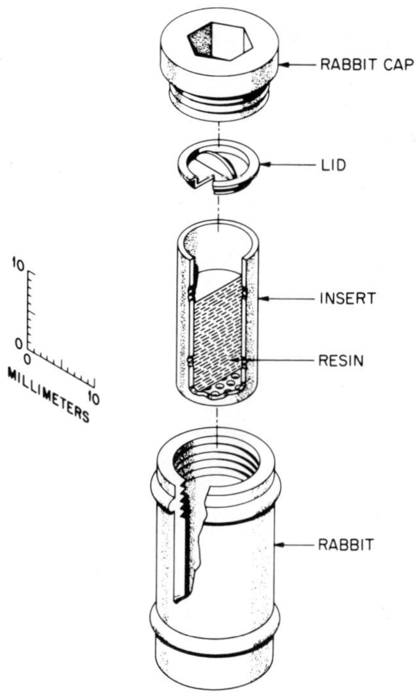

Figure 1. Irradiation rabbit and
insert with ion exchange resin.

Because of the small size of the column, special
apparatus was constructed to aid in loading the ^{99}Tc on the
resin column. The apparatus is shown in Figure 2. Vacuum
is applied to the botton of the insert from a 250 ml filtering flask and is controlled by a stopcock to adjust the flow
rate through the column.

Bio-Rad Dowex 1X1 (50-100 mesh) anion exchange resin in
the chloride form is converted to the nitrate form by adding
20-30 gram of the slurried resin to a 100 ml funnel.
Entrained air bubbles are removed by gentle stirring. The
foam which collects at the liquid surface is carefully
removed after the resin has settled. Stirring is repeated
to ensure that all the foam is released and removed. The

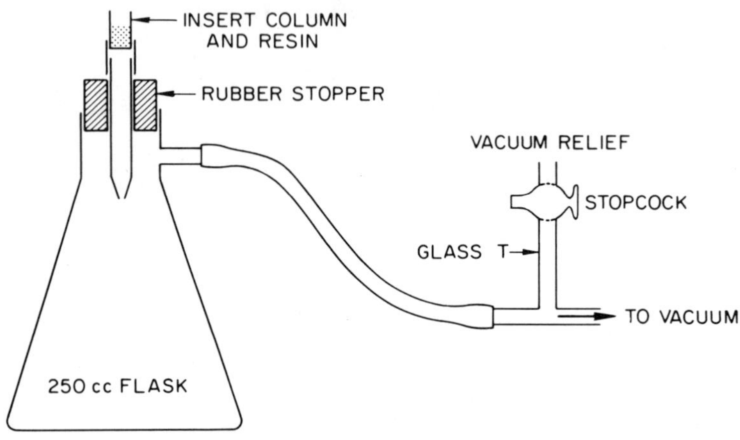

Figure 2. Apparatus for loading of
solution on ion exchange resin column.

resin is washed with 40 ml of 2 $\underline{M}$ NaNO$_3$ and then with water
until the volumn of the resin has attained the prewashed
condition. Washing of the resin with 2 $\underline{M}$ NaNO$_3$ and water
is repeated twice, and the resin is transferred as a slurry
to a bottle for storage. The chloride content of the resin
should be less than 50 ppm.

The ^{99m}Tc isotope used in this work was obtained by
irradiating a molybdenum salt and making a ^{99}Mo-^{99m}Tc
generator (7). Technetium-99m was separated from ^{99}Mo
immediately before use. Technetium-99 standards were made
by dissolving technetium metal in nitric acid and stabilizing
as ammonium pertechnetate by adding NH$_4$OH and H$_2$O$_2$. The
technetium content of the solution was determined by liquid
scintillation measurements using a ^{99}Tc standard obtained from
the Radiochemical Centre, Amersham, England. The technetium
standard was solution number S5/17/23 and contained 13.40
microcuries of ^{99}Tc per gram of solution with an overall
uncertainty of $\pm$1.2% (RSD).

The gamma-ray spectrometer system used in this work was
a Nuclear Data PDP-15 series 50-50 system coupled to a Ge(Li)
detector of 10% efficiency. Irradiations were performed in

the NAA facilities at the High Flux Isotope Reactor (HFIR) at
Oak Ridge National Laboratory. This facility has a high and
stable neutron flux (thermal: 5×10^{14}; resonance: 1.6×10^{13}
$n/cm^2/sec$).

<u>Recommended Procedure</u>

The procedure for the separation of the ^{99}Tc from other
fission products is given below along with information on the
irradiation and radioactivity measurements.

One milliliter of ^{99m}Tc or ^{95m}Tc internal standard for
yield determination and a predetermined volume of sample
solution are added to a 50 ml separatory funnel. If the
sample is basic, nitric acid is added until the sample is
acidic. Five milliliters of 5 $\underline{M}$ K_2CO_3 are added slowly to
neutralize the sample. The contents of the funnel are mixed,
1 ml of 30% H_2O_2 is added, and the solution is mixed. This
solution is allowed to oxidize for 30 minutes with occasional
shaking. Ten milliliters of cyclohexanone are added, and the
solution is shaken for two minutes. The solution is then
centrifuged to separate the layers, and the cyclohexanone is
transferred to a clean separatory funnel. Ten milliliters
of cyclohexanone are added to the original sample for a
second extraction of 2 minutes duration. This cyclohexanone
is added to the first extract. Four milliliters of distilled
water, 1 ml of 5 $\underline{M}$ K_2CO_3, and five drops of 30% H_2O_2 are
added to the combined cyclohexanone extractions, and the
separatory funnel is shaken for two minutes. The solution
is centrifuged to separate the layers, and the wash solution
is discarded. Five milliliters of water and 10 ml of CCl_4
are added to the separatory funnel. The solution is shaken
for two minutes and centrifuged, and the aqueous layer is
transferred to a clean separatory funnel. The mixture is
shaken for two minutes, centrifuged, and the aqueous layer
is transferred to the first strip solution. One half
milliliter of 1 $\underline{N}$ NaOH is added to the Tc strip solution to
ensure that the solution is basic prior to adsorption on
the ion exchange resin.

A cleaned resin column insert is approximately one-half
filled with slurried Bio-Rad 1X1 resin as shown in Figure 2.
Excess liquid is removed by adjusting the stopcock on the
vacuum line to maintain a flow rate of one drop per second.
The resin is washed with 2 ml of distilled water to ensure
that air bubbles are removed from the column. After the Tc
is adsorbed, the resin column is washed with 5 ml of 1 $\underline{M}$ $NaNO_2$
solution to remove the adsorbed iodide. The column is then
washed with 5 ml of distilled water to remove sodium. The
stopcock on the vacuum line is closed to remove the water

179

from the resin, and the resin column insert is capped and
placed in a labeled vial.

The chemical yield of technetium in the separation is
determined by adding a 1 ml aliquot of ^{99m}Tc or ^{95m}Tc internal
standard solution to a centrifuge cone with 1 ml of 1 $\underline{N}$ NaOH.
The solution is mixed and transferred to a column insert.
Technetium is quantitatively retained by the resin column.
The resin column is dried, capped, and placed in a counting
vial for gamma-ray spectrometry measurements. The internal
standard and samples are counted at the same geometry. Count
rates of the internal standard and samples are used to determine the chemical yield of the separation. After the yield
measurements are made, the resin column inserts are placed
on the multiple sample drying vacuum system (Figure 3) for
at least two hours of drying before irradiation.

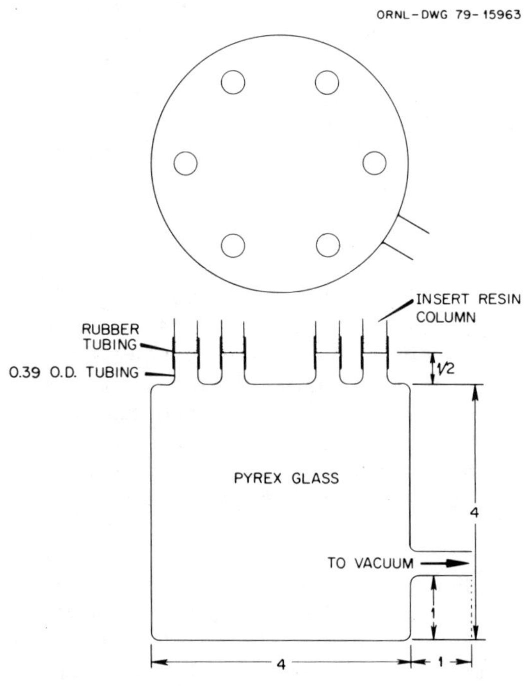

Figure 3. Apparatus for drying
of ion exchange columns.

At our laboratory, the insert resin column was placed
in a rabbit for an 11 second irradiation in the HFIR
pneumatic tube facility. After irradiation, the rabbit was
opened, and the insert resin column was transferred to a
clean vial for counting on a Ge(Li) detector. Induced ^{100}Tc
radioactivity was measured for 30 seconds. The decay time
from the end of irradiation to start of measurement ($\sim$ 30
seconds) was recorded and used when the gamma-ray spectrum
was processed by computer program MONSTR (8).

RESULTS AND DISCUSSION

<u>Solvent Extraction Separation of Technetium</u>

Solvent extraction techniques were chosen for separation
of ^{99}Tc in the recommended procedure because extractions are
in general easy to apply to highly radioactive samples and
do not require a chemical carrier. As reported by Boyd and
Larson (9), heptavalent technetium can be readily extracted
by ketones, amines, and organo-phosphorus compounds. High
selectivity is obtained with ketones, and cyclohexanone was
chosen for the procedure. Table I shows that ^{99m}Tc, as
pertechnetate, is efficiently extracted from several aqueous
media with cyclohexanone. Also, cyclohexanone has very
low solubility in aqueous media which provides good phase
separation and low volume changes during extraction.

Table I

<u>Extraction of ^{99m}Tc and ^{106}Ru with Cyclohexanone</u>

Solution	^{99m}Tc, % Ext.	^{106}Ru, % Ext.
3 <u>M</u> HNO$_3$	95.1	60.0
1 <u>M</u> HNO$_3$	96.8	–
3 <u>M</u> KOH, 1 ml H$_2$O$_2$ (30%)	85.6	18.8
3 <u>M</u> K$_2$CO$_3$,[1] 1 ml H$_2$O$_2$ (30%)	92	0.5
3 <u>M</u> K$_2$CO$_3$,[1,2] 1 ml H$_2$O$_2$ (30%)	90.3	0.0015

---

[1]3 <u>M</u> HNO$_3$ added prior to K$_2$CO$_3$ addition.

[2]30 min. oxidation time before extraction.

181

After the initial extraction with cyclohexanone, ^{99}Tc
is stripped from the organic phase into an aqueous phase for
subsequent loading on an anion exchange resin column. This
stripping step provides additional decontamination from
fission products and gives an aqueous solution which can be
applied directly to the resin column. Technetium was stripped
by addition of a non-polar, organic diluent to the cyclohexanone as described previously by Foti and coworkers (2).
Table II shows data on several organic diluents that were
tested to determine their effectiveness in removing ^{99m}Tc
from cyclohexanone. This data was obtained by addition of
10 ml of the diluent to 20 ml of cyclohexanone and stripping
with 5 ml of water for 1 minute. All diluents remove greater
than 80% of the ^{99m}Tc in a single stripping step. Carbon
tetrachloride was chosen as the diluent for the procedure
because it is advantageous to have an organic phase heavier
than water. Chloroform could also be used as a diluent in
the analysis. With carbon tetrachloride, greater than 99%
of the technetium can be stripped from the organic phase
using two successive contacts with 5 ml of distilled water
as specified in the procedure. Table III gives decontamination factors for the extraction and stripping steps for
several radionuclides present in old fission product mixtures.

Table II

Stripping of ^{99m}Tc from Cyclohexanone

Diluent	% Removal
Cyclohexane	84.0
Benzene	91.4
Xylene	82.3
Chloroform	83.0
Carbon Tetrachloride	81.8

Table III

Fission Product Separation

| | Radioactivity, dps/sample | | | |
Isotope	Original Solution (x 10^{-5})	Washed Cyclohexanone	Aqueous Strip	Decontamination Factor
^{106}Ru	2.78	<2	<2	>2x10^5
^{125}Sb	1.91	<4	<4	>1x10^4
^{134}Cs	26.6	<2	<2	>2x10^4
^{137}Cs	9.29	<2	<2	>9x10^5
^{154}Eu	2.54	<5	<5	>2x10^5
^{131}I	526	500	15	350

Table IV

Residual Impurities in Bio-Rad
Anion Exchange Resin in Nitrate Form

| | Concentration of Impurities, ppm | | | |
Resin	Cl	Br	I	Al
1x8	1900	4.7	<1	9.7
1x4	510	0.42	0.065	1.3
1x2	210	2.0	1.81	3.70
1x1	34	0.93	0.045	0.70

Selection of Anion Exchange Resin

The final step in the separation is adsorption of pertechnetate on a small anion exchange resin column. This step provides additional decontamination and concentrates the technetium for the subsequent NAA. It was desirable that the NAA be performed on the loaded resin, and this required that the resin be free of large quantities of contaminants which would give radioactive products during neutron irradiation. Several Bio-Rad anion exchange resins were irradiated to determine the impurity level. As shown on Table IV, aluminum, chlorine, bromine, and iodine are the contaminants which would cause difficulties in the analysis. The low cross-linked resin (1X1) showed the lowest level of contamination and is recommended for the analysis. Careful cleaning of the resin, as specified in the experimental section, is necessary to obtain the lowest possible contamination level.

Resin samples after thorough drying contained about 0.1% oxygen. This oxygen forms 29 sec ^{19}O during the NAA. Oxygen-19 radioactivity causes no difficulty in the gamma spectral measurements of ^{100}Tc if a Be absorber is used to attenuate the high energy beta particles from ^{19}O.

Fission Product Separation

Iodine and ruthenium are the only fission products that extract with technetium and could cause interference in the analysis. Up to 10% of the iodine in the sample can accompany technetium through the extraction and stripping steps. Since fission product mixtures may contain several isotopes of iodine, special provisions were made to eliminate iodine as a potential interference.

Selective elution of iodine from the anion exchange resin column with sodium nitrite was found to be effective for separation of iodine and technetium. Figure 4 shows the elution behavior of iodine (as iodide) and technetium (as pertechnetate) on the anion exchange resin column. Iodide is rapidly eluted from the column with 1 $\underline{M}$ NaNO$_2$ while technetium is retained. Elution with 5 ml of 1 $\underline{M}$ NaNO$_2$ removes greater than 90% of the iodide with negligible losses of technetium. The selective elution of iodine with sodium nitrite was tested on samples containing 150 µg/ml ^{129}I, and a decontamination factor of 5000 was obtained.

Ruthenium is partially extracted with cyclohexanone, and special pretreatment of the sample prior to extraction is necessary to prevent ruthenium isotopes from interfering in the analysis. As shown in Table I, ruthenium extraction

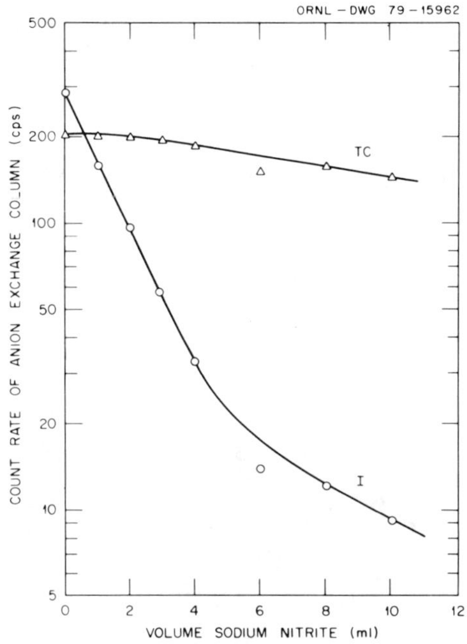

Figure 4.   Elution of technetium and
iodine from ion exchange column.

is reduced to an insignificant level if the sample is pretreated by adding 1 ml of 30% H_2O_2 and allowing the sample
to stand for 30 minutes before extraction.   Presumably,
ruthenium is oxidized by the peroxide to an inextractable
valence state and remains in the aqueous phase during the
extraction.

Results Obtained from Standards and Samples

Several standard ^{99}Tc samples were irradiated, and the
radioactivity was measured to determine the best counting
parameters for the [100]Tc radioactivity and the detection
limit of the method.   The results of this study are shown
in Table V.   The concentration of ^{99}Tc found agrees very well
with the known value of the ^{99}Tc standards except for samples
4 and 5 which were measured without a Be absorber.   As

Table V

Neutron Activation Analysis of ^{99}Tc Standards

Sample No.	^{99}Tc, μg	
	Added	Found
1	1.02	1.00
2	0.0665	0.0673
3	0.0665	0.0654
4	0.0665	0.0572*
5	0.0665	0.0552*
6	0.0133	0.0124

*No Be absorber used

mentioned previously, the Be absorber is necessary to attenuate the beta particles from ^{19}O.

The data in Table VI are the results of analysis of two dissolved reactor fuel solutions and three caustic gas scrubber traps. The result of the ^{99}Tc analysis for the dissolver solutions are in good agreement with the values obtained from the computer code ORIGEN (10). Data in Table VII are the results of replicate analyses on two samples; the maximum deviation in these results is less than 10%. Chemical yields for the separation procedure are also given in Table VII. The chemical yield was determined by measuring the ^{99m}Tc radioactivity in the separated product and comparing this with the original ^{99m}Tc radioactivity added to the sample. The yield of the ^{99m}Tc was lower than desired in two of the samples.

Discussion

The internal standard used in the development of this method was ^{99m}Tc but ^{95m}Tc could also be used for yield determination if a longer-lived isotope is needed and ^{95m}Tc is available. The half-life of ^{99m}Tc is 6 hours, and the ^{99m}Tc must be milked daily from ^{99}Mo which has a 66 hr. half-life. A new ^{99}Mo-^{99m}Tc isotope generator system can

Table VI

^{99}Tc in Reactor Fuel Material

Sample	Found ^{99}Tc, μg/ml	Estimated ^{99}Tc, μg/ml*
Dissolver Solution – A	237	200
" " – B	184	200
Caustic Scrub Trap – A	0.032	
" " " – B	0.0034	
" " " – C	0.00065	

*From computer code ORIGEN (10)

Table VII

Replicate Chemical Yield and Analysis Results on Two Samples

Sample	Chemical Yield, %	Found ^{99}Tc, μg/ml
Caustic Scrub Trap – A-1	68.7	0.030
" " " – A-2	71.0	0.030
" " " – A-3	84.5	0.034
" " " – A-4	83.1	0.030
Dissolver Solution – B-1	58.7	172
" " – B-2	95.5	177
" " – B-3	92.7	178

only be used for three weeks before it has decayed beyond its usefulness. Therefore, ^{95m}Tc would be preferred for yield determination if it is available.

The irradiation of technetium samples adsorbed on Bio-Rad 1X1 resin was normally performed for 11 seconds which results in the highest ^{100}Tc radioactivity in relation to the radioactivity resulting from resin impurities. The measurement of the ^{100}Tc radioactivity when adsorbed on the resin must be done with a Be absorber to eliminate the interference of high energy beta radioactivity in the sample. The counting time for the sample is 30 seconds which is about two half-lives of the ^{100}Tc. This short counting time also aides in reducing the interference of other radioactivities in the sample.

The detection limit for the NAA is about 5 nanograms using this separation and activation analysis method. Samples of 100 ml have been run by the method, giving a ^{99}Tc detection limit of approximately 0.05 nanogram per milliliter.

The limits of detection could be improved with resin that contained lower quantities of impurities. The resin used in this work could not be placed closer than 3 cm to the detector due to ^{28}Al, ^{38}Cl, ^{19}O, and ^{40}A radioactivities. If the contamination level of the resin and insert could be reduced, the samples could be measured closer to the detector for a higher counting efficiency. Also, the time from the end of the irradiation to the start of measurement was about 30 seconds, which allowed the ^{100}Tc to decay two half-lives before measurements were started. An improved method of transferring the samples from the reactor to the gamma-ray spectrometer would also enhance the sensitivity of the analysis.

ACKNOWLEDGEMENTS

Oak Ridge National Laboratory is operated by the Union Carbide Corporation under contract W-7405-eng-26 with the U.S. Department of Energy.

REFERENCES

1.   Edward Anders, "The Radiochemistry of Technetium, NAS-NS-3021, Feb. 1961.

2.   S. Foti, E. Delucchi, and V. Akamian, "Determination of Picogram Amounts of Technetium in Environmental Samples by Neutron Activation Analysis", Anal. Chem. Acta., <u>60</u>, p. 269-76 (1972).

3.   S. Foti, E. Delucchi, and V. Akamian, "Determination of
     Picogram Amounts of ^{99}Tc by Neutron Activation Analysis",
     Anal. Chem. Acta., $\underline{60}$, p. 261-68 (1972).

4.   J. H. Kay and N. E. Ballou, "Determination of Technetium
     by Graphite Furnace Atomic Absorption Spectrometry",
     Anal. Chem., $\underline{50}$, p. 2076-78 (1978).

5.   J. H. Kaye, M. S. Rapids, and N. E. Ballou, "Determination
     of Picogram Levels of Technetium-99 by Isotope Dilution
     Mass Spectrometry", Proc. of 3rd Inter. Conf. on Nuclear
     Method in Environmental and Energy Research, Univ. of
     Mo., Columbia, Mo, in press.

6.   C. M. Lederer, J. M. Hollander, and I. Perlman, Table
     of Isotopes", 6th Ed., John Wiley and Sons, Inc., New
     York, NY (1967).

7.   L. C. Bate, "A Novel ^{99}Mo-^{99m}Tc Generator System",
     report in preparation.

8.   J. F. Emery and F. F. Dyer, "Multielement Determination
     in Environmental NAA using MONSTR", Proc. of 2nd Conf.
     on Nucl. Methods in Environmental Research, Columbia,
     Mo, July 29-31, 1974, p. 123-130.

9.   G. E. Boyd and Q. V. Larson, "Solvent Extraction of
     Heptavalent Technetium", J. Phys. Chem., $\underline{64}$, p. 988-
     996 (1960).

10. J. M. Bell, "ORIGEN - The ORNL Isotope Generation and
     Depletion Code", Oak Ridge National Laboratory Report
     4628, May 1973.

MULTIELEMENT PROTON ACTIVATION ANALYSIS: APPLICATION
TO AIRBORNE PARTICULATE MATTER

P. Priest, M. Devillers, and G. Desaedeleer,
Université Catholique de Louvain, 2, chemin du
Cyclotron, 1348 Louvain-la-Neuve, Belgium

ABSTRACT

Proton activation analysis in the range of 25 to 30
MeV proton energies allows the determination of Na, Mg, Ca,
Ti, Fe, Zn, As, Sr, Sn and Pb in airborne particles collected
by 4-7 stage impactors. Under normal, not limitative irradiation and counting conditions, the determination is accurate for samples collected from 1-10 m^3 of air in rural
atmospheres.

INTRODUCTION

Nondestructive atomic or nuclear methods are successfully applied to the determination of many elements present
in airborne particles. Among nuclear methods, instrumental
neutron activation analysis (INAA) is largely used because of
the wide range of elements investigated; however a few
elements of great toxicological concern (i.e. Cd, Sn and Pb)
require alternative methods. Proton activation in the range
of 10-12 MeV energy has been used for a long time for the
determination of light elements or trace impurities in
refractory metals i.e., in thick targets[1-3]. The method
described here is aimed towards proton activation of thin
samples of air particles collected by cascade impactors.
The method, already used for the determination of lead in
aerosols[4], was developed because of the following advantages:

- many samples can be irradiated at a time together with
  standards.

- the proton beam can easily be focused on the aerosol spot,

- the high sensitivity of 30 MeV proton activation allows
  the determination of many heavy elements in the relatively
  small amount of matter collected by the impactors.

EXPERIMENTAL

Air particles are sampled by two cascade impactors:
Aries (flow = 1.1 1/min.) or Delron (flow = 12 1/min.); they
are deposited on Whatman 540 cellulose filter paper (thickness 10 mg/cm^2). Standards are made by punctual spot impreg-
nation on the same kind of support with a 2 microliter
calibrated standard solution.

Fifteen to twenty samples are stacked in a pile and
mounted in a water cooled target holder; heat dissipation in
the stack is ensured by a flow of helium.

The irradiated area corresponds to the spot of aerosol
and standard (0.8 cm^2) and is defined by appropriate colli-
mators; as a consequence, relatively high activities are
induced in the heavy elements to be determined, as compared
with activities induced in the support material (paper).

Irradiations are carried out in the external beam of
the isochronous cyclotron of Louvain-la-Neuve. In routine
analysis, 30 MeV incident protons are used, under beam
intensities of 300 nA. Irradiations last for 2 hours; this
leads to integrated currents of the order of 2 mC, as measured on a Faraday cup coupled to an integrator. Along the
stacked samples, the proton beam energy decreases to about
25 MeV, without any appreciable loss of intensity.

Radionuclides are identified by γ-spectrometry using
a Phillips APY 40 AIN Ge(Li) detector with a useful volume
of 80 cm^3, 16% relative efficiency, and a resolution of
2.09 keV at 1333 keV. The detector signal is fed into a
Northern Econ II multichannel analyzer through a Canberra
2011 amplifier. The samples are counted after at least
4 hours cooling time to allow sufficient decay of the short-
lived β^+ emitters induced in the filter paper.

ANALYTICAL RADIOISOTOPES - INTERFERENCE PROBLEMS

In a previous investigation[5], analytical properties of
the following elements were studied: Na, Mg, Cl, Ca, Ti,
Cr, Mn, Fe, Ni, Cu, Zn, As, Br, Sr, Cd, Sn, Sb, and Pb.
Since elements contiguous in the periodic table can lead to
the production of the same radionuclide, other elements were
also considered on an interference basis: S, Sc, V, Co, Ga
and Se. This preliminary study made it possible to select

the best analytical radionuclides for each element considered,
as well as counting and decay times. This is summarized in
Table I; this procedure is the best compromise when a rather
large number of samples are to be counted routinely. The
interference appearing in Table I causes no serious problem,
except for Mn and Fe -ecause it appears that, in the peculiar
case of airborne particles, elements likely to add parasitic
activities in an analytical peak of a given element are
present in a much lower concentration than this element:
see for example the interference of Sc on Ca (Table I) at
372 keV which is reduced to a negligible amount because of
the very low content of Sc as compared with Ca (typical
concentrations: Ca 1.300 ng/m^3 – Sc 0.43 ng/m^3. See
Table II).

Interferences in the γ–spectrum are few when using the
selected γ peaks of Table I: at 159 keV, a contribution from
^{47}Sc (T = 82 h arising from Ti) and at 372 keV, a contribution
from ^{43}K (T = 22 h arising from Ca) and from ^{204}Bi.

Finally, since the proton energy decreases along the
pile of samples, we determined the production rate of each
of the analytical radionuclide at various proton energies in
the range 30–25 MeV; the results are presented elsewhere[5].

RESULTS

The sampling site is located at Louvain-la-Neuve, 30 km
S-E of Brussels in a semi-rural area. Air pollution episodes
from industrial areas of the country are detected on the
sampling site by means of a sulfur dioxide monitor. A
previous study[7] has shown that heavy elements such as Br, As,
Sb, Se, Cr, ... are closely associated with SO_2 during the
episodes.

Twenty-eight impactor samples have been collected
between May '78 and May '79; 19 samples correspond to "high"
SO_2 levels (80–520 μg/m^3); 9 samples have been collected in
"clean air" conditions.

The sensitivity of the analytical method allows one to
restrict the sampling time to a few hours. As a consequence,
it is possible to strictly synchronize granulometric sampling
with SO_2 immissions eqisodes which typically last for 2 to
6 hours. Such a procedure markedly differs from the usual
procedures based on a regular time basis sampling.

Table I

30 MeV Proton Activation

Analytical characteristics of the determined elements

Decay and counting time	Analytical radioisotope (half-life)	γ-energy (keV)	Observed activity counts. $\mu g^{-1} X m C^{-1}$ in the element			
			To be determined		Likely to interfere	
(o = interferences in the γ-spectrum (see text)).						
t_d=4h	^{34m}Cl (32 min)	146	Cl	360	S	3.6
	^{43}Sc (3.9 h)	372 o	Ca	21	Sc	320
	^{44}Sc (3.9 h)	1157	Ti	720	Ca	10
					Sc	6900
	^{61}Cu (3.3 h)	283	Cu	720		
	^{66}Ga (9.4 h)	1039	Zn	130		
	^{73}Se (7.2 h)	361	As	5600	Se	100
	^{109}In (4.3 h)	204	Cd	3800		
	^{117}Sb (2.8 h)	159 o	Sn	6900		
	^{118m}Sb (5.1 h)	254	Sn	2500		
t_d=15h	^{24}Na (15 h)	1368	Mg	220	Na	40
t_c=2000s	^{55}Co (17.9 h)	931	Fe	480		
	^{57}Ni (36 h)	1378	Ni	840	Co	26
	^{66}Ga (9.4 h)	1039	Zn	240		
	^{67}Ga (78 h)	93	Zn	490	Ga	3200
	^{73}Se (7.2 h)	361	As	7800	Se	140
	^{77}Br (56 h)	239	Br	400	Se	420
	^{87}Y (80 h)	388	Sr	3200		
	^{87m}Y (14 h)	381	Sr	7200		
	^{111}In (67 h)	171	Cd	3000		
	^{119m}Te (113 h)	153	Sb	5400		
	^{204}Bi (11.3 h)	899	Pb	1400		
t_d=10d	^{22}Na (2.6 yr)	1274	Na	38		
t_c=10000s	^{51}Cr (27.7 d)	320	Cr	650		
	^{54}Mn (312 d)	835	Mn	280	Fe	80
	^{119m}Te (4.7 d)	153	Sb	6900		
	^{206}Bi (6.2 d)	803	Pb	820		

Table II

Mean Composition of Various Urban Aerosols (ng/m^3) [6]

Na	620	(x)	Cu	100	(x)
Mg	510	x	Zn	260	x
S	1500		Ga	0.5	
Cl	480	(x)	As	3.1	(x)
Ca	1300	(x)	Se	1.2	
Sc	0.43		Br	60	(x)
Ti	85	(x)	Sr	3.9	(x)
Cr	7.5	(x)	Cd	2.3	(x)
Mn	37	(x)	Sn	10	(x)
Fe	1350	x	Sb	4.6	(x)
Co	1.4		Pb	600	x
Ni	10	(x)			

x: detectable in 1 m^3 of air (by proton activation)

(x): detectable in 10^3 of air (idem).

Table III

Average Concentrations in Air
at Louvain-La-Neuve (ng/m^3)

Element	Low SO_2 Conc.	High SO_2 Conc.	MMED (μm)
Na	170	330	3.5
Mg	290	590	3.5
Ca	2500	3800	4.5
Ti	30	140	3
Fe	340	1350	2.5
Zn	320	560	1.5
As	1.8	8.5	1.2
Sr	1.1	7.8	3
Sn	4.6	32	0.25
Pb	110	370	.65

Table II gives a list of the elements usually detected in the air particles. In Table III are presented the mean concentrations of the elements easily detectable during "high" or "low" SO_2 levels episodes. The last column of the Table gives the observed mass-median equivalent diameter (MMED).

The elements under study can be roughly classified into three groups:

- elements of MMED greater or equal to 3 μm: Na, Mg, Ca, Ti and Sr: these are mainly soil-derived.

- Fe and Zn, with MMED around 2 μm: both are soil and combustion derived.

- elements with MMED around or lower than 1 μm: As, Sn, and Pb which are emitted by high temperature processes as volatile compounds and clearly of human origin. It has recently been assumed that municipal incinerators are major sources of airborne Zn, and possibly Sn in many areas[8], while the automative origin of Pb is well established.

As a conclusion, it can be mentioned that proton activation analysis allows the determination of many toxic elements likely to be deposited in the human lungs[9]: Cr, Ni, As, Cd, Sn, Sb and Pb. The method is thus of interest in toxicological studies.

REFERENCES

1. V. Krivan, Anal. Chem., _47_, 469 (1975).
2. J. N. Barrandon, S. Benaben, and J. L. Debrun, Anal. Chim. Acta, _83_, 157 (1976).
3. K. Strijckmans, C. Vandecasteele, and J. Hoste, Anal. Chim. Acta, _89_, 255 (1977).
4. G. Desaedeleer, C. Ronneau, and D. Apers, Anal. Chem., _48_, 572 (1976).
5. P. Priest and G. Desaedeleer, Anal. Chem., (1979), in press.
6. K. A. Rahn, Technical Report, "The Chemical Composition of the Atmospheric Aerosol," (ed. by Graduate School of Oceanography, Univ. Rhode Island), (1976).
7. C. Ronneau, J. L. Navarre, P. Priest, and J. Cara, Atm. Env., _12_, 877 (1978).
8. R. R. Greenberg, W. H. Zoller, and G. E. Gordon, Env. Sci. Tech., _12_, 566 (1978).
9. H. Schroeder, Environment, _13_, 18 (1971).

ENVIRONMENTAL ANALYSIS

SPECIAL PROBLEMS IN THE ANALYSES FOR PLUTONIUM AND
NEPTUNIUM IN GROUND WATERS

T. F. Rees, Transuranium Research Project, Water
Resources Division, U.S. Geological Survey, Lakewood,
Colorado, USA.

ABSTRACT

The wide variety of chemical constituents associated
with waters found in commercial low-level radioactive waste
facilities present special problems in the analyses for
plutonium and neptunium. Samples collected from Maxey
Flats, Kentucky provide data that illustrate the sometimes
subtle problems which affect analytical results. These
samples contain very high dissolved carbon, representing a
large variety of organic compounds. Routine organic destruction steps are not always effective and often give results
with low precision. A procedure is given for neptunium
using a ferrous sulfamate reduction step. The procedure for
plutonium uses a perchloric acid organic destruction followed
by a permanganate oxidation.

INTRODUCTION

The United States Geological Survey (USGS) is presently
conducting aqueous speciation experiments to determine the
geochemical behavior of the transuranic elements under
prevailing conditions at existing low level waste burial
sites in an attempt to develop favorable geologic and
hydrologic criteria for future sites. To normalize the
results of these laboratory studies to _in situ_ ground waters
at the facilities, it is necessary to determine how much
neptunium and plutonium are present in the ground waters.
Two procedures will be presented which allow determining
these two elements in the complex matricies found at the
present sites.

Commercial radioactive waste disposal sites began
operation in 1962 at Beatty, Nevada. Since that time, six
disposal sites have been operated by commercial concerns:
Maxey Flats, Kentucky; Beatty, Nevada; Sheffield, Illinois;
Barnwell, South Carolina; West Valley, New York; and Richland,
Washington. Although managed by private industry, all sites
are located on Federal or State-owned land in remote, sparsely
populated areas.

Maxey Flats was selected for initial study because
waters present in the waste trenches contained plutonium in
concentrations sufficiently high to be readily detectable.
Typically, the solid wastes were buried in shallow trenches
60-150 meters (m) long, 13.5 m wide and 6 m deep. The
trench floors were sloped to one or more sumps, which were
fitted with standpipes to allow removal of collected water
either for monitoring or disposal. The wastes were placed
into the excavated trenches in an assortment of containers,
such as cardboard and wooden boxes, plastic bags, and steel
drums. The filled trenches were then covered with soil and
planted with shallow-rooted ground cover to prevent erosion.

Maxey Flats receives approximately 1 m of rain per
year. The backfilled trenches are more permeable to water
than the rock into which they were dug; as a result, rain
water percolates and collects in the trenches. Excess water
is removed from the trenches using the standpipes. The
presence of water in the trenches promoted premature failure
of many containers, which allows wastes to leach into the
water.

Various wastes buried at Maxey Flats resulted in
leachates, the term describing the solutions present in the
trenches, of extremely complex composition: alkalinity
ranges up to 2,720 mg/L as $CaCO_3$; hardness (Ca + Mg) up to
5,700 mg/L as $CaCO_3$; pH from 2.0 to 12.3; and total solids
up to 11,200 mg/L (see Table I). In addition, many organic
compounds are present; a listing from Weiss, Francis and
Colombo is given in Table II. Considerable care must be
exercised to assure good analytical results from these
samples.

PROCEDURES

Standardization

Two standardization techniques are available to the
alpha radiochemist to determine the chemical yield of a
purification procedure. In the method or technique of
isotopic dilution, an isotope of the analyte element is
added directly to the analyte sample prior to any chemistry
being performed. This procedure requires an isotope which:
(1) is not present in the original sample; and (2) emits

Table I.--Range of measurements of water samples
from trenches at Maxey Flats[a]

Measurement	Range
Alkalinity	125 − 2,720 mg/L as $CaCO_3$
Hardness (Ca + Mg)	398 − 5,700 mg/L as $CaCO_3$
pH	2.0 − 12.3
Residue	1,400 − 11,200 mg/L

[a]Weiss, Francis, and Columbo (1979).[1]

---

alpha particles at energies different from any isotope being
measured. The technique or method of standard additions is
used when no isotope meets these two criteria. Several
aliquots of sample are analyzed with no tracer added along
with several aliquots containing differing quantities of
added analyte isotope. It is assumed that if all samples
are run identically and simultaneously, the amount of added
tracer recovered in the spiked samples will be the same as
the chemical recovery of the unspiked samples. Based on
recovery of the added tracer in either technique, chemical
yield of the separation scheme can be determined and used to
normalize the results to the original sample.

Neptunium Analysis

Since the laboratory did not have access to a suitable
diluent tracer to use in the neptunium analysis, the standard
additions technique was used. Five aliquots of sample were
measured into Teflon beakers, two of which contained a known
quantity of added neptunium-237. All five aliquots were run
simultaneously to insure that all were treated identically.
An equal volume of 16 $\underline{N}$ nitric acid was added to each of the
samples; then the samples were evaporated to dryness. The
residues were then treated with nitric-perchloric acids to
destroy organics. After evaporating once again with concentrated nitric acid, the residues were dissolved in 4 $\underline{N}$ nitric
acid, and 0.1 mL 2.5 $\underline{M}$ ferrous sulfamate was added to
reduce the neptunium to the +4 oxidation state. After
sitting one hour, the solution was made 8 $\underline{N}$ in nitric acid.
When gas emanation ceased, the samples were loaded onto
8 $\underline{N}$ nitric acid Biorad AG 1X4 anion exchange columns. After
rinsing with 8 $\underline{N}$ nitric acid, neptunium was eluted using

201

Table II.--Organic compounds identified in
trench water samples at Maxey Flats[a]

Compound	Compound
Acetic acid	Formic acid
Acetovanillon	4-Hydroxy-3-methoxy benzoic
Arachidic acid	acid
Benzoic acid	Heptanoic acid
Biphenyl	3-Hexanol
Bis (2-chloroethyl) ether	Hexanoic acid
Bis (2-ethyl hexyl) adipate	Iso-butyric acid
Borenol	Iso-octylphthalate
Butanoic acid	Iso-valeric acid
2-Butanol	2-Mercaptobenzoic acid
o-Cresol	2-Methyl-2-butanol
p-Cresol	2-Methyl butanoic acid
Cyclohexane	Methylcyclohexanol
Cyclohexanol	Methyl isobutyl ketone
Diacetone alcohol	2-Methoxy phenyl acetic acid
Dibutylphthalate	Octanoic acid
1, 1-Diethoxyethane-2-	Octanol
chloroethane	Oleic acid
1, 1-Diethoxyethane	Oxalic acid
Diethylphthalate	Palmitic acid
Di-2-ethylhexylphthalate	Paraldehyde
2, 3-Dimethyl-2-hexene	Phenylacetic acid
Dioctyladipate	2-Phenylpropionic acid
Di-(isooctyl) adipate	Phenylpropionic acid
Dioctylphthalate	Stearic acid
p-Dioxane	Tetrahydrofuran
2-Ethylheptanoic acid	Tributylphosphate
2-Ethylhexanol	Triethyl phosphate
3-Ethylhexanol	3, 3, 5-Trimethylcyclohexanol
2-Ethylhexanoic acid	Toluene
Fenchone	Toluic acid
	Vanillin

[a]Weiss, Francis, and Columbo (1979).

1 $\underline{N}$ hydrochloric acid. The eluate was evaporated, treated
with perchloric acid, dissolved in 12 $\underline{N}$ hydrochloric acid,
and evaporated to about 0.25 mL. This residue was diluted
with 4 $\underline{M}$ ammonium chloride; the pH was adjusted to 4 using
dilute ammonium hydroxide and hydrochloric acid; and the
resulting solution was electroplated onto a stainless steel
planchet for one hour at 0.6 ampere. The electroplated disc
was counted on an alpha spectrometer.

Plutonium Analysis

     Four replicate samples were transferred to Teflon
beakers and a known quantity of plutonium-236 isotopic
diluent added to each. An equal volume of 16 $\underline{N}$ nitric acid
was added and the samples were evaporated to dryness. The
residues were treated with nitric-perchloric acids to destroy
organics. This residue was then treated for one hour with
sufficient acid 5 percent (weight/volume) potassium permanganate to maintain the purple color. After 1 hour the samples
were cooled, and the remaining permanganate decolorized with
dropwise additon of 5 percent sodium bisulfite. Additional
bisulfite was added until the brown manganese dioxide just
dissolved. The resulting clear solution was evaporated to
dryness, evaporated twice with 16 $\underline{N}$ nitric acid to dryness,
and finally dissolved in 8 $\underline{N}$ nitric acid. A small amount of
sodium nitrite was added to adjust the plutonium to the
tetravalent oxidation state, and the samples allowed to sit
overnight. The samples were then loaded onto Biorad AG 1X4
anion columns that had been pretreated with 8 $\underline{N}$ nitric acid.
After rinsing the columns with 8 $\underline{N}$ nitric acid, the plutonium
was eluted with 0.4 $\underline{N}$ HNO_3--0.04 $\underline{N}$ HF. The eluate was
evaporated and treated with nitric-perchloric acid to destroy
any organics which might have eluted from the column. The
ashed residue was dissolved in 12 $\underline{N}$ hydrochloric acid,
evaporated to 0.25 mL and electrodeposited by the same
procedure as the neptunium. The dried disc was then counted
on an alpha spectrometer.

DISCUSSION

Neptunium Analysis

     Neptunium forms an analytically useful anionic Np(IV)
nitrate complex. To form this complex, it is first necessary
to reduce neptunium from its stable +5 oxidation state.
Sodium bisulfite is a useful reductant in soil analyses, but
more stringent ferrous sulfamate reduction was necessary in
the leachate samples. The problem was to decide at which
acidity to add the ferrous sulfamate. The nitrate complex
forms more efficiently in 8 $\underline{N}$ acid; however, at that concentration, ferrous sulfamate is destroyed before neptunium is
quantitatively reduced. Reduction, therefore, was done in
4 $\underline{N}$ acid, where the kinetics favor reaction with neptunium.
Acidity was then increased to 8 $\underline{N}$ for the ion exchange step.
This increased acidity destroyed the ferrous sulfamate with
the evolution of gas, which inhibited contact with the resin
bed. A waiting period was necessary to allow this gas to
dissipate prior to ion exchange.

Precision of the technique was studied two different
ways. In the first, twenty samples were analyzed and chemical
recoveries determined. This yielded Result 1 in Table III:
91 $\pm$ 7 percent recovery. In the second technique, replicates
of different samples were analyzed and the relative standard
deviation (RSD) between aliquots was calculated; this yielded
Result 2 in Table III: 16 percent RSD.

Table III.--Neptunium results

[1 Becquerel (Bq) = 1 decay per second]

Precision	
Result 1----------	91 $\pm$ 7 percent recovery
Result 2----------	16 percent RSD
Accuracy	
Calculated value (C)----	1870 Bq/L
Measured value (M)------	1830 $\pm$ 100 Bq/L
M/C--------------------	0.98

Accuracy was determined by generating a blind sample:
mixing volumes of two different samples together and calculating the resulting concentration. This blind sample was
analyzed and gave the final result in Table III: 98 percent
of actual value.

Plutonium Analysis

Because the organic material in the leachates would
interfere with the ion exchange, an organic destruction step
was needed prior to analysis. The obvious choice of perchloric acid by itself proved unusable in most samples, giving
recoveries generally less than ten percent of the tracer
added. This suggested two things might be occurring in the
samples. First, some classes of organic compounds are known
to be oxidized only to oxalates by perchloric acid. Any
oxalate would then react with the plutonium preventing it
from behaving as expected. The other occurrence would
involve the oxidation of some inorganic species such as
phosphorus compounds to phosphates, which might then tie up

the plutonium. This did not appear to be a factor, however,
since the samples which gave the highest recovery also had
the most phosphorus. Clearly, another approach was needed
for those samples where perchloric acid was ineffective.

Addition of an acidic permanganate step was tried.
At first, erratic results were obtained, as well as inconsistent coloration prior to neutralization with bisulfite. By
making sure there was excess permanganate after digestion,
more consistent results were obtained.

Precision and accuracy of the plutonium method were
determined by the same techniques as for the neptunium;
these values are given in Table IV. Result 1 was
95 ± 5 percent recovery, and Result 2 was 3 percent RSD.
Accuracy was 97 percent of actual value.

Table IV.—Plutonium results

Precision	
Result 1----------	95 ± 5 percent recovery
Result 2----------	3 percent RSD

Accuracy	
Calculated value (C)---------	220 Bq/L
Measured value (M)----------	210 ± 5 Bq/L
M/C-------------------------	0.97

SUMMARY

These techniques have been found useful for determining
neptunium and plutonium in ground waters associated with
Maxey Flats; some analytical pitfalls have been suggested.
Other more or less stringent techniques might be needed at
other locations, depending on the water composition.

REFERENCE

1. A. J. Weiss, A. J. Francis, and P. Columbo, Characteriza-
    tion of trench water at the Maxey Flats low-level radio-
    active waste disposal site, *in* Management of low-level
    radioactive waste, Pergamon Press, New York, 2, 747–762
    (1979).

PROBLEMS ASSOCIATED WITH TRANSURANIUM DETERMINATION
OF SUSPENDED SOLIDS IN SEAWATER SAMPLES

K. M. Wong, T. A. Jokela, and V. E. Noshkin,
Environmental Sciences Division, Lawrence Livermore
Laboratory, Livermore, California, USA

ABSTRACT

Particulate material collected by filtration from the
north Equatorial Pacific Ocean has been analyzed for
plutonium and other radionuclides. Different filter pore
size, types of filter substrates, flow rates, and sample
volumes were evaluated.
Retention of $^{239+240}$Pu was found to vary with the
sample volume filtered and was not greatly effected by the
type of filter substrate, flow rate, or porosity tested.
About 7 $\pm$ 3% of the $^{239+240}$Pu activity in north Equatorial Pacific surface water is found with the particulate
material filtered from 120 liters of water, while less
than 1% is retained on samples with volume of 20,000–
70,000 liters.
Clearly an understanding of these results is necessary to correctly assess the quantity of plutonium, other
radionuclides and trace elements associated with and transported by particulate material in the marine environment.

INTRODUCTION

The quest to decipher the complex interaction of
suspended particulates in seawater has been the object of
many oceanographic investigations and numerous findings,
theories, and reviews have appeared in the literature.
(1-4) The introduction of the transuranium elements into
the ocean, mainly from atmospheric nuclear weapon testing
in the 50's and 60's, and more recently from nuclear waste
disposal, created an even greater interest in this subject
as these potentially toxic elements are found to associate

with suspended particulates.(5) The preconcentration of
transuranium elements onto suspended particulates, in
turn, provided a unique tracer for the study of particle
movement in the seas.

For example, Miyake, et al, were able to deduce a
sinking rate of particulates in the western north Pacific
and Japan Sea from the measurement of plutonium.(6)Noshkin
and Bowen developed models describing the sinking rate of
plutonium associated with particles in the Atlantic.(7)
Krishnaswami, et al, Lal, and Hodge, et al, have also
applied plutonium and other radionuclide measurements to
the studies of suspended particulated in the ocean.(8-9)

With few exceptions, the analysis of suspended particulates or transuranium elements in the oceans invariably
requires the processing of large volumes of water. The
collection of particulate materials from seawater samples
as large as 10^5 liters has been reported.(8) Separation
of the suspended particulates from seawater is generally
done by filtration or centrifugation. Certain inherent
drawbacks in the centrifugation method, as well as filtration, have been discussed elsewhere and the method of
choice has been by filtration.(4,10)

In this paper we will present our observations on the
filtration of large volume of seawater for the determination of plutonium and americium radionuclides in suspended
particulates. It should be emphasized at this point that
this paper is a by-product of several studies involved
with the analysis of transuranium elements in seawater
samples and it was not the specific aim of these studies
to evaluate the best available filtering systems for the
collection of suspended particulates.(11-13)

EXPERIMENTAL CONDITIONS AND RESULTS

The dimension of the cartridge is 6.5 cm diameter by
25 cm in length. The filter material is made of cotton
reinforced with acrylic fibers and supported with a poly-
propylene core. Filter cartridges made with different
materials, size, or porosity are also available from a
number of manufacturers at a cost of about $2.50 to $45.00
each. Different cartridge holders from transparent
plastic to stainless steel are also available from various
manufacturers that can be used interchangeably with these
or different cartridge filters with or without adaptors.
We find holders equipped with "O" rings are more convenient and leak proof than those equipped with flat gas-
kets. Some of the criteria for selection of the cotton

cartridge were: high filtration rate, high retention
($\geq$ 95% of 0.45 μm membrane filter), convenience and the
availability from many commercial sources. We have used
this type of cartridge filter for filtration of suspended
particulates, as well as for other radiochemical separations involving large volume samples.(14)

Within our original experimental constraints, we have
tested several collection variables in the field. Both
membrane and woven cotton cartridges have been compared;
the filter porosity of 0.2 μm (membrane), and 1.0 μm in
cotton cartridge filters were tested; and flow rate between 3 to 30 1/min. were also tried. Samples from about
120 to 70,000 1 were collected. Except for sample size,
we find no significant differences in any of the variables
mentioned above.

The filtration time was less than 15 minutes for a
120 1 or smaller sample. Larger samples, 500-5000 1, took
several hours and as long as 2 days for samples up to
70,000 1.

The samples tabulated in Table I and Figure 1 were
collected in the northern Equatorial Pacific between 0°
to 16° N latitude and 157° E to 177° W longitude in
a joint cruise with the Woods Hole Oceanographic Institution in June-August 1978. Data from other reports (8,9,
12) on plutonium concentration in suspended particulates
from the northern Pacific and northern Equatorial Atlantic
are also plotted in Figure 1 for comparison. Some of the
parameters mentioned above were tested in Enewetak Lagoon
in an area of relatively high activity in order to minimize the effect of blank and counting error. The results
of the Enewetak samples are shown in Table II.

DISCUSSION

At first glance, the plutonium concentration data in
Table I appear to vary by a factor of about 30. We have
not seen this large of a variation in fallout $^{239+240}$Pu
concentration in any surface seawater that we have ever
sampled in the north Equatorial Pacific. The $^{239+240}$Pu
concentration of the suspended particulates does not seem
to correlate with filtration rate or with latitudinal differences within the area sampled. Krishnaswami, et al, (8)
observed a factor of 3 variation at these latitudes in the
Atlantic.

However, there seems to be a correlation of plutonium concentration with sample volume filtered, as shown in
Figure 1. The same trend is also noted for samples collected from the north Equatorial Atlantic (8), southern and
northern California Coast (9,12) and samples from Enewetak
Lagoon (see Table II).

TABLE I

Plutonium Concentrations in Suspended Particulates from the
North Equatorial Pacific Collected July–August 1978

Sample No.	Start	Stop	Flow Rate (1/m)	Volume Filtered (1)	$^{239+240}$Pu fCi/1000 1
5674	12N 165W	12N 165W	30	125	4.0(60)[a]
5684	12N 163W	12N 163W	30	120	17.0(30)
5696	12N 161W	12N 161W	30	115	29.0(26)
5707	13N 159W	13N 159W	30	125	34.0(25)
5719	14N 165W	14N 165W	30	120	19.0(33)
5732	10N 165W	10N 165W	30	120	2.0(30)
6044	14N 170W	12N 167W	8	2150	3.0(24)
6056	16N 157W	16N 157W	10	7080	2.4(33)
6059	16N 157W	15N 162W	7	7500	3.7(21)
6062	15N 162W	14N 165W	6	6400	4.3(17)
6084	7N 165E	5N 165E	19	28600	1.8(13)
6085	5N 165E	5N 165E	25	48000	1.2(12)
6086	5N 165E	8N 167E	30	36200	3.3(9)
6087	8N 167E	8N 169E	31	44700	2.9(7)
6088	8N 169E	8N 171E	29	42760	1.5(8)
6089	8N 171E	7N 175E	28	70920	1.4(19)
6091	7N 177W	6N 176W	29	41000	2.7(13)
6092	6N 176W	5N 175W	25	52050	2.6(11)
6093	5N 175W	5N 175W	29	35130	3.5(15)
6096	1N 173W	0 170W	27	60840	2.8(8)

[a] The value in parentheses is counting error expressed
as % of activity. Activity has been corrected for
reagent blank (7 ± 2 fCi/sample).

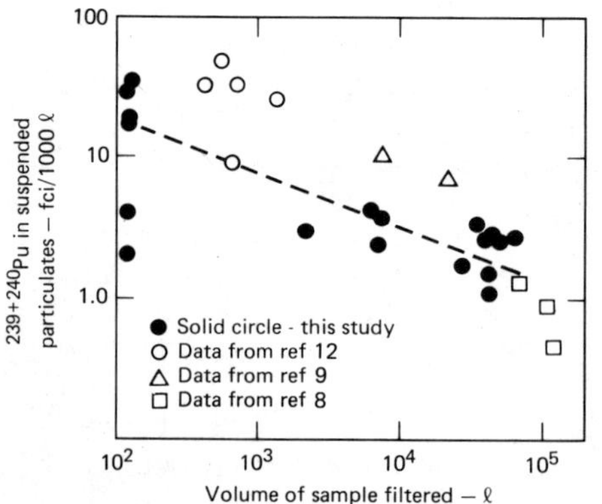

Fig. 1. $^{239+240}$Pu variation with sample volume.

210

TABLE II
Plutonium and ^{241}Am Concentrations in Suspended
Particulates from Enewetak Lagoon

Station No.		Filter Size (μm)	Volume Filtered (1)	fCi/1000 1 (a)		
				$^{239+240}$Pu	^{238}Pu	^{241}Am
C	26	1.0	343	4440	1240	960±120
	27	1.0	1110	3920	1180	750
	29	1.0	1640	3520	1200	
	28	1.0	3823	3040	840	
	30	1.0	4397	3260	950	540
H	37(b)	1.0	100	3430	1030±103	
	38(c)	1.0	100	190±23	--	
	39(b)	1.0	855	2650	770	850
	40(c)	1.0	855	6±2	--	
I	41	1.0	108	6450	2330±190	
	43	1.0	980	2010	583	
J	32	1.0	1909	1570	209±15	
	33	0.2(d)	1744	1510	217±13	
	35	0.2(d)	105	4070	650±80	

(a) Activity due to reagent blank and counting error is less than 5%, except as indicated.

(b&c) Samples filtered in series.

(d) Membrane cartridge.

It seems tempting at first to attribute the high plutonium values found in the small samples to the uncertainty in the reagent blank (7 $\pm$ 2 fCi/sample).

In order not to under estimate the blank correction, the reagent blank was based on two sets of independent analysis -- one was measured from a series of 120 1 of distilled water. The range of these water blanks was 4.5 to 9.5 fCi/sample with an average of 7 $\pm$ 2 fCi/sample. The second estimate was made from a series of actual collections of suspended particulates from the water column where the plutonium concentration profile is known to be minimal (e.g., at 50-150 m). The average value from the plutonium minimum should represent the maximum reagent blank activity one could expect from these analysis, which was also 7 $\pm$ 2 fCi/sample. In view of these, and the fact that the normal reagent blank in our laboratory for this

type of analysis has been 4 ± 2 fCi/sample, we cannot
justify attributing the high plutonium values in these
samples to contamination.

Furthermore, since a similar trend was observed in the
Enewetak Lagoon water where the blank correction was
insignificant for these samples; therefore, other processes
related to the sample collection must be responsible for
this decrease in plutonium concentration with sample volume.

Another approach in examining the validity of these
data is to compare them on the basis of concentration
factors of the suspended particulates. Using an average
measured distribution coefficient of 2.3×10^5 tabulated
by Noshkin (15) for sediments in seawater, a suspended
particulate concentration of 20 μg/l, and an average
$^{239+240}Pu$ concentration in filtered surface seawater from
the northern Equatorial Pacific of 0.26 fCi/l, the calculated $^{239+240}Pu$ in the suspended particulates should have
been 1.2×10^{-3} fCi/l, which would be in reasonable
agreement with the measured values in samples larger than
20,000 l.

On the other hand, if the original plutonium is
predominately in the (+4) state, as shown by Nelson and
Lovett in the Irish Sea samples, the applicable distribution coefficient would be 2.5×10^6 and the calculated
$^{239+240}Pu$ concentration of 1.3×10^{-2} fCi/l would be
closer in agreement with the small volume samples.(16) It
is not totally inconceivable that Pu (+4) could be oxidized
to Pu (+6) in the presence of an increasing concentration
of organic detritus, as is the case with filtration of
large samples.

It has also been known that plutonium forms strong
complexes with certain anionic constituents in seawater.
Under certain conditions, disproportionation of Pu (+4) to
Pu (+3) and Pu (+6) is possible.(17,18) In addition, the
chelation of plutonium with natural organic compounds, such
as humic acids from seawater, has been measured.(19) All
these processes could remove plutonium from suspended
particulates and during prolonged filtration the desorption of plutonium from suspended particulates would be
favored.

Another variable may yet be introduced if we wish to
consider adsorption by the filter media. Samples from
Station H, shown in Table II, were collected in series.
One set of samples was collected for 100 l and the second
set for 855 l. In the smaller sample, 19 fCi of
$^{239+240}Pu$ was found in the second filter which represented 5.5% of the activity in the first cartridge, but
less than 6 fCi or 0.3% was found in the second cartridge
of the 855 l sample. If we are to speculate, we may attribute the 5.5% to ion exchange adsorption by the filter
matrix since only filtered water is being passed through

the second cartridge. As for the 855 1 sample, the
smaller amount of plutonium found on the second cartridge
may be an indication of the net result of adsorption and
desorption processes previously mentioned.

At this time we do not have sufficient data on other
transuranium radionuclides to make similar comparison.
However, the few available ^{241}Am data from the Enewetak
Lagoon tabulated in Table II seem to indicate the same
trend as plutonium.

In summary, the present data show that the accuracy
of determining the actual plutonium concentration in the
particulate materials in the natural marine environment is
subject to large error. This uncertainty is due partly to
the difficulty associated with analysis of extremely low
level samples and also partly due to obscure interactions
in the filtration processes which could add or remove
transuranium radionulcides from the filtered material.

Unfortunately, most of these interactions cannot be
directly measured or verified in the natural environment
at this time. This may or may not apply to other
elements, but it would be prudent to take this factor into
consideration for other trace element analysis.

We gratefully acknowledge the helps from Dr. V. T.
Bowen for arranging the joint cruise on the RV KNORR;
Brenda Olson for collecting many of the filters; Learh
Nelson and Jim Brunk for analyzing most of the samples.

REFERENCES

1. R.J. Giubs, <u>Suspended Solids in Water</u>, Plenum Press,
    New York, NY (1974).
2. T.R. Parson, <u>Chemical Oceanography</u>, J.P. Riley and G.
    Skirrow, Eds., Academic Press, London, 2nd Ed., vol.
    <u>2</u>, 365 (1975).
3. D. Lal, Science, <u>198</u>, 997 (1977).
4. W.M. Sackett, <u>Chemical Oceanography</u>, J.P. Riley and
    R. Chester, Eds., Academic Press, London, 2nd Ed.,
    vol. <u>7</u>, 127 (1978).
5. V.T. Bowen, K.M. Wong, and V.E. Noshkin, J. Mar.
    Res., <u>29</u>, 1 (1971).
6. Y. Miyake, Y. Katsuragi, and Y. Sugimura, J. Geophys.
    Res., <u>75</u>, 2329 (1970).
7. V.E. Noshkin and V.T. Bowen, <u>Radioactive Contamina-
    tion of the Marine Environment</u>, Internat. Atomic
    Energy Agency, Vienna, 671 (1973).
8. S. Krishnaswani, D. Lal, and B.L.K. Somayajulu, Earth
    Plant. Sci. Lett., <u>32</u>, 403 (1976).

9. V.F. Hodge, M. Koide, and E.D. Goldberg, Nature, <u>277</u>, 206 (1979).

10. P.S. Lisa, et al, <u>The Nature of Seawater</u>, E.D. Goldberg, Ed., Dahlem Konferenzen, Berlin, 453 (1975).

11. V.E. Noshkin, K.M. Wong, R.J. Eagle, and C. Gatrousis, Lawrence Livermore Laboratory Rept. UCRL-51612 (1974).

12. V.E. Noshkin, K.M. Wong, T.A. Jokela, R.J. Eagle, and J. Brunk, Lawrence Livermore Laboratory Rept. UCRL-52381 (1978).

13. V.E. Noshkin, R.J. Eagle, and K.M. Wong, Nature, <u>261</u>, 745 (1976).

14. K.M. Wong, G.S. Brown, and V.E. Noshkin, J. Radioanal. Chem., <u>42</u>, 7 (1978).

15. V.E. Noshkin, Lawrence Livermore Laboratory Rept. UCRL-80587 (1978).

16. D.M. Nelson and M.B. Lovett, Nature, <u>276</u>, 599 (1979).

17. G.H. Coleman, <u>The Radiochemistry of Plutonium</u>, NAS-NS-3058, National Academy of Science (1965).

18. J.M. Cleveland, <u>The Chemistry of Plutonium</u>, Am. Nuc. Soc., La Grange Park, Ill. (1979).

19. F. Harrison, Lawrence Livermore Laboratory, private communication, (1979).

"Work performed under the auspices of the U.S. Department of Energy by the Lawrence Livermore Laboratory under contract number W-7405-ENG-48."

DETERMINATION OF ACTINIDES IN SOIL

N. W. Golchert, F. S. Iwami, and J. Sedlet, Argonne
National Laboratory, Argonne, Illinois, USA

ABSTRACT

A method is described for the sequential separation of
environmental levels of plutonium, thorium, uranium, and depending on conditions, neptunium and americium from up to
100 grams of soil. A leach procedure is used to solubilize
the nuclides of interest from soil, bottom sediment, ashed
vegetation or food materials, or air-filter residues. The
leach procedure consists of treatment of these materials
with combinations of hydrochloric, nitric, and hydrofluoric
acids, and results in a final solution which is $8M$ in nitric
acid. The separations involve ion exchange and solvent extractions followed by electrodeposition. The nuclides
present in the electrodeposited samples are determined by
alpha spectrometry.

INTRODUCTION

The described method was developed to provide flexibility in the analysis of such materials as: soil, bottom sediment, ashed vegetation or foodstuffs, or air-filter residues.
The dissolution of these materials and subsequent chemical
separations allows the individual identification of various
radionuclides of plutonium, thorium, uranium, neptunium, and/or
americium. The specific analysis of these radionuclides at
low or environmental levels is essential for: a facility to
demonstrate compliance with relevant emission standards,
provide data to make decisions on the extent of decontamination
and decommissioning of excess sites, determine the magnitude
and distribution of radioactivity from an incident, or examine
the migration of radionuclides from a radioactive waste burial
area.

It is assumed that adequate consideration has been given
to the proper method of collecting samples. Two frequently
used techniques are the DOE Environmental Measurements
Laboratory[1] site selection and sampling criteria and the
Nevada Applied Ecology Group (NAEG)[2] sampling procedure.
These are presented in <u>AEC Regulatory Guide 4.5</u>[3] and extensively reviewed by Bernhardt.[4] After collection, the soil
is dried at 110°C, ground, mixed, and usually 100 g aliquots
are taken for analysis. Bottom sediments are treated in the
same manner as soil samples, while vegetation and foodstuffs
are washed, dried, and ignited at 600°C before beginning the
chemistry. Air particulate samples are collected on Delbag,
a polystyrene filter media, and are ignited at 600°C before
processing. Appropriate tracers are added for each element
to be determined. They are: thorium-234, uranium-232,
neptunium-239, plutonium-236 or plutonium-242, and americium-
243.

PROCEDURE

The dissolution procedure is as follows: Weigh 100 g of
soil into a one liter heavy duty beaker and add tracers as
needed. Add a 5 cm long magnetic stirring bar to the sample
and place the sample on a stirring hotplate. Add slowly
400 ml of conc. HNO_3, and turn the stirrer on. (Some soils
react more vigorously in HNO_3, others more vigorously in
HCl.) After the initial reaction has subsided and the slurry
is mixing well, add 100 ml of conc. HCl even more slowly,
~5-10 ml at a time. Turn the heat on, raise to moderate heat
within 1/2 hour, and maintain a near boiling temperature for
about two hours. Turn off the heat and stirrer, add 200 ml
of water, stir with a glass rod, and cool. Filter the mixture through Whatman #2 fluted filter paper, collecting the
solution in an 800 ml or one liter beaker. Wash and clean
the beaker with $8N$ HNO_3 using a rubber policeman. Filter the
washings and add to the solution. After the HNO_3 has drained
through the paper, rinse the beaker with ~100 ml of water,
filter, and add to the solution. Drain until the undissolved
soil looks dry. Place the undissolved soil and paper in
~500 ml platinum deep dish and place in muffle furnace. Heat
to 300°C for 1/2 hour, then turn up to 600°C and ash for four
hours. Take the original filtrate solution and evaporate the
volume down to about 300 ml. Do not boil vigorously. (Put
aside until the ashed solids have been treated.) Place the
cooled dish containing the ashed solids in a sand bath on a
hotplate. Add enough water to wet the sample, then add
100 ml conc. HNO_3. Stir with a platinum rod. Add slowly
~25 ml portions of conc. HF until 200 ml has been added.
Turn hotplate on to low heat and heat with occasional stirring

until the sample is dry. Add 100 ml $8N$ HNO$_3$, stir, and evaporate to dryness. Transfer the solids to the original
beaker and clean the platinum dish with 200–300 ml of $8N$
HNO$_3$. Boil the sample in the above beaker on a hotplate for
about one hour, add 100 ml of water, stir, and cool. Filter
the sample through fluted filter paper and drain the filtrate
into the original solution. Wash the beaker with $8N$ HNO$_3$ and
add to the sample. Drain until the solids are dry. Add 6–8
glass beads to the solution and boil on a hotplate until the
volume is ~250 ml. Cool, add 150 ml water and filter through
Whatman #2 paper. Wash the beaker and paper with 100 ml of
$8N$ HNO$_3$. Discard everything else.

The same procedure is used on bottom sediment samples,
while the volumes are scaled down by a factor of four for
most plants, foodstuffs, and air-filter residues. The final
volumes are 500 ml for soils and bottom sediments, either
100 ml or 200 ml for plants and foodstuffs, and 100 ml for
air-filter residues.

The separation of the nuclides of interest follows one
of two schemes depending on initial sample size. For samples
of five to one hundred grams, the procedure is illustrated in
Figure 1. Initially, an aliquot is removed for uranium
analysis, either by fluorometric uranium analysis or by extraction of the uranium, followed by electrodeposition and
alpha spectrometry. The only actinides retained by a nitrate-
form anion-exchange column are thorium, neptunium, and
plutonium. The uranium distribution coefficient under these
conditions is about ten, and the uranium must be removed from
the column by adequate washing with $8N$ HNO$_3$. For sample
volumes of 500 ml, the initial volume plus the amount of $8N$
HNO$_3$ needed to completely remove the uranium will also remove
some of the thorium since its distribution coefficient is
only several hundred. The remaining thorium is then removed
from the column with $12N$ HCl, since it does not form a
chloride complex, to separate it from neptunium and plutonium.
The plutonium is removed by reduction to Pu(III) with $12N$
HCl-0.1M NH$_4$I solution. The neptunium is then removed by
dilute acid.

For samples of less than five grams, the procedure is
outlined in Figure 2. In this case, the initial portion of
the separation is similar to the above case except that because of the much smaller volumes, essentially all of the
thorium is retained on the column. The removal of the
thorium, neptunium, and plutonium sequence is the same as
above, except that the neptunium and plutonium can be eluted
together with dilute acid, using the added plutonium tracer,
i.e., plutonium-236, to determine the neptunium content.

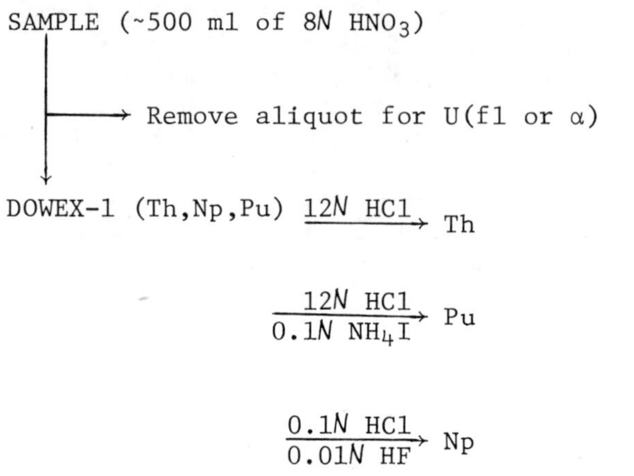

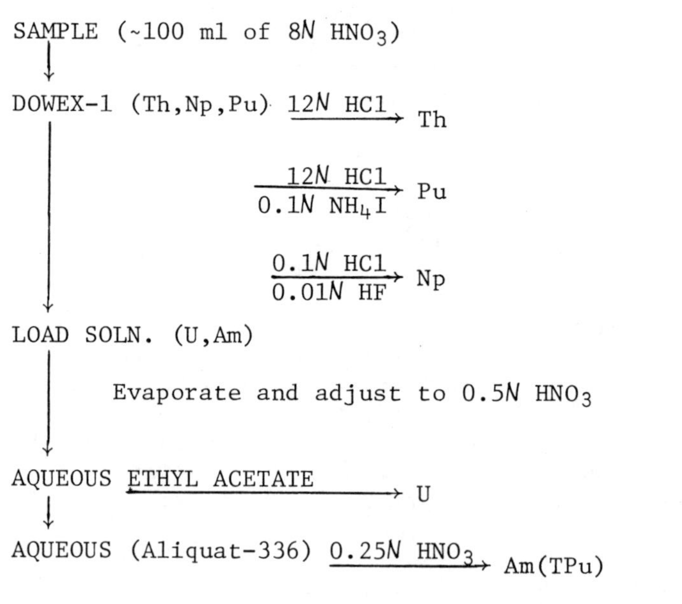

Figure 1. Separation Procedure for Samples of 5 to 100 Grams

SAMPLE (~100 ml of 8N HNO₃)
DOWEX-1 (Th,Np,Pu) 12N HCl → Th
12N HCl / 0.1N NH₄I → Pu
0.1N HCl / 0.01N HF → Np
LOAD SOLN. (U,Am)
Evaporate and adjust to 0.5N HNO₃
AQUEOUS ETHYL ACETATE → U
AQUEOUS (Aliquat-336) 0.25N HNO₃ → Am(TPu)

Figure 2. Separation Procedure for Samples < 5 g.

The column effluent solution, which contains the uranium
and americium, is evaporated to near dryness, taken up in 10M
NH_4NO_3, which is 0.5M in HNO_3, and the uranium is removed by
extraction with two successive equal volumes of ethyl acetate.
The americium is then extracted from the aqueous with 30%
Aliquat-336 in xylene. The Aliquat-336 is scrubbed with 10M
NH_4NO_3 to remove residual calcium and aluminum, and the
americium is stripped into dilute acid. The Aliquat-336
extraction will not only remove americium, but all trivalent
transplutonium elements.

Each actinide fraction is electrodeposited onto a silver
disc from a chloride solution. The individual radionuclides
are determined by alpha spectrometry. For environmental
level samples, typical measurement parameters for our alpha
spectrometry system are: counting efficiencies of 33%, counting times of 3000-4000 minutes, and background plus reagent
blanks of 3-7 counts per thousand minutes depending upon
which nuclide is considered.

DISCUSSION

The above techniques have been used by the ANL Environmental Monitoring Group, in some form, since 1971, to determine various actinides in environmental media samples. As an
example, the concentrations of several actinides measured in
soil collected from off-site locations are shown in Table I.
Two different sampling depths are indicated. While the
thorium and uranium concentrations are similar at all depths,
the plutonium concentrations are higher for the 0-5 cm
samples since most of the fallout plutonium still resides in
the top few cm of soil. The 0-30 cm cores have lower plutonium concentrations due to dilution by the deeper soil
which has little plutonium. In terms of deposition, the
0-5 cm samples contain 1.2 nCi/m^2 of plutonium-239, while the
0-30 cm samples contain 2.5 nCi/m^2. The analysis of rain
samples for plutonium since 1973 has allowed the calculation
of ground deposition and is a measure of the fallout from
recent atmospheric nuclear tests.

Similar data to that in Table I exists for bottom sediment, plants, and foodstuffs. The plutonium concentration of
bottom sediment samples vary widely between locations and is
strongly dependent on the retentiveness of the sediment.
Thorium and uranium concentrations are similar to soil samples.
Grass and foodstuffs are washed before analysis to remove
surface soil, which contains considerably more radioactivity
per gram than the samples. In terms of deposition, the
plutonium-239 concentration in grass was about a factor of

10^4 less than soil from the same location. Typical results of the analysis of samples from these various media is illustrated in the annual report of the program.[5]

TABLE I

Radionuclides in Soil

(concentrations in pCi/g)

Year	Depth (cm)	^{228}Th	^{230}Th	^{232}Th	Uranium (total)	^{238}Pu (x 10^3)	^{239}Pu (x 10^3)
1978	0-5	0.34	0.48	0.36	1.4	0.9	23.5
1977	0-5	0.19	0.40	0.24	1.7	0.7	14.1
1976	0-30	0.27	0.31	0.24	1.1	0.4	6.3
1975	0-30	0.28	0.26	0.24	1.4	0.5	6.0
1974	0-30	0.37	0.47	0.33	1.4	0.4	7.6
1973	0-30	0.43	0.49	0.37	1.5	0.5	6.3
1972	0-30	0.43	0.45	0.37	1.3	0.5	4.9

Another example of the application of the separation procedure is the analysis of air-filter residues. Plutonium, uranium, and thorium concentrations in air have been measured since 1973. The specific activity of the thorium and uranium in the air-filter residues is related to the specific activity of thorium and uranium in soil. It appears that the bulk of these elements in the air is due to resuspended soil. Using the uranium and thorium in air data, the contribution of plutonium in air from stratospheric fallout and from resuspended surface soil can be determined. Using this information, resuspension factors can be calculated for thorium, uranium, and resuspended plutonium. A comparison of the factors indicates quite good agreement.[6]

Another study involved the determination of plutonium-241 in soil. Separated plutonium fractions from soil samples analyzed in 1971 were leached from the plating disc and the americium-241 separated and determined. Conversion to plutonium-241 activity indicated that in 1971, the concentration of plutonium-241 in the top five cm of soil was about 100 fCi/g and the ^{241}Pu/^{239}Pu ratio was 3.2.

A pathway study was conducted to examine the distribution of plutonium and neptunium discharged into a creek. Samples were collected of water and suspended sediments and analyzed for plutonium and neptunium. The suspended sediment was separated into two fractions by filtration, coarse (particle

sizes greater than 2.5 μm) and fine (particle sizes between
2.5 and 0.45 μm). Water was collected at the point the
Laboratory discharges its waste water into the creek and at
various points downstream.

The water and sediment data[7] indicates that the loss by
sedimentation is greater for plutonium than for neptunium.
Because of the relatively high velocity of the creek to a
point about 200 meters below the outfall, loss of both
elements from the water is by association with the coarse
sediment fraction. Beyond the 200 meter point, the stream
velocity slows sufficiently so that the loss of plutonium
and neptunium with the fine sediment dominates.

CONCLUSION

This paper presents a method for solubilizing various
environmental media samples and sequentially separating some
of the actinides. The applicability of the method is demonstrated both in the routine monitoring program and use in
special studies.

REFERENCES

1. J. H. Harley (Ed.), HASL Procedure Manual, USAEC Report
   HASL-300 (1972).

2. E. B. Fowler, R. O. Gilbert, and E. H. Essington,
   "Sampling of Soils for Radioactivity: Philosophy,
   Experience, and Results, " in Atmosphere-Surface
   Exchange of Particulate and Gaseous Pollants, R. J.
   Engelman and G. A. Sehmel, Coordinators, ERDA Symposium
   Series 38, CONF-740921, pp. 709-726 (1974).

3. AEC Regulatory Guide 4.5, "Measurement of Radionuclides
   in the Environment, Sampling, and Analysis of Plutonium
   in Soil"(1974).

4. D. E. Bernhardt, Evaluation of Sample Collection and
   Analysis Techniques for Environmental Plutonium,
   ORP/LV-76-5 (1976).

5. N. W. Golchert, T. L. Duffy, and J. Sedlet, Environmental
   Monitoring at Argonne National Laboratory, Annual Report
   for 1978, USDOE Report ANL-79-24 (March 1979)

6. N. W. Golchert and J. Sedlet, "Resuspension Studies on
    Fallout Level Plutonium," in Selected Environmental
    Plutonium Research Reports of the NAEG, M. G. White and
    P. B. Dunaway (Eds.), USDOE Report NVO-192, p. 723
    (June 1978).

7. N. W. Golchert, T. L. Duffy, and J. Sedlet, Environmental
    Monitoring at Argonne National Laboratory, Annual Report
    for 1977, USDOE Report ANL-78-26 (March 1978).

ANALYSIS OF ROCKY FLATS SOILS FOR SIX RADIONUCLIDES
BY A SEQUENTIAL METHOD

J. R. Stevens, D. R. Guhlow,[*] T. F. Rees,[**]
and K. J. Grossaint. Analytical Laboratories,
Rockwell International, Golden, Colorado, USA.

ABSTRACT

A dramatic increase in soils analysis for actinides
has occurred at the Rocky Flats nuclear facility in recent
years. Intensive investigation of the extent of potential
contamination of soils was needed to generate a valid data
base for evaluating the true environmental impact of plant
operations.

A method for the sequential analysis of Rocky Flats
soils for six actinides of interest was developed to
facilitate the large number of radionuclide analyses. The
isotopes determined were: U-238, U-235, U-234, Th-232, Th-
230, Th-228, Np-237, Pu-238, Pu-239/240, Am-241, and Cm-244.
The method included addition of tracer isotopes for each
actinide, pyrosulfate fusion to help solutize minerals and
decompose organic matter, sequential separation by cation
and anion exchange techniques, extraction into thenoyltrifluoroacetone (TTA)/organic phase, and subsequent counting
of alpha activity by pulse height analysis. The efficieny
of recovery for tracer isotopes ranged from 40% for curium,
to 95% for plutonium. The detection limits for the method
were determined at 0.2 disintegrations per minute per gram
(DPM/g) of soil for thorium, uranium, neptunium, and plutonium. The limits were a factor of three higher (0.6 DPM/g)
for americium and curium, primarily because of inefficient
extraction of these components by TTA. The sequential
analysis was developed into a routine procedure to be used
when multi-element soil analyses were requested.

*Present Address: Hach Chemical Company, Loveland, CO, USA.
**Present Address: US Geological Survey, Denver, CO, USA.

INTRODUCTION

The emphasis on "Low as Practicable" pollutant emission limits for nuclear facilities such as Rocky Flats sometimes places a strain on laboratories performing the analyses required for any monitoring effort.

Where on-site contamination of soils has occurred, the monitoring effort has a great significance. Analyses performed to determine the extent of contamination contributed by plant operations are complicated by the presence of low levels of fall-out activity, and also by the natural radioactivity prevalent in the plant environs. Natural radioactivity in soils surrounding Rocky Flats is very significant, as illustrated by the presence of several high-grade ore Uranium mines in the front range, within 30 miles from the plant.

The responsibility for the soil monitoring program at Rocky Flats lies with the Environmental Studies Group, which is a part of the Health, Safety, and Environment Organization. That group defines the sampling program and sampling techniques for soils, as well as the other myriad analyses required by Environmental Protection Agency (EPA), state regulatory agencies, Department of Energy (DOE), and plant self-imposed, regulations or emission limits.

The Analytical Laboratory provides a significant portion of the analytical support for the soil monitoring program. We also support an extensive research and development effort at Rocky Flats, which is developing processes for decontaminating soils.

We found that although not requested for every sample, analysis for the radionuclides of plutonium, americium, uranium, thorium, neptunium, and curium, was requested often enough to put a severe strain on our ability to meet a reasonable time scale for reporting results. As a response to this realization, a method was developed to permit sequential separation of those six elements from only four aliquots of each soil sample.

EXPERIMENTAL

The chemical form of environmental radionuclides cannot be known with certainty, in most cases. It is therefore necessary to rigorously treat the sample for dissolution, both to be certain of complete chemical dissolution, and to convert to the expected soluble species, prior to the separation steps. The procedure should use a minimum of hydro-

fluoric acid, because of its influence on the ion exchange
separation of certain actinides.[1,2,3] Our studies have indicated that prior treatment of the sample by fusion with
potassium pyrosulfate greatly aided in attaining complete
dissolution of radionuclide contaminants in soil. An added
advantage may be the chemical breakdown of humus, vegetative
debris, or other organic material present in the soil sample.
This is definitely an advantage to us, considering the
hazards of perchloric acid digestion, or other possible
digestion techniques.

Preparation of the Sample

     The procedure followed to obtain the sequential sepa-
ration of the six elements was as follows:

     Sieve the sample through a standard 10-mesh screen.
Determine the weight fraction of fines to coarse particles.
Dry the less-than-10-mesh fines at 100°C, to constant weight.
(Determine the wet-to-dry weight ratio). Grind the dry
fines to a homogeneous powder, using an analytical mill.

     Muffle the soil at 500°C for 24 hours. Cool, re-weigh
to determine weight loss. Homogenize by tumbling 10 minutes.

     It should be noted that Rocky Flats experience indi-
cates the preponderance of the activity is associated with
the less-than-10-mesh fines (particularly contaminant plutonium). If a more complete inventory is desired, the greater
than 10-mesh fraction should be analyzed alongside the fines.

     Weigh separate, appropriately sized, sample aliquots
into four clean platinum beakers or Pyrex^R glass beakers.
Thoroughly mix each aliquot with potassium pyrosulfate
powder, in a ratio of one part sample to three parts pyrosulfate, by weight. Add known amounts of tracer isotopes
U-232, Pu-236, and Am-243 activity to three of the aliquots.
If possible, the tracer count rate should be approximately
equal to the expected sample activity for the corresponding
element.

     Add known amounts of tracer isotopes Np-237, Th-232
and/or Th-228, and Cm-244 to the fourth aliquot.

     Dry, and fuse each aliquot over the flame of a labora-
tory burner, for a period of 10-15 minutes. Remove from
heat and cool. Take care that the sample is not lost by
breakage of the glass beaker during the fusion and cooling
steps.

Separation of Americium and Curium

Dissolve the fused sample in a minimum of 0.5 $\underline{N}$ HNO_3.
Note that an insoluble residue of approximately 1–10% by
weight may remain. Filter through Whatman 540 or 541
filter paper, rinsing with 0.5 $\underline{N}$ HNO_3, until the filtrate
volume is about 25 ml.

Prepare a 50W–X8, 50–100 mesh, hydrogen form resin
column, about 6 cm high. Condition the column with one
column volume (CV) of 0.5 $\underline{N}$ HNO_3.

Load the sample filtrate on the column, collecting
the effluent in a clean beaker. Rinse the column with
two CV of 0.5 N HNO_3, adding the rinse effluent to the
original. Set the collected effluent aside for later
separations of Pu, Th, Np, and U.

Elute the **americium and curium** from the column with
one CV of 8 $\underline{N}$ HNO_3, one CV of 16 $\underline{N}$ HNO_3, then another CV of
8 $\underline{N}$ HNO_3, collecting the eluate in a clean beaker.
Evaporate the eluate to dryness, dissolve in 12 $\underline{N}$ HCl,
evaporate to dryness, and re-dissolve in 12 $\underline{N}$ HCl a second
time.

Prepare an ion exchange column containing 3 cm of 1–X4
Bio-Rad^R resin, rinsing with 12 $\underline{N}$ HCl. Pour the sample
through the column, collecting the effluent in a clean
beaker. Evaporate the effluent to dryness, add 16 $\underline{N}$ HNO_3,
evaporate to dryness, and repeat. On the third evaporation,
use caution to take the sample to just past incipient dryness. Dissolve the residue in a minimum of 0.1 $\underline{N}$ HNO_3, and
transfer to a 4-dram vial, using a total of about 5 ml of
0.1 $\underline{N}$ HNO_3. Add 0.5 ml of pH 5 ammonium acetate buffer,
and 5 ml of 0.5 $\underline{M}$ TTA in xylene.

Shake the mixture on a wrist action shaker for approximately 15 minutes. Let stand to separate phases. Transfer
the (top) organic phase as quantitatively as possible onto
a 3-cm stainless steel planchet and evaporate to dryness.
Flame at red heat, and count the Am and Cm alpha activity
on a pulse height analyzer. Multiple extractions may be
useful to increase the chemical recovery of the americium
and curium. Typical first extraction recoveries range from
near 10 to about 50%. It should be noted that transfer of
$\underline{any}$ of the aqueous phase seriously degrades the quality of
the planchet.

Separation of Thorium and Plutonium

Add an equal volume of 16 $\underline{N}$ HNO$_3$ to the effluent
collected from the Am and Cm separation, to make that
solution approximately 8 $\underline{N}$ HNO$_3$. Add 0.1 gm sodium nitrite.
Prepare an ion exchange column containing 3 cm of 1-X4,
100-200 mesh chloride form resin, rinsing with 1 CV of 8 $\underline{N}$
HNO$_3$. Load the sample on the column, collecting the effluent containing uranium and neptunium in a clean beaker.
Set aside for further separation. Rinse with one additional
CV 8 $\underline{N}$ HNO$_3$. Elute the thorium from the column with 3 CV of
9 $\underline{N}$ HCl, collecting in a clean beaker.

Elute the plutonium from the column with 5 CV of 0.4 $\underline{N}$
HNO$_3$/0.1 $\underline{N}$ HF, collecting in a clean beaker. Evaporate the
thorium eluate to incipient dryness, and convert to the
nitrate form with 16 $\underline{N}$ HNO$_3$. Evaporate to dryness a second
time, and redissolve in 5 ml of 0.1 $\underline{N}$ HNO$_3$.

Transfer the redissolved thorium eluate to a 4 dram
vial containing 5 ml of 0.5 $\underline{M}$ TTA in xylene, and shake on a
Burrell wrist action shaker for 15 minutes. Let stand to
separate phases. Quantitatively transfer the organic phase
to a planchet, evaporate, flame, and count the thorium
alpha activity by pulse-height analysis.

Evaporate the plutonium strip solution to incipient
dryness, re-dissolve in 1 $\underline{N}$ HNO$_3$, and transfer to a 4-dram
vial. Add 5 ml of 0.5 $\underline{M}$ TTA in xylene and 0.1 g of sodium
nitrite. Shake for 15 minutes, and transfer the organic
phase as before, prior to counting the plutonium alpha
activity by pulse-height analysis.

Separation of Uranium and Neptunium

Evaporate the effluent containing neptunium and urani-
um to dryness, and convert to the chlorides with 12 $\underline{N}$ HCl.
Evaporate to incipient dryness a second time and redissolve
in 12 $\underline{N}$ HCl. Add ammonium iodide and hydroxylamine hydrochloride. Set aside for 10 minutes.

Prepare an ion exchange column containing 3 cm of 1- X4
100-200 mesh, chloride form resin. Condition the column
with 12 $\underline{N}$ HCl. Load the sample onto the column, discarding
the effluent. Rinse the column with 12 $\underline{N}$ HCl. Elute the
neptunium from the column with 3 CV of 4 $\underline{N}$ HCl/0.1 $\underline{N}$ HF,
collecting in a clean beaker.

Elute the uranium from the column with 2 CV of water,
collecting in a clean beaker.

Evaporate the neptunium eluate to dryness, redissolve
in 16 $\underline{N}$ HNO3, and evaporate to incipient dryness. Redissolve

in 0.5 $\underline{N}$ HNO$_3$. Add 0.1 g ferrous chloride (FeCl$_2$) and 0,1 g hydroxylamine hydrochloride. Let stand for 10 minutes. Transfer to a 4 dram glass vial, add 5 ml of 0.5 $\underline{M}$ TTA in xylene, and shake for 15 minutes on a wrist action shaker. Quantitatively transfer the organic phase to a planchet as before, prior to counting the neptunium alpha activity by pulse-height analysis.

Evaporate the uranium eluate to dryness and redissolve in 8 $\underline{N}$ HNO$_3$. Evaporate to incipient dryness and redissolve in 0.1 $\underline{N}$ HNO$_3$. Transfer to a 4 dram vial with approximately 5 ml of 0.1 $\underline{N}$ HNO$_3$. Add 1 ml of pH 5 ammonium acetate buffer. Add 5 ml of 0.5 $\underline{M}$ TTA, and shake for 15 minutes on a wrist action shaker. Let stand to separate the phases. Quantitatively transfer the organic phase to a stainless steel planchet, evaporate, flame, and count the uranium alpha activity by pulse-height analysis.

DISCUSSION

The sequential procedure requires four aliquots of each soil sample because of the necessity of adding activity of the same isotope analyzed for, in the case of curium and neptunium. We had no supply of alpha-emitting isotopes of these two elements, except Cm-244 and Np-237. The other four elements' chemical recovery was monitored with isotopes known to be absent in Rocky Flats soils. For 75 test samples of Rocky Flats soils, the chemical recoveries were as listed in Table I.

Table I

Sequential Method Chemical Recoveries, and Agreement Between Replicate Separations, for 75 Test Samples of Rocky Flats Soils

| Actinide | Chemical Recovery | | Sample Count Agreement |
	Ave(%)	Range(d)	(3 Planchets)
Americium	50(b)	8–64	$\pm$ 12%
Curium	40(b)	11–55	$\pm$ 47%(c)
Thorium	80	16–90	$\pm$ 11%
Plutonium	95	12–104	$\pm$ 9%
Neptunium	60	6–88	$\pm$ 26%
Uranium	60	21–87	$\pm$ 14%

(a) Lowest recovery observed, and highest recovery observed, for the set of samples.

(b) May include one, two or three successive extractions.

(c) Large value because of the counting statistic error for actinides with activity near the detector background.

Although not spectacular, these chemical recoveries were achieved on analyses performed on a routine basis, by non-degreed technicians, on soils of variable composition, and under pressures of time.

The manhours of effort to perform the sequential analyses was approximately 24 hours per sample. This compares with approximately 43 hours per sample average estimated for doing the analyses on separate aliquots.

The calculated limits of detection for the radionuclides, based upon these chemical recoveries, the aliquot size (approximately one quarter gram), and the counting efficiencies of our detector system were 0.2 disintegrations per minute per gram for thorium, plutonium, neptunium and uranium.

The limit of detection for americium and curium were calculated at 0.6 disintegrations per minute per gram, primarily because of the difficulties encountered in the extraction step for these elements.

These limits of detection approximate the level expected for plutonium in soils surrounding the Rocky Flats facility.[4]

The procedure for sampling soils at Rocky Flats is specified by an internal document.[5] The criteria for sampling include:

A. Area chosen
   1. open, relatively flat area
   2. undisturbed location during the time interval of interest
   3. no excessive run-off during heavy rains
   4. no mechanical disturbance of soil evident
   5. only light to moderate vegetation

B. Collection procedure
   Normal samples – five samples (10 x 10 x 5 cm). One at center, one at four compass points a specified distance from center composited.

Resuspendible samples – 1 cm depth, all soil particles less than two millimeters in size analyzed.

Special samples – use an orchard auger, 8.3 cm diameter barrel.

Profile samples – motorized backhoe to dig trench,

(60 to 120 cm) samples taken from walls after all loose
soil is removed.

ACKNOWLEDGEMENTS

     This work was supported by Department of Energy Con-
tract number DE-AC04-76DP03433.

REFERENCES

1.   IPST Cat. No. 2057, Analytical Chemistry of Elements
     Series, Analytical Chemistry of Uranium, translated from
     the Russian, N. Kramer, 1963 Israel Program for Scienti-
     fic Translations Ltd., S. Monson, Jerusalem, Israel, and
     Daniel Davey and Co., Inc., New York, (1963).

2.   Complex Compounds of Transuranium Elements, translated
     from the Russian, C. N. Turton and T. I. Turton, Con-
     sultants Bureau Enterprises, Inc., New York, (1962).

3.   Anion Exchange Studies of the Fission Products, K. A.
     Kraus and F. Nelson, Proceedings of the International
     Conference on the Peaceful Uses of Atomic Energy, Vol.
     7, pp. 113-125, United Nations, New York, (1956).

4.   Plutonium Determination in Soil by Leaching and Ion-
     Exchange Separation, N.Y. Chu, Analytical Chemistry,
     Vol. 43, pp. 449, (1971).

5.   D. D. Hornbacher, Private Communication, Environmental
     Control and Analysis, Energy Systems Group, Rockwell
     International, Rocky Flats Plant, Golden, Colorado.
     (August, 1979).

METHOD FOR DETERMINATION OF ^{237}Np IN LOW LEVEL ENVIRONMENTAL SAMPLES

E. Holm and M. Nilsson, Radiation Physics Department,
University of Lund, S-221 85 Lund, Sweden

ABSTRACT

A method for determination of ^{237}Np in environmental
samples was developed. A radiochemical procedure based on
isolation of neptunium from other interfering α-particle
emitters by precipitations and ion exchange was employed.
The radiochemical yield was determined by use of ^{239}Np and
Ge(Li) spectrometry. ^{237}Np was counted with silicon surface
barrier detectors. Biological samples up to 200 g dry weight
were processed with a minimal detectable activity in the
order of 10 fCi at 10^6 s counting time.

INTRODUCTION

Neptunium-237 ($t_{1/2}$= 2.1x10^6 y) is the remaining transuranium element in global fallout from nuclear detonation
tests, which has not been investigated.

This is certainly due to analytical problems consisting
of low activity concentrations in environmental samples and
low concentration factors in food chains. To measure very
low activities, α-spectrometry must be employed. Peak interferences from other α-particle emitters such as ^{234}U, ^{230}Th,
^{231}Pa and $^{239+240}$Pu present at much higher concentrations in
our samples must be considered. By mass ^{237}Np can be expected to be present in the environment in the same order as ^{239}Pu
since the (n, 2n)/(n, γ) reaction radio for ^{238}U at a thermo-
nuclear device can be as high as 0.5-1[1]. At equal oxidation
state neptunium and plutonium chemically resemble each other
but neptunium is frequently present in one oxidation state
higher. Being the daughter product of ^{241}Am, ^{237}Np is considered as a serious problem for long time storage of wastes
from nuclear industry.

Several methods have been described for the radiochemical separation of neptunium: anion exchange, cation exchange, liquid-liquid extraction, neutron activation etc.[2] The method described here was developed in order to investigate ^{237}Np in the environment. Problems especially considered were yield determination and interfering α-particle emitters.

METHOD

Use of Radiochemical Yield Determinant

As radiochemical yield determinant about 9 Bq (250 pCi) of ^{239}Np from a stock solution of ^{243}Am was used. Since plutonium determination is carried out simultaneously $\sim$180 mBq ($\sim$5 pCi) of ^{242}Pu was used for this purpose as yield determinant. If americium is in the +3 state and neptunium kept in the +4 state they both precipitate quantitatively at the conditions described below. It was thus not necessary to separate the ^{239}Np from the ^{243}Am prior to adding to the sample. Neptunium and americium will be separated at the first ion exchange procedure. For example Figure 1 shows the low energy γ-ray spectrum of the ^{243}Am-^{239}Np standard. Figure 2 shows the spectrum of a sample plus standard after separation. The 227 and 278 keV γ-ray peaks of ^{239}Np are seen, but the ^{243}Am is now gone.

Separation Procedure

The dried samples were ashed at 550°C until no traces of carbon remained. After leaching with aqua regia and hydrogen peroxide (30%), neptunium (+4) was precipitated with 20 mg lanthanum as fluoride. The precipitate from the centrifuged sample was dissolved by heating in nitric acid and a small amount of boric acid (added as solid). A second precipitation was carried out by adding ammonia, until precipitation started, then adding saturated ammonium carbonate solution. The supernate was again discarded and the precipitate dried at 105°C. Neptunium was kept in the +4 state with ferro ammonium sulphate during the procedure. From 9 $\underline{M}$ hydrochloric acid, 0.1 $\underline{M}$ ammonium iodine solution, neptunium was sorbed on an anion exchange resin (8 cm^3, BIORAD 1x8, 100-200 mesh). The effluent and 30 cm^3 9 $\underline{M}$ hydrochloric acid washing contain plutonium, americium, thorium and the lanthanum added. Remaining impurities were removed from the column with 30 cm^3 of 1 $\underline{M}$ nitric acid - 93% methanol (v/v). Neptunium was eluted with 4.5 $\underline{M}$ hydrochloric acid -0.05 $\underline{M}$ hydrogen fluoride. After evaporation with boric acid and sulphuric acid (1 cm^3) the samples were electrodeposited onto stainless steel discs from an ammonium sulphate solution as described by Talvitie.[3] For most biological samples a neptunium yield of about 50% can be obtained by this method. However, the radiochemical yield

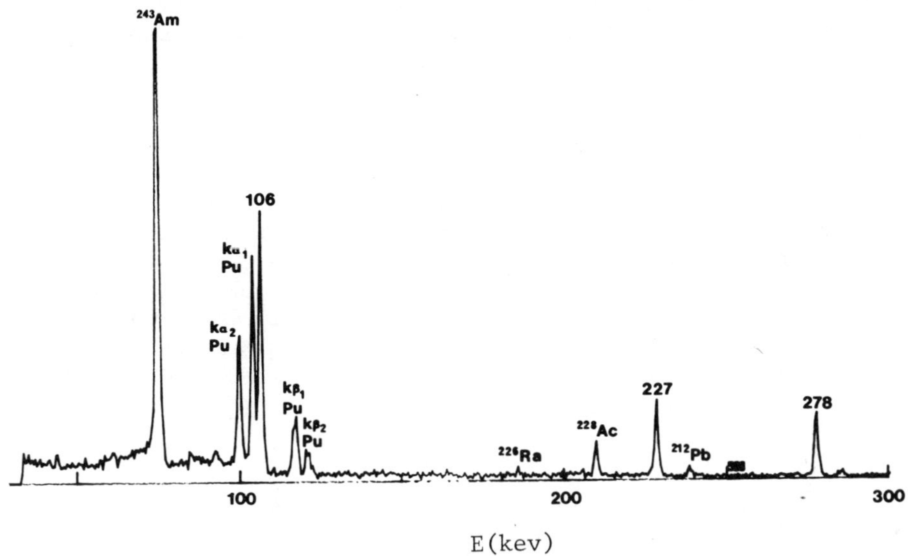

Figure 1. Ge(Li) spectrum of ^{243}Am–^{239}Np-standard.

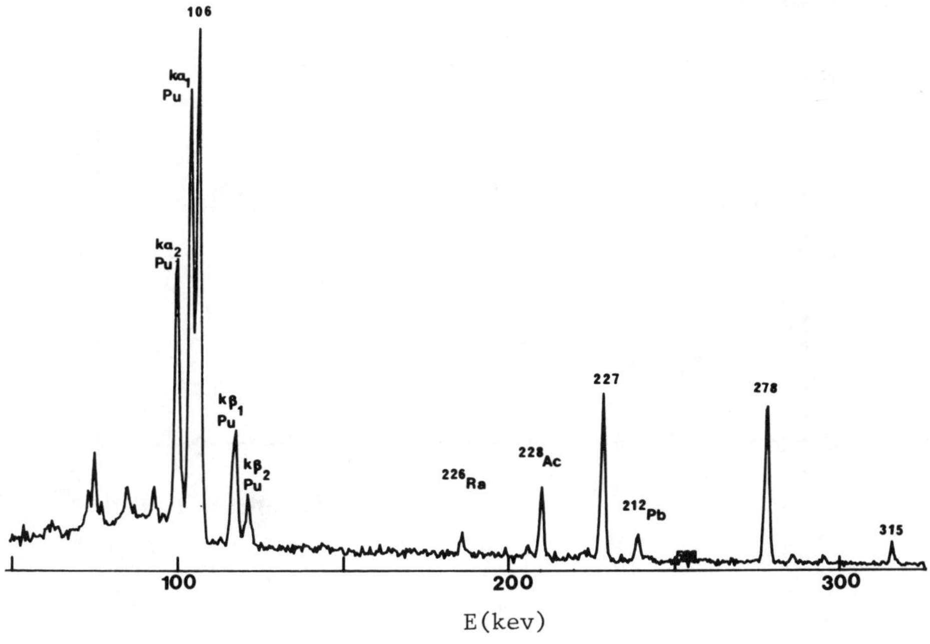

Figure 2. Ge(Li) spectrum of separated sample plus standard.

tended to be much lower when applying the method to soil and
sediment samples. This is probably due to the presence of
aluminum preventing americium and neptunium from precipitating as fluorides. Figure 3 shows results for a lichen sample.

The problem was solved by avoiding this fluoride precipitation. After loading the column neptunium was directly
eluted with 4.5 $\underline{M}$ hydrochloric acid -0.05 $\underline{M}$ hydrogenfluoride.
After evaporation with nitric acid - boric acid, neptunium
was extracted into 0.5 $\underline{M}$ TTA (thenoylfrifluoroacetone) -
xylene and back extracted with 8 $\underline{M}$ nitric acid. The drawback
to this procedure is that we do not get the effective reduction of sample sizes, as by using fluoride precipitation,
larger volumes at the column operation must be used.

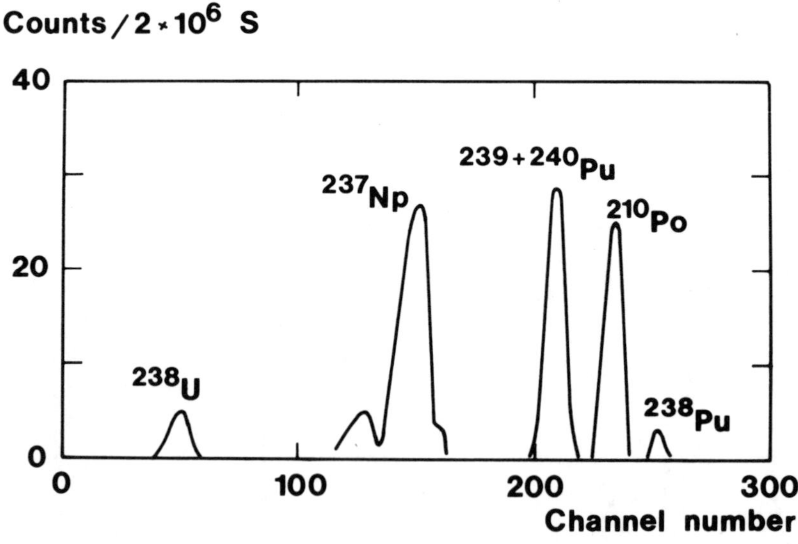

Figure 3. Alpha-spectrogram of lichen sample after
separation of ^{237}Np (150 g sample, 2 · 10⁶ S counting).

Measurements

The yield was determined by Ge(Li) spectrometry of the
sample compared to a ^{243}Am–^{239}Np standard calibrated by alpha
spectrometry. The peaks from kα1 Pu, kα2 Pu and 106, 227
and 278 keV from ^{239}Np were integrated. In Figure 1 and 2
the Ge(Li) spectra of the standard and of a sample are shown.
Note the absence of ^{243}Am after separation. Correction for
decay of ^{239}Np ($t_{1/2}$= 2.35 d) during separation from ^{243}Am
and yield measurement has to be performed. The samples were
counted for ^{237}Np with surface barrier detectors for about
10^6 seconds.

DISCUSSION

The actinides to take into consideration and their
interference at ^{237}Np measurements are shown in Figure 4.

The major interfering nuclide is ^{234}U present in large
amounts in the environment and having the same main α-energy
as ^{237}Np. The activity ratio ^{234}U/^{237}Np can be as high as
$5 \cdot 10^5$ in certain environmental samples. Remaining traces
of ^{234}U is corrected for by observing ^{238}U then also present.
(See Figure 3). The procedure described gives excellent
separation from uranium. Uranium does not precipitate as
fluoride or carbonate and is not complexed by the hydrogen
fluoride as neptunium at the elution stage from the column.

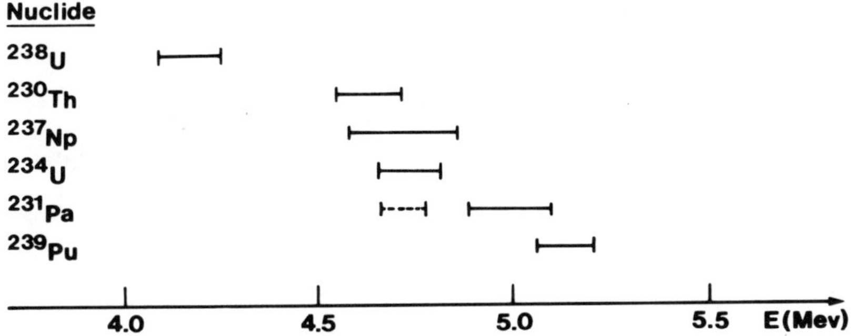

Figure 4. Actinides of importance in alpha-spectrometry of
^{237}Np. The energy region which they occupy is indicated.
(Surface barrier detector, ORTEC BA-18-300-100, 512 channels
per detector, 6 keV per channel). Minor branching dotted.

Plutonium-239 and 240 might interfere by tailing of the peak into the ^{237}Np energy region. At the column procedure plutonium is kept in the +3 valence state and will not be sorbed. Only a minor not interfering fraction of $^{239+240}$Pu will appear in the spectra.

Protactinium-231 has two major energies of which one (12% branching) will interfere into the ^{237}Np energy region (Figure 4). In addition ^{231}Pa could possibly follow in the separation procedure described. This problem was solved by integrating ^{237}Np over 4.70-4.83 MeV. This will contain 81% of the ^{237}Np peak and almost no ^{231}Pa. This energy integration also avoids interference from ^{230}Th. In spite of that thorium should not be sorbed together with neptunium at the column separation.

REFERENCES

1.   L. E. de Geer. Personal communication (1979).

2.   G. A. Burney and R. M. Harbour. The radiochemistry of neptunium NAS-NS-3060 (1974).

3.   N. A. Talvitie, Anal. Chem., _44_, 280 (1972).

LOW-LEVEL AIRBORNE MONITORING TECHNIQUES
UTILIZED IN A FIELD MEASUREMENT PROGRAM AT
OPERATING PRESSURIZED WATER REACTORS

D. W. Akers, N. K. Bihl, S. W. Duce, J. W. Tkachyk,
B. G. Motes, Exxon Nuclear Idaho Company, Inc.,
Idaho Falls, Idaho 83401

ABSTRACT

An in-plant measurement program, sponsored by the Nuclear
Regulatory Commission's Office of Nuclear Regulatory Research
is underway at commercial nuclear power stations. The objective of this program, the in-plant source term measurement
program, is to provide the Nuclear Regulatory Commission with
experimental data that can be used in the evaluation of plant
designs for liquid and gaseous effluent treatment systems.
As a part of this program, measurements have been made of low-
level airborne radioactivity concentrations and release rates.
A discussion of these measurement methods, which have been
either utilized or developed specifically in support of this
measurement program, is given. The monitoring methods discussed include several types of radioiodine species samplers
in addition to an atmospheric ^{14}C and tritium sampler. Each
method is individually evaluated with emphasis upon: system
description, observed detection limits, sample preparation,
analysis, problems encountered in field use, and appropriate
usage in a nuclear power plant ventilation system.

INTRODUCTION

Since June 1976, a field measurement program known as the
Source Term Measurement Program (STP) has been conducted at
commercial nuclear power stations. As a part of this program,
a large number of measurements have been made of low-level
radionuclide concentrations in the gaseous effluent streams
and at specific locations within the plants. In conjunction
with these measurements several methods have been used to

determine exhaust duct flow rates and the validity of the
sample locations used. From this data, the sources of airborne radioactivity and their release rates to the environment were determined.

The radionuclide of principle interest has been ^{131}I due
to its postulated impact, via the air-grass-milk pathway, to
man. Several types of samplers have been used for the monitoring of gaseous radioiodine in this project. The samplers
used for radiodine measurements is the Particulate-Iodine (PI)
sampler developed by <u>Keller, Duce, and Maeck</u>.[1] In special
cases, where greater sensitivity is required, a high capacity
model of the PI sampler[2] or a modified commercial air sampling
system[3] is used. As indicated, all iodine samplers are designed for collecting particulate radionuclides.

Other gaseous radionuclides which are normally sampled
in the STP due to their possible contribution to the environmental dose commitment are ^{14}C and ^{3}H. A sampler was specifically developed in support of the STP to collect these
radionuclides for periods of up to two weeks and to determine
their species.[4] A portion of this sampler is also used for
making measurements of only the oxidized forms of ^{3}H.

The subjects which will be discussed are, in order: the
sampling systems; proper usage and sample preparation; application in typical nuclear power plant ventilation systems;
and analytical methodologies, including the sensitivities
associated with sampling and analysis.

SAMPLING SYSTEMS

The basic PI sampler is shown in Figure 1. It consists
of five components. The components are: (1) a particulate
filter, (2) CdI_2 on a matrix of chromosorb-P support to retain I_2, (3) iodophenol on activated alumina to retain
hypoiodous acid, (4) silver exchanged 13X molecular sieve to
retain organic iodides, and (5) a TEDA impregnated charcoal
backup filter to check for penetration through the silver
xeolite bed. The normal flow rate is approximately 0.1 L/s.

The higher capacity model of the PI sampler is similiar to
the smaller sampler but to larger scale. This sampler has a
particulate filter and three media beds. The first two beds
are the same as in the smaller PI sampler, however; instead
of using silver xeolite to retain the organic iodides, a
charcoal filter bed is used due to the high cost of silver
zeolite.   The maximum flow rate for this model of PI sampler
is nominally 0.9 L/s., approximately eight times the flow rate
of the smaller sampler.

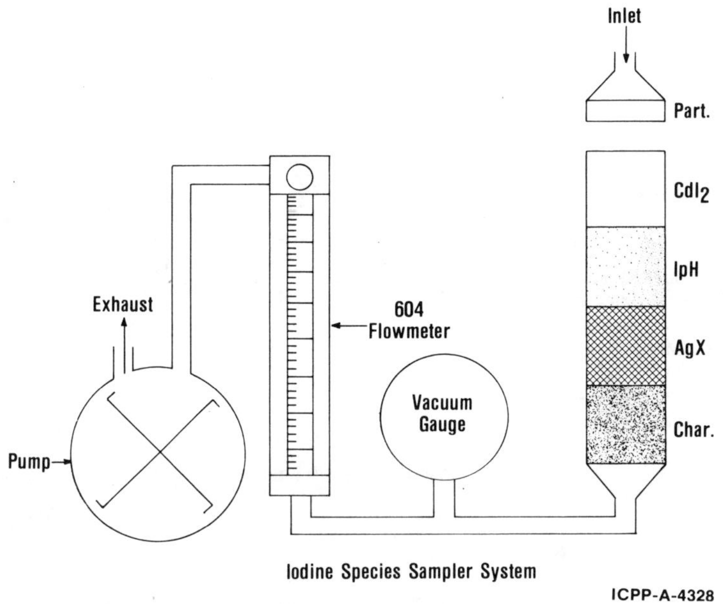

Figure 1. Basic sampler

As indicated in Figure 1, there is a vacuum gauge between
the sampler and the flowmeter. This component is required
for determination flow rate corrections as the flow rates may
vary somewhat due to particulate loading and the vacuum conditions in the duct being sampled.

The third type of iodine sampler, which is used generally
for environmental level field measurements, is a Misco (1825
Eastshove Highway, Berkley, CA) Model 680 high volume air
sampler with a ten inch diameter circular head. This sampler
has a particulate filter and two media beds. The normal media
used are CdI_2 on chromosorb-P for the retention of I_2 and
TEDA impregnated charcoal for the retention of all other
species. The normal flow rate for this sampler is approximately 14 L/s.

With regard to sample preparation for analysis, the small
PI sampler media cups are analyzed intact, whereas, the media
from both the high volume samplers is removed from the sample
cups. The different media are then analyzed in suitable geometries by gamma ray spectroscopy either at the Idaho
National Engineering Laboratory (INEL) or at the STP mobile
analysis trailer.

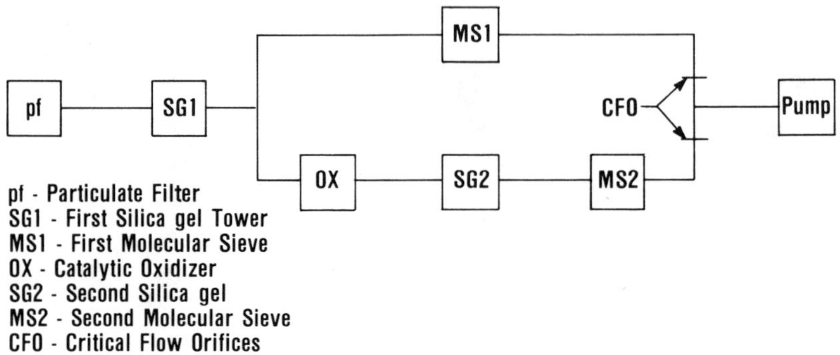

Figure 2.   Schematic for ^{3}H and ^{14}C sampler

Figure 2 schematically details the ^{14}C and ^{3}H gaseous
sampler used for the collection of the oxidized and non-
oxidized species of these radionuclides. The sample flow
(.003L/s.) is split after passing through the first silica
gel collection bed (Davidson-6X16 mesh), which removes H_2O
and HTO from the gaseous stream. Then, one aliquot goes
through a molecular sieve collection bed (4A) which adsorbs
CO_2, CO, and light hydrocarbons. The other portion of the
gaseous stream passes, in turn, through a heated catalytic
oxidizer (Pt.-Pd.), which converts all oxidizable species to
H_2O and CO_2; a silica gel collection bed; and a molecular
sieve collection bed. At this point, the gaseous streams enter critical flow orifices which are 1 cm stainless steel
discs with approximately 100 micron diameter laser-produced
holes. These orifices are calibrated and matched for a particular sampler, and require fourteen inches of vacuum to
maintain a constant flow rate.

At the end of the sample period, the molecular sieve
columns are capped in the original sample containers and returned to the INEL for analyses, and the silica gel is placed
in polyethylene bottles, sealed, and returned to the INEL
for analyses.

A portion of the ^{14}C and ^{3}H sampler is used for the
measurement of oxidized tritium. This sampler is a standard

silica gel tower with a flowmeter and vacuum gauge.

An integral part of the source term measurement program is measurements made of duct flows and the verification of sample locations. Two methods are used: (1) the standard pitot tube traverse method and (2) a helium dilution technique.[3] The pitot tube method consists of measuring the gas velocity pressure at the center of a specific number of equal areas of the duct perpendicular to the gas flow and calculating the duct flow rate. Sample validity of sampling locations is found by a helium dilution method. In this technique, pure helium is released into the duct at several locations upstream of the sample point at a known flow rate. The helium concentration at the sample location is determined using a helium leak detector, a mass spectrometer which is specific for helium. The degree to which the measured helium concentration at the sample location agrees with the concentrations calculated from the known injection rate is a measure of the validity of the sample location and/or a check of the duct flow rate.

NUCLEAR PLANT FIELD USAGE

The standard procedure upon entering a new reactor is to first identify major potential release points and then perform pitot tube traverse and helium dilution measurements on the release points and on the major feeder ducts. If it is found that the sample locations are indeed valid the ^{14}C, ^3H, and small PI samplers are installed at the release point sample locations and in the major ducts. Depending upon the results of a number of measurements, additional sample locations are added upstream of the first samplers in an effort to identify the major sources of radionuclides. Since many in-plant ducts contain very low concentrations of radionuclides, some or all of the iodine species present are not detected with the low volume PI samplers. Consequently, the higher volume samplers will be added to increase the sensitivity. For normal release rate calculations, these low concentrations of radionuclides are not significant, but for the purposes of precise measurements of the decontamination factors for filter banks and the calculation of a mass balance for the ventilation system, the measurement of the low-level radionuclide concentrations is critical.

The normal sampling period is two weeks making the total flow through a small PI sampler approximately 1.2(5)* liters

---

* 1.2(5) denotes 1.2×10^5

(4400 ft.^{3}). This compares with a total flow of approximately
9.9(5) liters (35000 ft.^{3}) for a large volume PI sampler and
1.71(7) liters (6.04(5) ft.^{3}) for the high volume Misco
sampler. Therefore, the sensitivity may be increased at a
particular sample location through the use of progressively
higher volume samplers.

The ^{14}C and ^{3}H monitors are normally used at the release
points only, but in some circumstances it is necessary to determine tritium mass balances within the plant, as the release
of tritium may not be directly related to the releases of
I^{131} within the plant. In this case, the oxidized tritium
portion of the ^{14}C and ^{3}H monitor is used.

In addition to the previously mentioned ventilation
studies, samples are taken of the containment and waste gas
decay tank (WGDT) systems. For containment, measurements are
made of the purge exhaust duct and of the containment air
during operation. These measurements are related to release
rates during purges and reactor coolant leak rates to the containment building. Samples from the WGDT's are used to calculate release rates via that pathway. Further special measurements include charcoal and HEPA filter banks decontamination
factor measurements.

ANALYSIS

Samples are either analyzed in the STP mobile laboratory
or returned to the INEL for analysis. The time delay for
samples returned to the INEL is between five and seven days,
which eliminates the need for analyses for nuclides with half-
life's on the order of a day, but requires that ^{131}I analyses
are performed quickly, especially for samples with marginal
radioiodine concentrations. The ^{14}C and ^{3}H samples are analyzed as time permits.

The small PI sampler media cups are analyzed in the individual media cups by gamma ray spectroscopy. Standards
traceable to the National Bureau of Standards (NBS) are prepared for each geometry. Due to the high adsorption of the
media,[1,2] the standards were prepared assuming a 0.3 cm average penetration of the iodine into the I$_2$ and hypoiodous
acid retaining media, i.e., a fiberglass filter containing an
NBS standard was placed 0.3 cm deep in the media. The silver
zeolite was assumed to have a penetration of 0.6 cm by the
organic iodides.[1] The standard is prepared in the same manner
as the I$_2$ and HOI adsorption media. Particulate filters were
standardized by placing the standard directly on the filter
medium. As some I$_2$ is retained on this filter in actual
sampling practice, corrections to the measured value have been

proposed to estimate the amount of retention occuring. However, the corrections are generally considered to be insignificant.

TABLE I

**SAMPLING SYSTEM SENSITIVITIES
AND COUNTING GEOMETRIES**

Sampling System	Species sampled	Geometry for analysis	Sensitivity - (uCi/cm.^{3}) ^{131}I	^{137}Cs	^{60}Co
P.I. sampler (low flow rate)	Particulate	Sample cup	1.3(-14)	3.5(-14)	2.6(-14)
	I$_2$	Sample cup	1.7(-14)		
	HOI	Sample cup	1.7(-14)		
	Organic	Sample cup	1.7(-14)		
P.I. sampler (high flow rate)	Particulate	filter in 40 ml. vial	1.6(-15)	4.1(-15)	3.0(-15)
	I$_2$	media in vials(2)	3.5(-15)		
	HOI	media in vials(2)	3.5(-15)		
	Organic	media in vials(3)	3.8(-15)		
High volume air sampler	Particulate	folded (1/8)	3.3(-16)	8.2(-16)	5.7(-16)
	Other media	Marinelli beaker	2.5(-16)		

ICPP-A-4330

Table 1 presents the sensitivities and counting geometries
for several radionuclides and the species of iodine. As noted,
the small PI sampler media is counted in the media cups. High
volume PI sampler media is transferred to forty dram vials
for gamma ray analysis due to the impracticality of counting
the large species cups directly.

The high volume sampler particulate filter is approximately ten inches in diameter. Standards and samples are
folded into eighths for a more reproducible geometry and analyzed. The iodine adsorbing media for the high volume sampler
is mixed; placed in a one liter Marinelli beaker; and analyzed.
This geometry allows the greatest sensitivity for the volume
of media used in these samplers. The sensitivities are calculated based upon the background at the energies of the
major gamma rays of the listed nuclides, and the standard
geometry calibrations for each media. The limits are listed
in μCi per cm^3 air. The sensitivities are calculated for a
standard two week sample period.

243

The listed sensitivities are the absolute sensitivities
for the samplers, that is the lowest possible airborne concentration which could be measured using the current INEL
gamma ray spectroscopy system. The actual sensitivity limits
are influenced by the counting periods used, changes in background, and Compton effects from higher energy nuclides.
These effects can reduce the sensitivities approximately one
order of magnitude. The large PI sampler is approximately
eight times as sensitive as the small PI sampler and the high
volume Misco sampler is approximately forty times as sensitive as the small PI sampler.

The analysis of the ^{14}C and ^{3}H samples is performed by
liquid scintillation techniques. The analyses for the two
radionuclides differ considerably. For ^{3}H there are three
options available for the analysis of the silica gel. The
first, which is normally used and has the least sensitivity
involves taking replicate 5-gram portions from the weighed
(approximately 250g) and well mixed sample. These portions
are placed in glass counting vials, treated with 3 mL of low
^{3}H content water, 14 mL of Ready-Solv^{R} HP liquid scintillation
cocktail, and counted directly. This method has the advantage
that it is the least time consuming of the three methods and
has a sensitivity of approximately $6.7(-11)\mu Ci/cm^{3}$ air. The
second method involves washing the silica gel and analyzing
an aliquot of the wash water. This method is more sensitive
than the first, but less sensitive than a third method. The
third method involves baking the water from the silica gel at
temperatures above $200^{\circ}C$ and analyzing the entire sample.
This method is the most sensitive as the tritium is in the
more concentrated form, HTO.

When the molecular sieve columns are returned to the INEL,
they are placed in the bakeout apparatus diagramed in Figure 3.
The cylinder is heated at $350^{\circ}C$ for four hours with a dry He
purge to remove the adsorbed CO_2 from the molecular sieve.
The gases from the cylinder pass through the ice-cooled trap
to remove adsorbed water vapor and then through the spiral
liquid nitrogen cooled traps which remove the CO_2. The CO_2 is
then allowed to expand into a calibrated volume and its
pressure measured, allowing the calculation of the quantity
of CO_2 captured. A portion of the sample is then transferred
to a glass scintillation vial containing ethanolamine which
reacts with the CO_2 to form a carbamate. Complete reaction
is indicated by the system pressure drop. Then 7 mL of
methanol and 12 mL of scintillation cocktail (Insta-gel^{R},
Packard Instruments) is added and the sample is counted. The
sensitivity for airborne ^{14}C is approximately $4.0X10^{-12}\mu Ci/cm^{3}$.

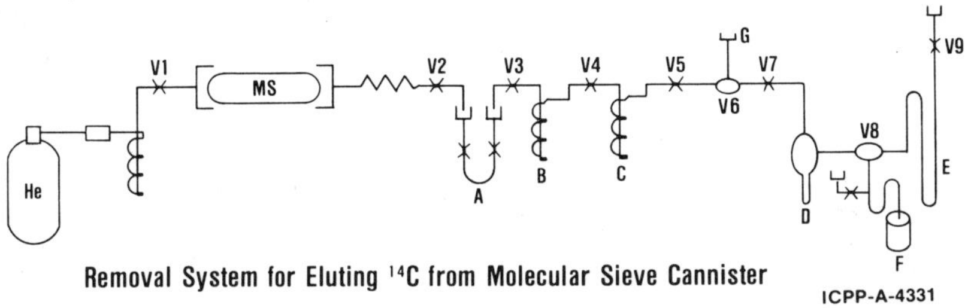

Figure 3. Removal system for eluting
^{14}C from molecular sieve cannister

SUMMARY

The samplers which have been discussed have been used at five reactor sites. These techniques can be effectively utilized in determing plant releases as well as mass balance and effluent treatment system's performance. In addition, the sampling methods have proven to have utilization in environmental monitoring studies. These sampling methods and analysis techniques used in the Source Term program are well suited to monitoring programs which require that large numbers of samples be collected and analyzed.

REFERENCES

1. J. H. Keller, F. A. Duce, W. J. Maeck, "A Selective Adsorbent Sampling System for Differentiating Airborne Iodine Species, Proc. Eleventh USAEC Air Cleaning Conference, CONF-700816, Vol.2, pg 621-634, (1970).

2. W. A. Emel, et. al.,"An Airborne Radioiodine Species Sampler and Its Application for Measuring Removal Efficiencies of Large Adsorbers for Ventilation Exhaust Air", Proc. Fourteenth E.R.D.A. Air Cleaning Conference, CONF-650822, Vol.2, (1976).

3.   N. C. Dyer, et. al., "Procedures Source Term Measurement Program", TREE-1178, Idaho National Engineering Laboratory, (1977).

4.   J. L. Thompson, S. W. Duce, J. H. Keller, "An Atmospheric Tritium and Carbon-14 Monitoring System", NUREG-1 CR-0386, (1977).

5.   C. A. Pelletier, R. T. Hemphill, "Surface Effects in the Transport of Airborne Radioiodine at Light Water Nuclear Power Plants", EPRI NP-876, pg B.2 (1978).

EXPERIMENTS INVOLVING ^{238}PuO$_2$ AND THE ENVIRONMENT

F. J. Steinkruger, G. B. Nelson, J. H. Patterson,
G. M. Matlack, and G. R. Waterbury. Chemistry-Materials
Science Division, Los Alamos Scientific Laboratory,
Los Alamos, New Mexico, USA.

ABSTRACT

The interactions of ^{238}PuO$_2$ with terrestrial and
aquatic environments have been examined. Samples were
removed periodically and analyzed for plutonium to determine
dispersion. The samples from terrestrial experiments were
acidified, reduced in volume, and analyzed by liquid scintillation counting while aquatic experiment samples were
diluted and counted by liquid scintillation. Results from
the aquatic experiments indicated that the release rate
was independent of the thermal power of the source. More
plutonium was released in cold fresh water systems than in
warm or cold seawater. Plutonium migration through soils
appeared to be independent of soil type. Seasonal variations in plutonium migration appeared to result from the
different quantity of water passing through the soil.
Large pieces of source material released more plutonium to
the environment which may be a thermal spallation effect.
A simulated tidal experiment in which the source cycled
into and out of the seawater twice a day resulted in severe
fracturing of the source.

INTRODUCTION

Plutonium enriched 80% in ^{238}Pu, an alpha emitter with
a half-life of 87.7 years, is used in the dioxide form as
the heat source in radioisotope thermoelectric generators
that provide power for instruments and data transmission
in many space missions. Because of the relatively short
half-life and the high alpha energy the thermal specific
power is 0.56 W/g of ^{238}Pu. Although the heat source container is designed with a large safety factor to withstand

both reentry from orbit and impact with the earth, additional information is continuously sought concerning the interaction of the heat source with the environment.

Since there are two general environmental conditions into which a heat source may be deposited on return from orbit, namely, on land or in water, several variations of each condition have been utilized in experiments at Los Alamos Scientific Laboratory. The discussion of these variations will include the conditions, the samples obtained, the radiochemical processing, and results.

EXPERIMENTAL

Terrestrial experiments are conducted in sealed environmental chambers. These chambers, with an internal volume of 1.8 m^3, contain a soil tray that is 0.3 meters deep by 0.9 by 0.9 meters in area. The chambers are equipped with environment control equipment including heaters and refrigeration for temperature control, and humidifier and dehumidifier coils for humidity control. Each chamber is equipped with a spray head that can provide a relatively uniform rain over the surface of the soil tray.

The samples obtained from a terrestrial experiment include rainwater that has percolated down through the soil, condensate from the dehumidifier coils, and soil core samples. Rainwater and condensate samples are collected in 20 liter polyethylene bottles. Core samples are obtained by driving a hollow tube down through the soil to the bottom of the tray. A core is approximately 27 cm long by 13 mm in diameter.

Two types of soils and two source configurations have been used in these experiments. A sandy soil containing approximately 90% sand and a loam soil containing 50% sand and 38% silt are the types used. Each soil is used in conjunction with two configurations from a particular source, large pieces and fines (from 10 μm to 6000 μm in diameter). This material is obtained from heat sources that have undergone impact testing as part of the overall source containment safety evaluation program.

Aquatic experiments are conducted in aquariums that have been outfitted with sealed lids and filter systems. These aquariums have heaters and refrigeration systems for temperature control. Fresh water has been utilized in experiments at 10°C and 37°C. Sources used in aquatic

experiments were usually cylindrical in shape and weighed
about 6 g for low source power experiments to 60 g for
high source power experiments. Samples obtained from these
experiments were aliquots of the aqueous phase.

A variation of the aquatic experiments involved placing the source on a tray equipped to cycle into and out of
the water twice a day. This experiment, conducted in
simulated seawater represents a source lying in a tidal
basin. In addition to aliquots of the aqueous phase,
samples of sand from the tray on the bottom of the aquarium
and a piece of the source were removed for analyses.

Other experiments being conducted include aquatic conditions in glass chambers, a high pressure seawater experiment, particle size solubility studies in $1\underline{M}$ $HClO_4$, ionic
and particulate plutonium on soil columns, and tests to
determine ionic versus particulate form.

Radiochemical processing and analyses of the samples
are rather straightforward. Rainwater samples are quantitatively transferred to beakers, acidified with HNO_3 and
a trace of HF, and reduced in volume by heating. The resulting small sample is quantitatively transferred to a volumetric flask and diluted to volume with $0.5\underline{M}$ HNO_3. An
aliquot is then counted by liquid scintillation using a
commercial instrument. A commercial cocktail, to which
0.1% D-2-EHPA is added, is used in a routine manner. A
condensate sample is treated in the same manner except that
only a 4 liter aliquot of the total 20 liter sample is
processed. Soil cores are cut into segments, dried, weighed,
and dissolved using HNO_3-HF followed by $HNO_3 - HClO_4$. The
final solution is treated in the same manner as a rainwater
or condensate. Aliquots of aquarium waters are pipetted
directly into the cocktail and counted.

RESULTS

Terrestrial Experiments

Two types of soils, sand and loam, were used in the
terrestrial studies. Large pieces and fines of the same
source were placed on each type of soil. Experiments using
loam were conducted on a humid winter-summer cycle, whereas
experiments using sand were on an arid winter-summer cycle.
These experiments are still in progress but some preliminary results are available at this time.

There appears to be a significant release of plutonium
from the large pieces of source material upon contact with

rain. These releases are attributed to thermal spallation
because the surface temperature of a large piece is higher
than that of fines. Several air filter tests have been
conducted covering the time interval before, during, and
after a rain. In a chamber containing large pieces the
airborne plutonium concentration increased by four orders
of magnitude during the first 5 minutes of the rain and
then decreased until it reached prerain levels about 1 to
2 h after the rain. In a chamber containing fines there
was no significant increase in airborne plutonium when the
rain commenced, but there was a reduction of a factor of
10 after the rain, which attributed to a cleansing
effect of the rain.

In all cases plutonium was seen in the percolated
rainwater collected during and after the first rain subsequent to source implanation. This result suggests that
plutonium migrates in the particulate form and the migration rate is independent of soil type.

In a majority of the core samples most of the plutonium was present in the top layer of the soil (0-5 cm).
The other layers contained measurable amounts of plutonium
but usually significantly less than the top layer. There
were a number of cores that contained a substantial amount
of plutonium at some intermediate depth, but no consistent
trend could be discerned.

Three soil columns, containing silt loam (840 μm to
1650 μm in diameter) with approximately 3 mg of ^{238}PuO$_2$
on the surface of each soil, have been in operation for
about 1000 days. These columns are being eluated with
distilled water. There is a continuous low level release
of plutonium, 20 pg/l, but almost 90% of the released
plutonium was collected during the first 5 days of the
experiment. These results again suggest that plutonium
migrates in the particulate form.

Aquatic Experiments

Table I is a summary of the aquatic environment
experiments. Several observations can be made concerning
the data. First, the lowest release rate is seen in warm
seawater, a condition that may occur in an isolated tidal
pool or in the tropics. The highest release rates are
observed in the cold fresh water systems. Third, at this
time no significance is attached to the variation in release
rates resulting from the different thermal power of the
sources. The substantial difference in the release rates
between fresh water and seawater experiments indicates
that removal reactions are occurring in the seawater

TABLE I

SUMMARY OF RELEASE RATES FROM PPO IN AQUATIC ENVIRONMENTS

WATER	POWER (W)	TIME (days)	TEMP. (°C)	RELEASE RATE (nCi/m^2-s)
FRESH	2.5	1633	10	171
FRESH	25	1610	10	460
SEA	2.5	1660	10	7.7
SEA	25	1615	10	12
SEA	2.5	993	37	2.7
SEA	25	968	37	2.3

experiments that do not occur in fresh water, or are occurring at a faster rate, or that the release rate is substantially reduced in seawater, or a combination of all three.

A variation of the aquatic experiment was the tidal basin. In this aquarium the source was placed on a tray that was cycled into and out of the water twice a day. Figure 1 is a graph of the plutonium concentration in the water. Note the initial increase of plutonium to a peak at about day 40. After that time there appears to be a gradual decrease. On day 103 a crack was noted in the face of the source and on day 393 the source had split into two pieces. By day 404 there were 7 pieces, and at the present time there are over 30 pieces. The sharp increase in plutonium concentration starting about day 400 is attributed to the exposure of new surface. Again a peak followed by a decrease is seen. On day 451 the elevator motor failed and the source was left submerged. The cause of the increasing plutonium level to the present time is not known.

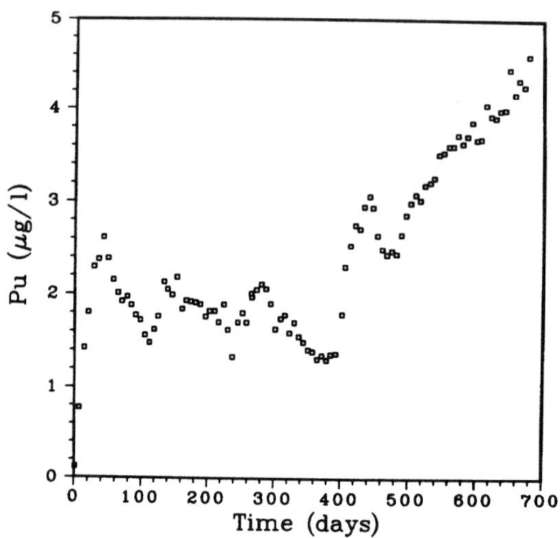

Figure 1. Plutonium concentration in the tidal experiment.

     Because of the possibility of plutonium removal by
plating out on the walls or settling to the floor of the
aquarium, a glass chamber was designed which would permit
total recovery of plutonium after termination of the
experiment. An indication of the magnitude of this problem
may be seen in the tidal experiment. Sand samples from
the tray on the bottom of the aquarium were removed and
analyzed for plutonium. The results, averaged over the
area of the tray, imply that there are 28 mg of plutonium
on the bottom of the aquarium. A total of 0.35 mg is
estimated to be in the aqueous phase.

     An initial experiment in the glass chamber designed
to measure a magnitude of the plate-out error, revealed
that a determination of plutonium released based on a
measurement of the plutonium level in the aqueous phase
would be low by 38%. This error is not as large as expected
on the basis of the tidal experiment, but the source configurations differ significantly.

     Additional experiments to determine a "total source"
term have been conducted in the glass chambers. In one
case a bare source was sealed in a dry chamber for 454
days. The total plutonium recovered from the chamber,
excluding the source itself, was 1.41 μg. This source was
then submerged in deionized water for 30 min. Again the
chamber was processed to determine the total plutonium
released. In this case 49 μg were recovered. The first

experiment indicated that a small amount of plutonium is
released from the surface of a dry source by recoil. The
second experiment was in agreement with a proposal by
R. L. Fleischer. This proposal suggests that the application of a solvent to the surface will remove a significant
quantity of plutonium that has been loosened by alpha decay
process near the surface.[1] The next experiment in the
sequence would be to allow the source to reach equilibrium
in air and then resubmerge it. This test would permit a
determination of the contribution of thermal spallation
to the total release. Before this test could be accomplished the source split into three pieces. Since there was
no longer a good estimate of surface area the experiment
was terminated.

SUMMARY

Many of the experiments described above and others not
described are still in progress at this time. The tenta-
tive conclusions mentioned are subject to change as more
data become available. Additional experiments in progress
are designed to determine the state of the plutonium in
the aqueous phase, ionic or particulate. Additional
terrestrial experiments are in the planning stages as are
experiments in the glass chambers.

REFERENCES

1.   R. L. Fleischer, Health Phys., 29, 69 (1975).
      R. L. Fleischer, and O. G. Raabe, Health Phys., 32,
      253 (1977).
      R. L. Fleischer, and O. G. Raabe, Geochim. Cosmochim.
      Acta, 42, 973 (1978).
      R. L. Fleischer, and O. G. Raabe, Health Phys., 35,
      545 (1978).

ANALYSIS OF LOW ENERGY BETA-EMITTERS †

D. L. Murphy, Energy and Environment Division,
Lawrence Berkeley Laboratory,
Berkeley, California, USA.

ABSTRACT

As part of an overall Survey of Instrumentation for
Environmental Monitoring, a survey was made of the instruments used for the determination of low energy beta radioactivity. Techniques commonly used are gas flow proportional counting, liquid scintillation counting, solid
scintillation counting, and internal ionization chamber
counting. Also used are solid state detector counting and
radiochemical separation followed by counting using one of
the preceeding techniques.

As a first step, the first four techniques were
examined and compared with each other. The sensitivities of
the techniques were compared on the basis of the detection
limits quoted for instruments described in the technical and
reviewed literature. The detection limits were then related
to the occupational and public individual maximum levels for
air and water given in Appendix B, Part 20 (Standards for
Protection Against Radiation) of Title 10 of the Code of
Federal Regulations. Attention is focussed primarily on the
continuous monitoring of air for ^{3}H and ^{85}Kr, a medium
energy β-emitter.

From a survey of U. S. commercial sources, it is clear
that several continuous air monitoring instruments are
readily available for measuring low energy β concentrations,
even in presence of certain other activity, at occupational
levels. However, these instruments do not typically have
sensitivities comparable to the public individual levels.
Moreover, their capabilities for giving results in real
time and for differentiating among the radionuclides
actually present is limited.

INTRODUCTION

Concurrent with an increasing use of energy, the realization has grown that environmental quality control is important. As a basis for such control, numerous monitoring programs have been developed, including a variety of instruments, some of which are elaborate and highly sophisticated. To aid monitoring organizations and analytical laboratories in choosing among the techniques and instruments available, Lawrence Berkeley Laboratory (funded by DOE and NSF) has for several years conducted a Survey of Instrumentation for Environmental Monitoring.[1] Instrumental methods covered are those suitable for monitoring and analyzing the conventional pollutants in air and water, as well as those for radiation measurements and analysis of biomedical samples.

As part of this work, a study was made of the instruments used for the measurement of low energy beta-emitters. The necessity for counting β particles lies in the fact that a number of natural or manmade radionuclides are β-emitters and either emit no gamma or X-rays, or emit them at so low a probability per decay as to make assay at low levels by photon detection difficult or impossible. Several such radionuclides with low β energies are listed in Table I, along with pertinent properties and sources of production. Not included are nuclides that decay by electron capture with the emission of conversion electrons. For this application semiconductor detectors can be particularly effective because of the low background attainable under the sharp conversion electron peaks.[2]

TABLE I. SELECTED LOW ENERGY β-EMITTERS AND PERTINENT PROPERTIES. DATA FROM REF. 3 EXCEPT WHERE NOTED.

Nuclide	Half-Life $t\ 1/2$	$\beta_{max.}$	$\beta_{ave.}$	Sources
^{3}H	12.33 yr	0.0186 MeV	0.0057 MeV	naturally occurring ternary fission[4] ^{6}Li (n,α)
^{14}C	5,730 yr	0.156 MeV	0.0467 MeV	naturally occurring ^{14}N (n,p)
^{35}S	87.4 d	0.167 MeV	0.061 MeV	^{34}S (n,γ)
^{79}Se	$\leq$6.5 x 10^4 yr	0.159 MeV	0.058 MeV[5]	fission
^{85}Kr^{a}	10.7 yr	0.687 MeV	0.249 MeV[5]	fission ^{85}Kr (n,γ)
^{99}Tc	2.14 x 10^5 yr	0.294 MeV	0.085 MeV[5]	fission
^{129}I	1.6 x 10^7 yr	0.192 MeV	0.040 MeV[5]	fission
^{135}Cs	3 x 10^6 yr	0.205 MeV	0.057 MeV[5]	fission daughter ^{135}Xe
^{147}Pm	2.6234 yr	0.225 MeV	0.0650 MeV	fission
^{151}Sm	90 yr	0.076 MeV	0.019 MeV[5]	fission
^{241}Pu	14.4 yr	0.0208 MeV	0.005 MeV[5]	multiple n-capture ^{238}U, ^{239}Pu, etc.

[a]Technically, ^{85}Kr is a moderate energy β-emitter,[6] but is included due to its widespread interest.

Of the nuclides in Table I the volatile and semi-volatile long-lived manmade radionuclides ^{3}H, ^{14}C, ^{79}Se, ^{85}Kr, ^{99}Tc, ^{129}I and ^{135}Cs are of concern in connection with the operation of certain nuclear fuel cycle facilities. They are difficult and expensive to contain and once emitted to the environment many become permanent ecological constituents with both local and global distributions. Since volatile and semivolatile species are the most difficult to trap, they are the most likely to be released and transported long distances. Control of the release of these radionuclides to the environment and assessment of their long-term effects require sensitive sampling and measurement methods.[7] Measurements of both radionuclide levels <u>and</u> their chemical forms are needed. However, attention in this paper will focus on the measurement of the former.

Low energy β radioactivity, due to its limited penetration, tends to be an internal rather than external hazard. Internal exposure is usually assessed through estimation of the body burden based on measurements of the activity concentrations in biological samples (bioassay), or on measurements of activity in the body by use of external counters. Alternatively, the potential body burden or the potential dose equivalent in body organs is assessed by comparing measurements of the concentrations of radionuclides in air and water taken in by an individual with the maximum permissible concentrations that have been specified. NCRP Report No. 57 says, "In areas where frequent or continuous air contamination is likely, the air should be sampled continually during periods of personnel occupancy.* The potential for nonoccupational exposure to airborne contamination should be assessed by sampling the gaseous exhaust stream from the facility."[7]

Commonly used techniques for the measurement of β-emitters are gas flow proportional counting, liquid scintillation counting, solid scintillation counting, and internal ionization chamber counting. For the purposes of this paper, it is assumed that the reader is familiar with the details of the instrumentation and procedures involved in these counting techniques, and I will simply compare the different techniques in the measurement situations outlined later. For detailed information on techniques the reader is referred to the literature.[1, 2, 8-12] In this paper attention is focussed primarily on continuous air monitoring,

---

*In an alternative philosophy, the environment of personnel is controlled to levels well below the MPC and air sampling is done primarily to indicate when control is lost and remedial actions are needed.

although some mention will be made when a technique is
applicable to continuous water monitoring as well.

DISCUSSION

First, a compilation of the detection limits quoted for
instruments in the technical and reviewed literature will be
presented to give an idea of the sensitivities obtainable
with the aforementioned techniques. Then the results of a
survey of U. S. commercial sources for off-the-shelf instruments will be presented to determine the extent to which
commercially available equipment is able to measure low
energy β-emitters at the various maximum permissible concentrations allowed by Appendix B of Title 10, Part 20 of
the Code of Federal Regulations which are summarized in
Table II.

TABLE II. MAXIMUM PERMISSIBLE CONCENTRATIONS IN AIR AND WATER.
           (FROM 10CFR20, APPENDIX B).

Radionuclide		Occupational MPC		Unrestricted MPC	
		Air[a,b]	Water[a,b]	Air[a,b]	Water[a,b]
^{3}H	S	5×10^{-6} µCi/ml	1×10^{-1} µCi/ml	2×10^{-7} µCi/ml	3×10^{-3} µCi/ml
	I	5×10^{-6} µCi/ml	1×10^{-1} µCi/ml	2×10^{-7} µCi/ml	3×10^{-3} µCi/ml
	Sub	2×10^{-3}		4×10^{-5}	
^{14}C	S	4×10^{-6}	2×10^{-2}	1×10^{-7}	8×10^{-4}
(CO_2)	Sub	5×10^{-5}		1×10^{-6}	
^{35}S	S	3×10^{-7}	2×10^{-3}	9×10^{-9}	6×10^{-5}
	I	3×10^{-7}	8×10^{-3}	9×10^{-9}	3×10^{-4}
^{85}Kr[c]	Sub	1×10^{-5}		3×10^{-7}	
^{99}Tc	S	2×10^{-6}	8×10^{-2}	7×10^{-8}	3×10^{-4}
	I	6×10^{-8}	1×10^{-2}	2×10^{-9}	2×10^{-4}
^{129}I	S	2×10^{-9}	1×10^{-5}	2×10^{-11}	6×10^{-8}
	I	7×10^{-8}	6×10^{-3}	2×10^{-9}	2×10^{-4}
^{135}Cs	S	5×10^{-7}	3×10^{-3}	2×10^{-8}	1×10^{-4}
	I	9×10^{-8}	7×10^{-3}	3×10^{-9}	2×10^{-4}
^{147}Pm	S	6×10^{-8}	6×10^{-3}	2×10^{-9}	2×10^{-4}
	I	1×10^{-7}	6×10^{-3}	3×10^{-9}	2×10^{-4}
^{151}Sm	S	6×10^{-8}	1×10^{-2}	2×10^{-9}	4×10^{-4}
	I	1×10^{-7}	1×10^{-2}	5×10^{-9}	4×10^{-4}
^{241}Pu	S	9×10^{-11}	7×10^{-3}	3×10^{-12}	2×10^{-4}
	I	4×10^{-8}	4×10^{-2}	1×10^{-9}	1×10^{-3}

[a]To convert to pCi/cm^3 divide numerical value by 10^{-6}.

[b]To convert to $µCi/m^3$ divide numerical value by 10^{-6}.

[c]Technically <u>not</u> a low energy β-emitter.

The results of the literature survey, as summarized in Table III, indicate that attention has focussed on instruments that determine ^{3}H and ^{85}Kr concentrations.

Since occupational monitoring is generally required when there is a probability that individuals will receive a dose in excess of 10% of the standard,[30] then it is desirable that available monitoring techniques measure as low as 0.1 MPC (occupational). Although this would be greater than that set forth by ANSI N 13.10 - 1974, it says itself, "These values represent current minimum standards. Improved sensitivities are always encouraged and should be used when improved state-of-the-art and commercial availability are realized."[31] In any case instruments should be available to measure at the 0.1 MPC (occupational) level, since the public MPC is even lower.

From perusing Table III it seems clear that all of the techniques are sensitive enough to at least be able to measure down to the occupational MPC_a. Only two of the techniques (without special sampling procedures) are adequate for measuring 0.1 occupational MPC_a. These are liquid scintillation and gas flow proportional counting. Using special sampling techniques the solid scintillator may become adequate for measuring even lower than 0.1 public individual MPC_a, although a rapid response is sacrificed. From the results obtained by Gregory and Parnell,[18] it seems possible that a specially designed gas flow proportional chamber can also measure down to these levels, and with the advantage of a shorter response time. This particular instrument is based on technology used in experimental high energy physics and should be tested further in practical applications for low energy β-emitters to determine its true potential. The techniques that give reliable readings at occupational MPC_a even when other β-activity is present (>10 times) are internal ionization chamber counting with a compensating chamber, liquid scintillation coupled with a bubbler, and either solid scintillation counting or internal ionization counting using permselective membranes for sampling. The technique with the fastest response time at occupational MPC_a is internal ionization chamber counting.

The results of a survey of U. S. commercial sources for off-the-shelf instruments based on three of these techniques is summarized in Table IV. The instruments have been identified by letters and in any case the inference should not be drawn that we recommend one over the other, or that one is inherently superior to another.

TABLE III.   INSTRUMENTS APPEARING IN TECHNICAL AND REVIEWED LITERATURE

Reference Source	Detector Type	Detector Characteristics	Nuclide Detected	System Response Time	Detection Limits[a]	γ Response	Other Nuclide Response	Remarks
Howell et al[14]	gas flow proportional	250 cm^3 volume 250 cm^3/m flow	^3H	~5 min.	~2.5 nC/1		Decreased by lowering concentration	Semipermeable membrane used to enhance HTO over HT and Kr.
Ehret[15, 16b]	gas flow proportional	1 ℓ volume methanet + 20 – 30% air counting gas, flow 25 1/hr	^3H	few minutes	0.1 pCi/cm^3		Used anti-coincidence proportional chamber	Memory effects due to tritiated water vapor absorption in plastic tube
Block et al[16, 17b]	Two thin window gas flow proportional chambers	0.2 ℓ volume	^3H		1 pCi/m^3	Compensation such that detection limit held in 3MR/hr field	Compensated against ^{41}Ar and ^{85}Kr	Thin Formvar window allows detection ^3H; aluminum mylar window opaque to ^3H.
Gregory and Parnell[18]	proportional	multiwire multi-plane proportional chamber; 112.5 cm^2 active area	^{85}Kr	200 s count time	4 x 10^{-3} pCi/cm^3	anticoincidence between planes		Technology used in high energy physics
Osborne and Coveart[19]	internal ion chamber with sealed ion chamber inside	1330 cm^3 vol., flow rate 30 cm^3/s, solid state electronics	^3H	70 s.	5-10^7 μCi/m^3	>95%	2.2 times more sensitive to ^{41}Ar	Portable air monitor for tritiated water vapor; ion trap.

[a] Taken as stated by the author; for precise meaning the reader is referred to the source.
[b] The data actually quoted is taken from ref. 16.

TABLE III (Cont'd.)

Reference Source	Detector Type	Detector Characteristics	Nuclide Detected	System Response Time	Detection Limits[a]	γ Response	Other Nuclide Response	Remarks
Jalbert[20]	two concentric ion chambers	1.6 ℓ sampling volume	^{3}H	18 s.	10^{-6} µCi/ml to 10^{-4} µCi/ml		w/background β of 2×10^{-5} µCi/ml to 0.4 µCi/ml, respectively	Sensitive to temperature or pressure changes; iontrap.
Jalbert and Hiebert[21]	internal ion chamber and sealed ion chamber	1 ℓ volume flow rate 10 ℓ/m	^{3}H	50 s	25 µCi/m^3	in 50 mR/hr field	no selectivity	Ion trap; chambers orbit about common axis.
Osloond et al[22]	liquid scintilation	5 g. gel 2 ml H_2O 18 ml scintillator solution	^{3}H	167 hr- sampling 20 min.- counting	4×10^{-11} µCi/ml low humidity 6×10^{-10} µCi/ml high humidity	inherently none	noble gases pass sampling gel	250 g. silica gel, with sample flow rates 6 l/hr; result must be calculated.
Osborne[23]	liquid scintillation		^{3}H	10 hr at 10 cm^3/ min. flow rate, or flow at 100 cm^3/ min.	0.1 pCi/cm^3 minimum	inherently none	selectivity against noble gas to <.013% and also HT	Conjectured; uses bubbler system to sample air flow.
Osborne and Tepley[24]	liquid scintillation	LS flow $\sim$2 mm^3/sec.	^{3}H	several minutes	$\sim$.3 nCi/m^3 minimum	inherently none	those gases not soluble selected against	HTO collected directly from air in LS; water monitoring possible with detection limit 2 µCi/kg.
Colmenares et al[25]	CaF$_2$(Eu) scintillator and photomultiplier	6.35 cm. dia. x 0.37 cm.	^{3}H gas	60 s.[b]	minimum 48 pCi/cm^3			Uses amplifying pulse shape analyzer; $\sim$.05 efficiency.

[a]Taken as stated by the author; for precise meaning the reader is referred to the source.
[b]Sample & background counting time used to calculate minimum detectable activity.

TABLE III (Cont'd.)

Reference Source	Detector Type	Detector Characteristics	Nuclide Detected	System Response Time	Detection Limits[a]	γ Response	Other Nuclide Response	Remarks
Osborne[26]	plastic NE 102 scintillator flow counter	Sheets 0.125 mm apart w/300 cm^2 surface; water flows through	^{3}H water vapor	200 s.	1 μCi/m^3 minimum	1 mr/hr[27] corresponds to .5 μCi/m^3	<0.7% for same conc. ^{3}H	Air stream vapor collected in water stream, purged; .05% of ^{3}H disintegration detected.
Moghissi et al[28]	anthracene scintillator	ground scintillator coating 100 parallel 3 mm diameter plexiglas rods, 45 mm long viewed by two PMT.	^{3}H		minimum $\sim$6nCi/ℓ			Signal out of SCA grated coincident in 30–50 nsec. and in energy window of interest.
Takamatsu et al[29]	plastic scintillator	polyvinyl-toluene plate discs stacked 10 cm and thickness 0.5 mm viewed by 2 PMT	^{85}Kr	1 hr.	min. 10^{-10} μCi/ml	anti-coincidence shielding + shielding	selectively enriches Kr over Xe and eliminates Co_2 and H_2O.	Uses silicone rubber membrane and 2 molecular sieves.

262

[a]Taken as stated by the author; for precise meaning the reader is referred to the source.

TABLE IV. OFF THE SHELF U. S. COMMERCIAL INSTRUMENTS (COMPILED AS A RESULT OF
SURVEY CONDUCTED NOVEMBER – DECEMBER 1978)[32]

Indentifier	Detector Type	Detector Characteristics	Nuclide Detected	System Response Time	Detection Limits[a]	γ Response	Other Nuclide Response	Remarks
A	ion chamber	2 ℓ volume, other sizes available	^{3}H	$\sim$10 sec. time constant	1 μCi/m^3 to .2 Ci/m^3	compensated to as high as 10 mR/hr.	no inherent selectivity	ion trap; dust trap; α selectivity possible.
B	2 internal ion chambers and 2 sealed	air flow 2–10 ℓ/min.	^{3}H	15, 45 sec. time constant	1 μCi/m^3 to 10^4 μCi/m^3	compensated up to 5 mR/hr.	no inherent selectivity	electrostatic precipitator submicron filter
			^{14}C	15, 45 sec. time constant	0.2 μCi/m^3 to 2000 μCi/m^3	compensated up to 5 mR/hr	no inherent selectivity	
C	ionization chamber	$\sim$1 ℓ	^{3}H		5 pCi/ m^3 10 μCi/m^3		no inherent selectivity	portable; shock sensitive; available on contract basis only.
D	scintillation	sampling volume 192 in.^{3} flow rate 4 CFM	^{133}Xe ^{85}Kr		3 x 10^{-7} μCi/cc min. 2 x 10^{-7} μCi/cc min.		no selectivity	very spotty specs; unclear whether it has a response below .30 MeV.
E	2 ion chamber	2ℓ volume	^{3}H		3 x 10^{-6} μCi/cc to 3	compensated	no selectivity	dust and ion elimination portable
F	2 ion chambers		^{3}H		10 $\mu\mu$Ci/cm^3 to 15,000	compensated to few mr/hr	no selectivity	portable; dust filter and electrostatic precipitator

^{a}Taken directly from manufacturers quoted specifications

TABLE IV (Cont'd.)

Indentifier	Detector Type	Detector Characteristics	Nuclide Detected	System Response Time	Detection Limits[a]	γ Response	Other Nuclide Response	Remarks
G	GM detector	thin window pancake 2" dia.	^{133}Xe	40 sec. time constant	2×10^{-5} µCi/ml		no selectivity	a background subtract circuit, separate operation
H	scintillator and photo-multiplier	10 mil thick plastic, sampling volume 100 ml	noble gas	1 min	2×10^{-7} µCi/cc	heavily shielded	no selectivity	unclear whether it has a response below .30 MeV.
I	ion chamber	2 ℓ/m air flow, 2ℓ volume	^{3}H	15 sec. constant	7.5 µCi/m^3- 5000	comp. up to 5 mR/hr.	no selectivity	filter and ion trap portable
J	GM detector	wall thickness 30 mg/cm^2	noble gas		10^{-7} µCi/cc -10^{-4} µCi/cc		no selectivity	can't detect ^{3}H or ^{14}C

[a]Taken directly from manufacturers quoted specifications

CONCLUSION AND RECOMMENDATIONS

It seems clear from a study of Table IV that instruments
are readily available for measuring low energy β-emitters at
occupational MPC_a. However, these are typically not capable
of measuring concentrations at the public individual MPC_a
for low energy β-emitters. These conclusions agree in a
qualitative fashion with other work.[33-35] In order to change
this situation it seems clear that several sampling techniques
from research would have to be applied commercially. The one
with the easiest and most general application would be
permselective membranes. In addition, there are three
fundamental areas where improvement might be sought: 1)
reducing background radioactivity in materials used for
construction, 2) improving particular components in a
detection system, and 3) paying critical attention to proper
design of the system geometry. A specific example of the
second factor is improvements in photomultipliers used for
scintillation counting. The pulse height resolution capabilities of a photomultiplier are important for the detection
and measurement of low-level scintillations in which only a
few electrons are produced, as is particularly true for
scintillations produced by ^{3}H. The recent significant high
electron resolution improvements observed for prototype
high-gain microchannel plate photomultipliers permit the
elimination of almost all single-electron dark pulses that
accompany low-level scintillations.[36] An obvious potential
usage is in tritium counting. In order to make significant
advances in all three areas for the future, significant
development effort is required. One can only hope that
this is done.

ACKNOWLEDGEMENTS

The author wishes to express his gratitude to Anthony V.
Nero and George A. Morton for their review and comments.

† This work was performed for the Office of Health and Environmental Research of the Department of Energy under contract W-7405-ENG-48. Data in Table IV resulted from work
for Lawrence Livermore Laboratory.

REFERENCES

1. Environmental Instrumentation Group, _Instrumentation for Environmental Monitoring_, in four volumes:
    I) _Air_, Parts 1 and 1A, Gases; Part 2, Particulates,
    II) _Water_, Parts 2 and 2A
    III) _Radiation_
    IV) _Biomedical_,
    Lawrence Berkeley Laboratory report LBL-1, 1 Cyclotron Road, Berkeley, CA 94720, updated periodically.

2. National Council on Radiation Protection and Measurements _A Handbook of Radioactivity Measurements Procedures_, NCRP Report No. 58, 7910 Woodmont Avenue, Washington, DC 20014   (November 1978).

3. C.M. Lederer and V.S. Shirley (editors), _et al_, _Table of the Isotopes_, seventh edition, a Wiley- Interscience Publication, John Wiley and Sons, Inc., New York (1978).

4. I. Halpern, "Three Fragment Fission," Annu. Rev. Nucl. Sci. Volume _21_, pg. 245, Annual Reviews, Inc., 4139 El Camino Way, Palo Alto, CA 94306 (1971).

5. _Radiological Health Handbook_, revised edition, compiled and edited by Bureau of Radiological Health and the Training Institute Environmental Control Administration, Department of Health, Education, and Welfare, Public Health Service, Rockville, MD 20852 (January 1970).

6. J.A.B. Gibson and J.U. Ahmed, _Measurement of Short-Range Radiations_, Technical Report Series No. 150, International Atomic Energy Agency, Vienna (November 1973).

7. F.P. Brauer, R.W. Goles, J.H. Kaye and H.G. Rieck, "Sampling and Measurement of Long-Lived Radionuclides in Environmental Samples," conference proceedings of the _Fourth Joint Conference on Sensing of Environmental Pollutants_, held November 6-7, 1977 in New Orleans, LA, sponsored by ACS, AIAA, AICE, AMS, EPA, ERDA, IEEE, ISA, NASA NOAA, DHUD, DoI(GS), DoS, and DoT, published by American Chemical Society (1978).

8. National Council on Radiation Protection and Measurements _Instrumentation and Monitoring Methods for Radiation Protection_, NCRP Report No. 57 (May 1978).

9. G.F. Knoll, _Radiation Detection and Measurement_, John Wiley and Sons, New York (1979).

10.   L.A.J. Venverloo, <u>Practical Measuring Techniques for Beta Radiation</u>, Philips Technical Library, The MacMillan Press Ltd., London, distributed in the U.S. by Crane, Russak and Company, Inc., 52 Vanderbilt Avenue, New York N.Y. 10017 (1971).

11.   G. Friedlander, J.W. Kennedy, and J.M. Miller, <u>Nuclear and Radiochemistry</u>, second edition, John Wiley and Sons, Inc., New York (1964).

12.   B.G. Harvey, <u>Introduction to Nuclear Physics and Chemistry</u>, second edition, Prentice Hall International Series in Chemistry, Prentice-Hall, Inc., Englewood Cliff, New York (1969).

13.   <u>Code of Federal Regulations</u>, Title 10, Energy, revised as of January 1, 1978, Chapter 1, Nuclear Regulatory Commission, Part 20, Standards for Protection Against Radiation, Appendix B, Concentrations in Air and Water Above Natural Radiation Background, U.S. Government Printing Office, Washington, DC 20402 (1978).

14.   R.H. Howell, J.C. Cate and C. Wong, "Separation of HT, Noble Gases and HTO Vapor with Semipermeable Membranes," Nucl. Instrum. Methods, <u>124</u>, pp. 579-583 (1975).

15.   R. Ehret, "Proportional Flow Counters for Measurement of Tritium in Air," <u>Assessment of Airborne Radioactivity</u> p. 531, International Atomic Energy Agency, Vienna (1967).

16.   R.J. Budnitz, "Tritium Instrumentation for Environmental and Occupational Monitoring – A Review, " Health Phys., <u>26</u>, pp. 165-178 (February 1974).

17.   S. Block, D. Hodgekins and O. Barlow, "Recent Techniques in Tritium Monitoring by Proportional Counters, "UCRL-51131, Lawrence Livermore Laboratory, Livermore, CA (1971).

18.   J.C. Gregory and T.A. Parnell, "Proportional Counter for Monitoring Low Levels of Krypton-85 and Other Beta- and X-ray-Emitting Nuclides, "IEEE Trans. Nucl. Sci.<u>NS-23</u>, No. 1, pp. 715-718 (February 1976).

19.   R.V. Osborne and A.S. Coveart, "A Portable Monitor for Tritium in Air," Nucl. Instrum. Methods, <u>106</u>, pp. 181-187 (1973).

20. R.A. Jalbert, "A Monitor for Tritium in Air Containing Other Beta Emitters, "Proceedings of 23rd Conference on Remote Systems Technology,"pp. 89-93 (1975).

21. R.A. Jalbert and R.D. Hiebert, "Gamma Insensitive Air Monitor for Radioactive Gases," Nucl. Instrum. Methods, 96, pp. 61-66 (1971).

22. J.H. Osloond, J.B. Echo, W.L. Polzer and B.D. Johnson, "A Tritium in Air Sampling Method for Environmental and Nuclear Power Plant Monitoring, "Tritium, edited by A.A. Moghissi and M.W. Carter, CONF-710809, pp. 531 (May 1973).

23. R.V. Osborne, "Monitoring Reactor Effluents for Tritium: Problems and Possibilities, "Tritium, edited by A.A. Moghissi and M.W. Carter, CONF-710809, pp. 496 (May 1973).

24. R.V. Osborne and N.W. Tepley, "Monitoring Tritiated Water in Air and Water Effluents, "Radiation Instrumentation, edited by W.W. Wadman, proceedings of the Health Physics Society Eleventh Midyear Topical Symposium on Radiation Instrumentation, held in San Diego, CA, p. 203 (January 16-19, 1978).

25. C. Colmenares, E.G. Shapiro, P.E. Barry and C.T. Prevo, " A Europium-doped, Calcium Flouride Scintillator System for Low-Level Tritium Detection, "Nucl. Instrum. Methods 114, pp. 277-289 (1974).

26. R.V. Osborne, "Central Tritium Monitor for CANDU Nuclear Power Stations, "IEEE Transactions on Nuclear Science, Volume NS-22, pp. 676-680 (February 1975).

27. R.V. Osborne, "Detector for Tritium in Water, "Nucl. Instrum. Methods, 77, pp. 170-172 (1970).

28. A.A. Moghissi, H.L. Kelley, C.R. Phillips and J.E. Regnier, "A Tritium Monitor Based on Scintillation," Nucl. Instrum. Methods, 68, p.159 (1970).

29. S. Takamatsu, M. Ohno, T. Miyazawa, O. Ozaki,"A Krypton-85 Monitoring System with Permselective Membranes, Monitoring of Radioactive Effluents from Nuclear Facilities, proceedings of an International Symposium on the Monitoring of Radioactive Airborne and Liquid Releases from Nuclear Facilities held by IAEA in Portoroz, Yugoslavia 5-9 September 1977, IAEA, Vienna (1978).

30. Department of Energy, <u>Standards for Radiation Protection</u>,
    SAN Chapter 0524, ERDA Manual, Volume 0000 General Admin-
    istration, Part 0500 Health and Safety (March 30, 1977).

31. American National Standards Institute, <u>American National
    Standard Specification and Performance of On-Site Instru-
    mentation for Continuously Monitoring Radioactivity in</u>
    Effluents, ANSI N13.10 - 1974, IEEE, 345 East 45th St.,
    New York, N.Y. 10017 (1974).

32. D.L. Murphy, unpublished work (1978).

33. R.B. Hower, B. Hekkala and D.T. Pence, "Radioactive Air-
    borne Effluent Measurement and Monitoring Survey of Re-
    processing and Waste Treatment Facilities, "COO-3049-9,
    prepared for Harvard Air Cleaning Laboratory, Science
    Applications, 4060 Sorrento Valley Boulevard, San Diego,
    CA 92121 (September 1977).

34. S.J. Fernandez, G.D. Pierce, D.C. Hetzer and B.G. Mates,
    "Methods Evaluation for the Continuous Tritium Monitor,"
    ICP-1149, Idaho National Engineering Laboratory, Idaho
    Falls, ID (April 1978).

35. S.J. Fernandez, G.D. Pierce, D.C. Hetzer and B.G. Mates,
    "Methods Evaluation for the Continuous Monitoring of
    Carbon-14, Krypton-85 and Iodine-129 in Nuclear Fuel
    Reprocessing and Waste Solidification Facility Off-Gas,"
    ICP-1187, Idaho National Engineering Laboratory, Idaho
    Falls, ID (March 1979).

36. C.C. Lo and Branko Leskovar, "Studies of Prototype
    High-Gain Micro-Channel Plate Photomultipliers, "IEEE
    Trans. Nucl. Sci. <u>NS-26</u>, No. 1, pp. 388-394 (Feb. 1979).

RADIOCHEMICAL DETERMINATION OF URANIUM, THORIUM
AND LEAD-210 IN COAL AND COAL ASH

V. R. Casella, C. T. Bishop, A. A. Glosby,
and C. A. Phillips
Mound Facility, Miamisburg, Ohio   45342

ABSTRACT

A radiochemical procedure is presented for the sequential determination of uranium, thorium, and lead-210 in coal and coal ash. The procedure consists of dry ashing the sample, a nitric-hydrofluoric acid dissolution, removal of iron with ether extractions, and separation of the elements of interest by anion exchange chromatography. Uranium and thorium isotopes are measured by alpha spectrometry, while lead-210 is measured by beta counting its daughter, bismuth-210. For 10-g coal samples and 1-g coal ash samples, the chemical yields for the radioactivities measured were 70-80%, and the relative standard deviations for replicate analysis were generally less than 9%. The deviations of the means from the reference values were usually less than ±5%.

INTRODUCTION

In order to meet increasing electric power requirements in the United States, it will be necessary to utilize even more of our abundant coal reserves. This expanded production of energy will have an important environmental impact and, since coal contains naturally occurring radioactives, the radiological impact must be considered. A project has been initiated at Mound Facility to evaluate the potential radiological impact of coal utilization.[1] In order to help carry out this project, a method has been developed for the sequential determination of uranium isotopes, thorium isotopes, and lead-210 in coal and coal ash. A schematic representation of the radioanalytical separations is presented in Figure 1.

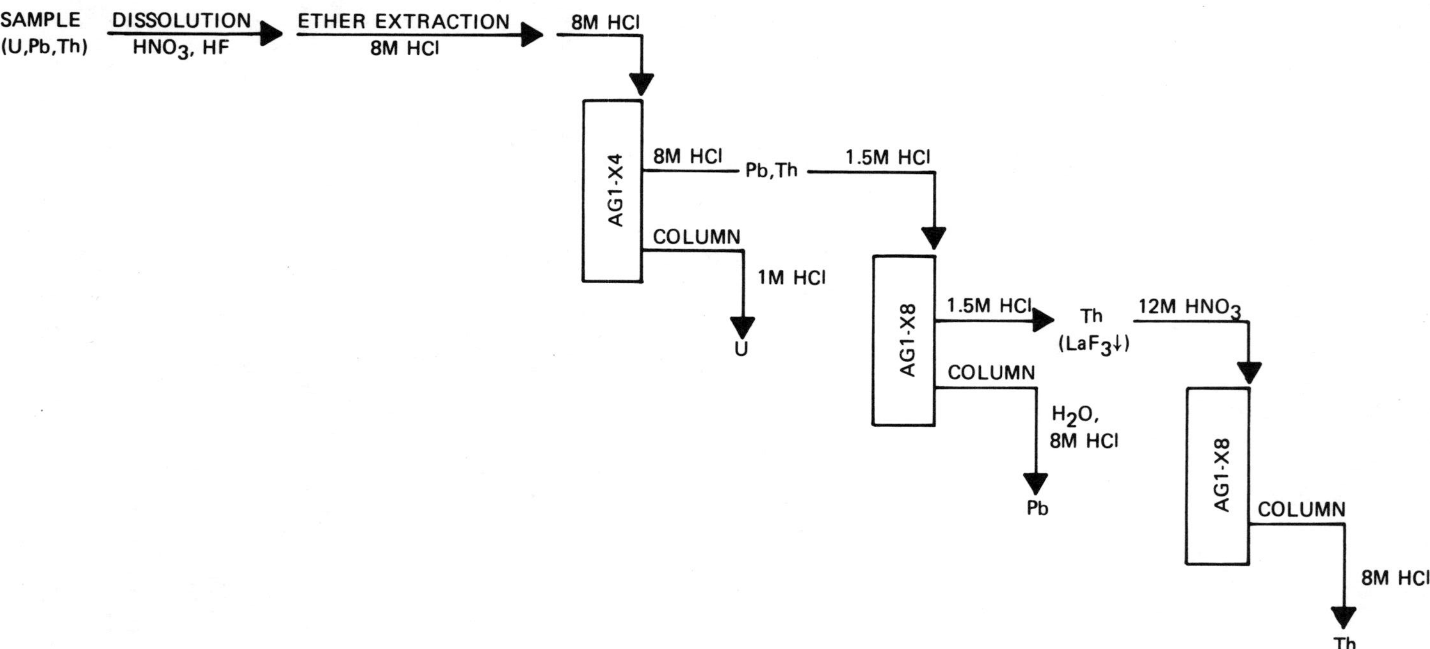

Figure 1. Sequential separation of uranium, lead, and thorium for radiochemical determination

EXPERIMENTAL

Reference Materials

Standard coal samples (SRM 1632a and SRM 1635) and a standard fly ash sample (SRM 1633), having known uranium and thorium concentrations, were obtained from the National Bureau of Standards (NBS).

Sample Decomposition

Add 1 ml each of uranium-232 and thorium-229 tracers ($\sim$ 10 dpm/ml) and 1 ml of a 2.5% lead nitrate solution to a 10-g coal sample or 1-g ash sample in a 150-ml porcelain evaporating dish and dry at 100°C. Heat the dish at 450-500°C overnight in a muffle furnace. Allow the sample to cool and transfer to a 250-ml Teflon beaker, rinsing the dish with 10 ml portions of 16 M nitric acid to a final volume of 60 ml. Add 30 ml of 29 M hydrofluoric acid, cover with a Teflon watchglass, and boil on a magnetic hotplate for 1 hr, while stirring continuously with a Teflon-coated stirring bar. Remove the beaker from the hotplate and allow to cool. Add 30 ml each of 16 M nitric acid and 29 M hydrofluoric acid and digest for an additional hour. Remove the beaker from the hotplate and allow to cool to room temperature. Add 20 ml of 12 M hydrochloric acid and boil on a hotplate until the solution has evaporated to a volume of about 30 ml. Add 50 ml of distilled water and digest on a hotplate, while stirring, for 10 min to dissolve the soluble salts.[2] Cool and transfer the total sample into a 200-ml centrifuge bottle using distilled water to rinse the beaker. Centrifuge the sample and decant the supernate into another 200-ml centrifuge bottle. Wash the residue with 10 ml of 0.1 M hydrochloric acid. Add this wash to the supernate and discard the residue.

Ether Extraction

Since large quantities of iron interfere with the subsequent anion exchange separation of uranium, the bulk of the iron is removed by ether extractions. Carefully add 15 M ammonium hydroxide to the sample solution to raise the pH to 9-10 as determined by pH paper. Centrifuge and discard the supernate. Dissolve the precipitate in a minimum of 12 M hydrochloric acid (add a few milliliters of distilled water if the precipitate does not readily dissolve), and then increase the volume of the solution to 50 ml with 8 M hydrochloric acid. Transfer the 8 M hydrochloric acid solution to a 250-ml separatory funnel using two 5-ml rinses of 8 M hydrochloric acid. Add 60 ml of isopropyl ether to the

separatory funnel and shake for 2 min. Transfer the aqueous
phase to a second separatory funnel and repeat the extraction.
If a deep yellow color persists, more extractions may be
necessary to remove additional iron present. The sample
solution should be nearly clear. Transfer the aqueous phase
to a 150-ml beaker and boil for 15 min to expel the ether.
Add 20 ml of 12 M hydrochloric acid to this solution and
proceed to the anion exchange separation of uranium.

Anion Exchange Separation of Uranium

     Prepare an anion exchange column of Bio-Rad AG1-X4
(100-200 mesh), height 8 cm, and inside diameter 1.3 cm
(volume ∿10 ml), and precondition the resin by rinsing the
column with 60 ml of 8 M hydrochloric acid. Pass the sample
solution through the column at a flow rate of 3 ml/min and
save the effluent containing the thorium and lead-210. Elute
any remaining iron from the column with six column volumes of
6 M hydrochloric acid containing 1 ml of 47% hydriodic acid
per 50 ml of 6 M hydrochloric acid (freshly prepared). Rinse
the column with an additional four column volumes of 6 M
hydrochloric acid and discard these solutions. Elute the
uranium with six column volumes of 1.0 M hydrochloric acid.
Evaporate the uranium fraction to about 20 ml and add 5 ml
of 16 M nitric acid. Evaporate the uranium fraction to near
dryness and proceed with the electrodeposition as described
previously.[3] Identify and quantify the uranium isotopes by
alpha spectrometry.[4]

Anion Exchange Separation of Lead-210

     Prepare an anion exchange column of Bio-Rad AG1-X8
(100-200 mesh), height 8 cm, and inside diameter 1.3 cm
(volume ∿10 ml), and precondition the column with three
column volumes of 1.5 M hydrochloric acid. Evaporate the
thorium and lead fraction from the uranium separation to
about 20 ml and dilute threefold with distilled water to a
final volume of about 80 ml. Transfer the sample to the
anion exchange column and allow to pass through the column
at a flow rate of 1 ml/min. Rinse the column with about
50 ml of 1.5 M hydrochloric acid and save the total effluents,
to this point, for the subsequent thorium determination. This
is the thorium fraction. Elute the lead from the column with
50 ml of distilled water and then with 50 ml of 8 M hydrochloric acid, noting the time of separation. Determine the
lead recovery of a 1 ml aliquot, diluted to 25 ml, by atomic
absorption spectroscopy. Evaporate the lead fraction to
about 10 ml, add 10 ml of 16 M nitric acid, and evaporate to
near dryness. Dissolve the residue in about 3 ml of 0.1 M
nitric acid (heating may be necessary). Transfer this

solution to a 5 cm diam. stainless steel disc and slowly
dry the disc using a heat lamp. Rinse the beaker with an
additional 3 ml of 0.1 M nitric acid, add to the disc and
dry. Cover the disc with an appropriate thickness ($\sim$5 mg/cm^2)
of aluminum foil to absorb any alpha and low-energy beta
activities and count the 1.6 MeV beta activity from the
ingrown bismuth-210 in a proportional counter.

Anion Exchange Separation of Thorium

Add 10 mg of La^{3+} (1 ml of a 2.5% lanthanum chloride
hexahydrate solution) to the thorium fraction from the anion
exchange separation of lead-210. Add about 5 ml of 29 M
hydrofluoric acid to precipitate lanthanum and thorium
fluorides, centrifuge, and discard the supernate. Dissolve
the precipitate in about 5 ml of 4 M nitric acid which was
previously saturated with boric acid. Add 25 ml of distilled
water and adjust the pH to 9 with 14 M ammonium hydroxide.
Centrifuge and again discard the supernate, and wash the
precipitate with 10 ml of 0.1 M ammonium hydroxide. Dissolve
the precipitate in 25 ml of 12 M nitric acid.

Prepare an anion exchange column of Bio-Rad AG1-X8 (100-200
mesh), height 8 cm, and inside diameter 1.3 cm (volume $\sim$10
ml), and precondition the resin by passing 10 column volumes
of 12 M nitric acid through the column. Pass the 12 M nitric
acid sample solution through the column which adsorbs the
thorium but not lanthanum, at a flow rate of 2 ml/min. Rinse
the column with about 60 ml of 12 M nitric acid at a flow
rate of 2 ml/min. Elute the thorium from the column with
80 ml of 8 M hydrochloric acid at a flow rate of 2 ml/min.
Evaporate this eluent to about 20 ml, add 10 ml of 16 M
nitric acid, and evaporate to near dryness. Proceed with
the electrodeposition as described previously.[3] Identify
and quantify the thorium isotopes by alpha spectrometry.
Since the thorium-229 tracer usually contains some thorium-
228 (<5%) and since thorium-228 is a decay product of the
uranium-232 tracer used in the present procedure, appropriate
corrections of the thorium-228 peak area must be made.

RESULTS AND DISCUSSION

Results for two coal samples and a fly ash sample,
obtained from the National Bureau of Standards, are given
in Table I. Each measurement shown in this table is an
average of three or four determinations.

Yields for uranium, thorium, and lead-210 in 10-g coal
samples averaged 73%, 69%, and 75%, respectively, while the
corresponding yields for 1-g ash samples were 81%, 72%, and

TABLE I. URANIUM, THORIUM, AND LEAD-210 RESULTS[a]

	SRM 1632[a] (Bituminous Coal) pCi/g	Reference Value[b] pCi/g	SRM 1635 (Subbituminous Coal) pCi/g	Reference Value[b] pCi/g	SRM 1633 (Fly Ash) pCi/g	Reference Value[b] pCi/g
U-238	0.444±0.016	0.430±0.007	0.0731±0.0046	0.080±0.007	4.01±0.04	3.90±0.07
U-235	0.0228±0.0019	0.0201±0.004	0.0049±0.0003	0.0037±0.004	0.179±0.012	0.182±0.003
U-234	0.448±0.012	0.430±0.007	0.0719±0.0044	0.080±0.007	4.07±0.12	3.90±0.07
Th-232	0.484±0.018	0.494±0.011	0.0619±0.0077	0.068±0.004	2.45±0.08	2.63
Th-230	0.452±0.017	0.430±0.007	0.0765±0.0079	0.080±0.007	3.74±0.17	
Th-228	0.499±0.011	0.494±0.011	0.0648±0.0041	0.068±0.004	2.23±0.05	2.63
Pb-210	0.449±0.024	0.430±0.007	0.0699±0.0013	0.080±0.007	3.37±0.13	

[a]Results are for 10-g coal sample and 1-g ash samples.

[b]Reference values are calculated from the NBS values. Secular equilibrium is assumed for U-238/daughter activities and Th-232/Th-228 in the coal and also for U-238/U-234 and Th-232/Th-228 in the fly ash.

81%. Relative standard deviations of replicate determinations
ranged from 1% to 6% for uranium, 3% to 12% for thorium, and
3% to 9% for lead-210. The deviations of the means from the
reference values were within the combined errors of the
reference values and the measured values. Excellent agreement with the NBS reference values was obtained for samples
SRM 1632 and SRM 1633. However, the measured values for
SRM 1635 were about 10% lower than that reported by the NBS,
although the NBS reports an uncertainty of about 10%.

ACKNOWLEDGEMENTS

     Mound Facility is operated by Monsanto Research
Corporation for the U.S. Department of Energy under
Contract No. DE-AC04-76-DP00053.

REFERENCES

1. C. E. Styron, V. R. Casella, B. M. Farmer, L. C. Hopkins,
    P. H. Jenkins, C. A. Phillips, and B. Robinson,
    "Assessment of the Radiological Impact of Coal Utiliz-
    ation I. Preliminary Studies on Western Coal," Mound
    Facility Report, MLM-2514, (1979).

2. U. S. Atomic Energy Commission, "Measurement of Radio-
    nuclides in the Environment - Sampling and Analysis of
    Plutonium in Soil," U.S. AEC Regulatory Guide 4.5,
    (1974).

3. C. T. Bishop, V. R. Casella, and A. A. Glosby, "Radio-
    metric Method for the Determination of Uranium in Water:
    Single-Laboratory Evaluation and Interlaboratory
    Collaborative Study," Office of Research and Development,
    U. S. Environmental Protection Agency, EPA-600/7-79-093,
    (1979).

4. C. W. Sill, Anal. Chem., _49_, 618 (1977).

ON LINE MONITORING
AND FACILITIES

ON-LINE MONITORING OF LOW-LEVEL PLUTONIUM CONCENTRA-
TIONS

K. J. Hofstetter, T. V. Rebagay, and G. A. Huff,
Allied-General Nuclear Services, Barnwell, South
Carolina, USA

Work supported by the U. S. Department of Energy, Fuel
Cycle Projects Office under Contract No. DE-AC09-78ET-
35900.

ABSTRACT

An on-line monitor has been developed at the Barnwell
Nuclear Fuel Plant (BNFP) to assay plutonium in nitric
acid solutions. The performance of the monitor has been
assessed by a laboratory experimentation program using
solutions with plutonium concentrations from 0.1 to
10 g/ℓ. These conditions are typical of the plutonium
solutions in an input stream to a plutonium-purification
cycle in a reprocessing plant following uranium/plutonium
partitioning. The monitoring system can be fully automated and shows great promise for detecting and quantify-
ing plutonium in situ, thus minimizing the reliance on
traditional sampling and laboratory-analysis techniques.
The total concentration and isotopic abundance of plutonium are determined by measuring the absolute intensities
of the low-energy gamma rays characteristics of ^{238}Pu,
^{239}Pu, and ^{240}Pu nuclides by direct gamma-ray spectroscopy
and computer analysis of the spectral data. The addition
of a monitoring system of this type to the input stream of
a plutonium-purification cycle along with other suitable
monitors on the waste streams and on the product stream
provides the basis for a near real-time materials control
and inventory system.

The results of the laboratory-evaluation program
employing plutonium in solutions with isotopic compositions typical of those involved in processing light water
reactor fuels are presented. The detailed design of a
monitoring cell and detection system is given. The

precision and accuracy of the results relative to those measured by mass spectrometry and controlled potential coulometry are also summarized.

INTRODUCTION

Effective materials accounting of special nuclear materials in a modern nuclear fuel reprocessing plant depends heavily on sophisticated data analysis techniques. The recent development of methods for studying on-line measurements of the total and isotopic plutonium concentrations by gamma-ray spectrometry[1,2] opens up new possibilities for monitoring not only high-level plutonium concentration streams but also low-level streams consistent with the plutonium abundances in the waste streams of nuclear fuel reprocessing plants.

The present study has been directed toward applying these techniques to the assay of dilute solutions containing from 0.1 to 10 g/ℓ. Because the method is nondestructive, it is directly applicable to the continuous, on-line analysis of dilute aqueous plutonium solutions.

EXPERIMENTAL

During the design phase of the BNFP, it was determined that many on-line monitors were required to insure adequate process control. Many monitors were installed which measure the physical properties of the system (e.g., tank level gauges, pressure indicators, flow meters, temperature probes, etc.). It was also desirous to measure several chemical properties of solutions flowing in key process streams. For example, it was found necessary to monitor the total plutonium content on the product stream (1BP) off the electropulse column. As designed, this stream would be monitored for plutonium content and would alarm when a plutonium concentration of 6.5 g/ℓ was reached. This boundary condition was determined by criticality restrictions on certain vessels in the plutonium purification cycle.

The primary partition step in the separation of uranium and plutonium occurs in the electropulse column at BNFP. Valence adjustment in this column permits the uranium to be retained in the plant organic stream (30% tri-n-butylphosphate in n-paraffin hydrocarbon

[TBP-NPH]) while the plutonium is stripped into the nitric acid phase. The majority of the fission products have been previously removed in an earlier step. Under normal operations, the composition of the 1BP stream is estimated to be 5 g Pu/ℓ and 10 g U/ℓ in approximately 2.9$\underline{M}$ nitric acid. The maximum fission product level in this stream is expected to be 40 μCi/ml, predominantly Ce, Ru, and Zr radioisotopes.

The purpose of this study was then to develop the techniques necessary to assay the 1BP stream for total plutonium content for process control purposes. After evaluation of alternate assay techniques, it was concluded that gamma-ray spectrometry was a possible candidate. In addition to determining the plutonium content, some isotopic information might be obtained by spectral analysis.

Discussion Of Techniques

It has been shown that the measurement of the plutonium content of a process stream can be accomplished by gamma-ray spectroscopic techniques.[1] The plutonium isotopes present in light water reactor fuel are ^{238}Pu, ^{239}Pu, ^{240}Pu, ^{241}Pu, and ^{242}Pu. Other nuclides almost always present in plutonium solutions are ^{241}Am and ^{237}U as decay products of ^{241}Pu. All these nuclides (except ^{242}Pu) emit characteristic low-energy gamma rays in sufficient abundance to be detected by high resolution germanium detectors. The energies and intensities of the gamma rays emitted in the decay of these nuclides have been carefully characterized.[3] Careful analyses of the gamma-ray spectra emitted by these nuclides permit the quantification of plutonium isotopic content in solutions containing mixtures of these nuclides.

It has been determined that the low-energy spectrum of plutonium nitrate solution of light water reactor grade isotopic composition is greatly influenced by the ^{241}Am content.[1] It is anticipated that the 1BP stream will contain solutions with low ^{241}Am content; the ^{241}Am is carried with the fission products during the initial organic solvent extraction separation. Consequently, the intensities of the characteristic low-energy gamma rays at 43.5, 51.6, and 45.2 keV can be measured to determine the ^{238}Pu, ^{239}Pu, and ^{240}Pu content. A portion of the low energy region of the spectrum is shown in Figure 1 for a solution of freshly separated plutonium nitrate with typical light water grade isotopic composition. The

38.7 keV gamma ray is unique to ^{239}Pu and can also be used to measure the ^{239}Pu content.

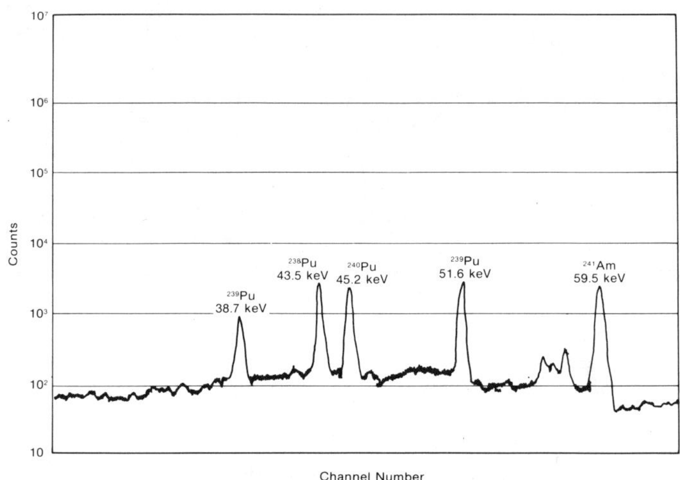

Figure 1.   Low energy spectrum

Experiments were designed to measure the accuracy and precision of gamma-ray spectroscopic techniques for determining the plutonium isotopic ratios for reprocessing type solutions. The experiments involved the direct measurement of the gamma rays emitted by plutonium nitrate solutions. A comparison of the total and isotopic plutonium assay with the laboratory plutonium analyses by controlled potential coulometry and mass spectrometry was conducted.

Detector-Assay Cell System

The detector chosen for these experiments was an intrinsic germanium low energy photon detector. The detector has an active area of 200 mm^2 with a sensitive depth of 7 mm and a 0.13 mm thick beryllium window on the cryostat. With suitable amplification, pulse pile-up rejection and digital stabilization, the spectrometer system is capable of maintaining a full-width at half-maximum resolution of 350 eV at 59.5 keV at a total count rate of 10K counts per second (CPS). Spectra were recorded with a computer-based multichannel analyzer with associated peripheral devices. This system has been

discribed previously.[2] A detailed discussion of the
operating system software and detailed data analysis
algorithms have also been published.[4]

The analysis cell has a 1-ml sample capacity with an
active area of 200 mm^2 and a thickness of 5 mm. The
dimensions were established by comparing the solution
self-absorption with the overall anticipated count rate.
The system was designed to limit the total system count
rate to 10K CPS for a solution containing 10 g Pu/ℓ. The
cell was mounted inside a Plexiglas secondary containment
box mounted external to an existing glovebox. Lines
containing plutonium nitrate solutions were run from the
glovebox to the secondary containment box. Stainless
steel (304L) was used in fabricating all cell components
and all fittings. Teflon tubing was used throughout the
liquid transport system. The cell is isolated from the
detector (and the room) by a 0.05-mm thick stainless steel
window. The cell and secondary containment box are
coupled to the detector with a mating flange and bolted
into place to ensure geometric reproducibility. The cell
to detector distance is maintained at 8 mm. Containment
of the solution is accomplished with a 3-mm thick Plexiglas window backed with a polycarbonate membrane to resist
the high concentration of nitric acid in the solutions. A
tantalum collimator is used to mask the edges of the cell
and minimize the effect of bubbles forming in the assay
cell. A schematic of this system is shown in Figure 2.

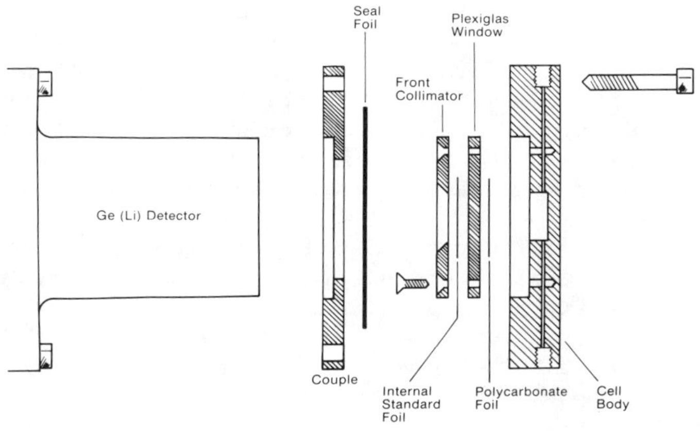

Figure 2.   Counting system

Plutonium nitrate solutions were prepared from reactor grade plutonium oxide dissolved in nitric acid and hydrofluoric acid. The resulting solution was then treated by ion exchange chromotography to separate ^{241}Am from the plutonium solution. The details of this separation have been reported by Gunnink, et al.[5] The stock solution was stabilized with sodium nitrite and suitable aliquots taken and diluted with 3<u>M</u> nitric acid to simulate the plutonium concentration expected in the plant stream. These solutions were drawn into the analysis cell and gamma-ray spectra were recorded. After removal from the system, the solutions were analyzed by controlled potential coulometry for plutonium content. The isotopic abundances were verified by mass spectrometry.

Experimental Procedure

The laboratory evaluation phase of the experimental program began with the design, fabrication and installation of the assay cell system. Prior to the assay of plutonium by the monitoring system described, a preliminary experiment was conducted to determine the relative response of the detector assembly to low-energy gamma rays. Solutions of ^{137}Cs of varying activities paralleling the expected activities of the plutonium solutions used in this study were circulated into the cell. From the observed intensities of the two low-energy x-ray lines having energies of 32.2- and 36.5-keV, the overall relative efficiency of the system including the attenuation by the cell windows, the detector efficiency, and solution self-absorption were determined. The characteristic barium Kα and Kβ x-rays emitted in the decay of ^{137}Cs not only simulated the low energy radiations from plutonium but also simulated the continuum background present in a typical fission product spectrum. Various thicknesses and types of windows were determined to optimize the system parameters as a result of this study.

It was found from this study that the cell dimensions and window thickness were sufficient to permit the acquisition of suitable pulse height spectra in reasonable counting times (approximately 10 minutes). The ratio of pulse height distributions between the gamma-ray peaks and the continuum background and the overall effect on the accuracy of peak integration were studied.

It was also determined as a result of this study that a source of ^{137}Cs could be fixed to the cell window and

serve as an internal standard to monitor short-term
detector efficiency drift and other system changes. This
addition would probably be included in the design of the
monitor installed in the plant.

The next series of experiments involved the introduction of plutonium nitrate solutions into the assay system.
The initial solutions studied were prepared directly from
the unseparated stock solution. The light water reactor-
grade plutonium had an 80% ^{239}Pu isotopic content. Dilutions were prepared from the stock solution by volume
aliquoting. Three ml of the solution were drawn into the
cell with a syringe in a typical experiment. The resulting spectra were recorded with the computer-based multichannel analyzer.

The first series of experiments were performed using
solutions with equilibrium quantities of ^{241}Am. Because
of the high gamma ray specific activity for ^{241}Am, it
contributes to the overall system dead time (approaching
50%) and changes the character of the resulting spectrum.
The low energy side of the 59.5 keV peak almost totally
obscures the 51.6 keV peak. A typical spectrum is shown
in Figure 3. It is evident that at equilibrium ^{241}Am
concentrations, the accuracy of this direct gamma ray
technique is suspect.

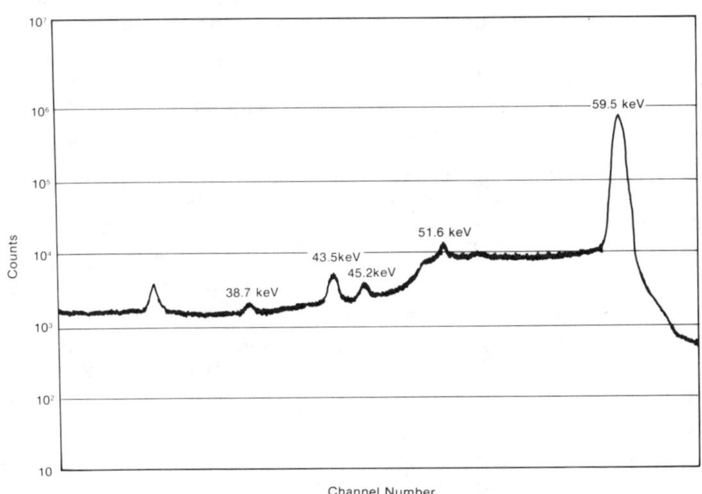

Figure 3. A portion of the low energy gamma spectrum of a
plutonium solution containing equilibrium quantities of Am-241

The next series of experiments involved the study of plutonium nitrate solutions that had been subjected to an americium separation. These solutions were introduced into the assay cell and the resulting spectra recorded and analyzed. A typical spectrum of ^{241}Am-free plutonium nitrate solution was shown in Figure 1. The low ^{241}Am content can be seen in the low intensity of the 59.5 keV gamma ray. Since the precision of a measurement is dependent on the ability of the cell system to replicate measurements made on samples over a given period, one sample was counted each day over a period of three weeks. The results showed that the reproducibility is excellent with no discernible interference from the ^{241}Am decay-daughter of ^{241}Pu. Over the three weeks period, the 59.5-keV gamma ray of ^{241}Am became an ever increasing component in the spectrum reaching an intensity of 40K counts per minute after 18 days. Beyond this level of gamma activity, the ^{241}Am content begins to affect the integrated area of the 51.6-keV gamma ray.

RESULTS

The results of the experiments on freshly separated plutonium nitrate solutions with concentrations up to 10 g/ℓ are shown in Figure 4. The gamma-ray intensities for each of the four radiations typical of ^{238}Pu, ^{239}Pu, and ^{240}Pu are plotted as a function of plutonium concentration as determined by controlled potential coulometry. The one sigma standard deviation fit to the data for each gamma ray was as follows: 43.5 keV (5.3%), 45.2 keV (4.8%), 51.6 keV (6.1%), and 38.7 keV (4.4%). Sufficiently long counting times were employed during these experiments that the uncertainties in the peak areas were less than one percent (one sigma relative standard deviation). With the calibration curves defined, the operational characteristics of the monitor could be deduced for plutonium solutions of this isotopic abundance. If the monitor was operated such that 1000 counts were recorded in the 51.6 keV gamma ray peak for example, the uncertainty in the concentration would be less than 6 percent. Unfortunately the isotopic composition of the plutonium entering the purification cycle will continually change due to recycle paths and differing isotopic input batches.

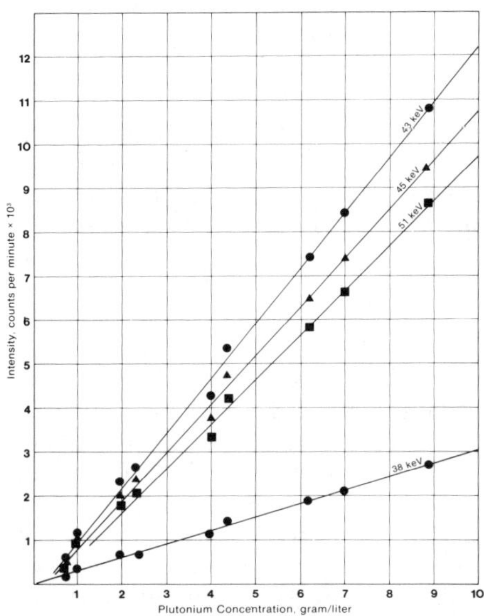

Figure 4. Calibration curves showing the count rate
response of the detector to plutonium
radiations at various plutonium concentrations

As the technique involves the direct counting of low
energy gamma rays emitted from the solution, significant
absorption of these radiations occurs within the sample.
The attenuation of the gamma rays traversing the solution
could be calculated provided the concentration is known.
As this is not the case, an estimate of the sample self
absorption can be obtained by measuring the ratio of the
51.6 keV gamma-ray to the 38.7 keV gamma ray both occurring in the decay of ^{239}Pu. As the bulk of the solution
is water, the effect of the addition of plutonium nitrate
and nitric acid is small compared to the attenuation of
the radiations due to water. The ratio of the 51.6 to
38.7 keV gamma ray intensities varied from 2.9 to 3.2 over
the plutonium concentration range studied. The ratio of
the intensities for these gamma rays being emitted from a
3M nitric acid solution found by extrapolating to zero
plutonium concentration was 2.88 ±0.02. The intensity
ratio for these radiations in an unattenuated source
should be 3.02. For solutions containing mixtures of
plutonium and uranium, the ratio technique should permit

the calculation of matrix self-attenuations and thereby
correct the efficiencies for all four radiations.

The isotopic ratios for the 238, 239, and 240 mass
plutonium nuclides found from these experiments were
within 2 percent of the mass spectrometer values with no
significant bias over the concentration range studied.
The deviations of the 239/238 isotopic ratios from the
mass spectrometer values were determined as a function of
plutonium concentration. The overall fit to the data was
1.15% one sigma relative standard deviation. The fit
could not be improved by including higher order terms and
thus the conclusion of no significant bias was justified.

As a monitor for the plutonium concentrations, cerium
activated Vycor scintillation detectors are installed on
all major waste streams in the plutonium purification
cycle to measure the alpha activity. Plutonium nitrate
solutions were circulated through an on-line alpha monitor
(OLAM) and the sensitivity, accuracy, reliability, and
extent of anticipated interferences for these monitors
determined. It was determined from this study that the
OLAMs installed on these low-level (less than 10^{-2} g/ℓ)
waste streams could be used to quantitatively assay the
plutonium content provided the isotopic composition were
known.

It is anticipated that the output of the 1BP monitor
would be used to determine the alpha specific activity of
plutonium. The alpha specific activity for plutonium is
dependent almost exclusively on the ^{238}Pu content. The
specific activity of plutonium is also a function of the
ratio of ^{238}Pu to ^{239}Pu + ^{240}Pu content. The specific
alpha activity of the plutonium used in this study as
determined by mass spectrometer isotopic abundances is
5.0 x 10^9 dps/gm. The specific activity determined by the
weight fraction of ^{238}Pu compared to ^{239}Pu and ^{240}Pu in
this study is 5.2 x 10^9 dps/gm. The technique should be
sufficient to determine the calibration factor in converting the alpha count rate from the OLAMs to grams per liter
plutonium in the plutonium purification cycle. Isotopic
correlation studies might be necessary to remove the
positive bias in this determination, by deducing the ^{241}Pu
and ^{242}Pu content from the relative abundances of the
other light-mass plutonium isotopes.

CONCLUSIONS

In the context of current safeguards requirements predicated on quantitative nondestructive analysis traceable measurements, the only practical solution to real-time safeguards testing and verification is one that can be proven to be accurate, sensitive, and reliable. The system reported here has these features and holds promise for essentially continuous near real-time determination of in-process plutonium. Additionally, it can serve as a redundant, independent verification system to confirm data measured by other chemical methods.

The development of an on-line monitor for plutonium content and isotopic ratios on the 1BP stream has proceeded with a laboratory experimentation and testing program. Using gamma ray techniques, the monitor can assay dilute plutonium solutions to within 3% of the reference values determined coulometrically. In addition, the monitor can determine the alpha specific activity of the plutonium in solution. On-line alpha monitors are strategically placed on the waste streams throughout the purification cycle. Since the monitor under study measures the properties of the material input to the cycle, the outputs of the OLAM's can be continuously updated to g Pu/ℓ concentration reflecting any changes in isotopic composition. With accurate measurements of the input and output of plutonium through the purification cycle along with measurements of the plutonium concentrations on the waste streams, near real-time inventory is possible. Modeling of the BNFP plutonium purification cycle[7] has indicated that for a 1% measurement precision in determining the plutonium concentration in the 1BP input and the product streams and a 10% precision on the waste streams, the average measurement error for one week will be about 1 kg.

Continued development of on-line plutonium monitors is progressing through laboratory experimental testing and in-plant installation and testing. Using gamma ray spectroscopic techniques, the assay methods are nondestructive and are easily computerized. These type monitors lend themselves to both safeguards and inventory monitoring operation. Continued testing of this monitor will be performed in the laboratory to determine the extent of uranium interferences and fission product levels which can be tolerated.

REFERENCES

1. R. Gunnink and J. F. Evans, "In-Line Measurement of
    Total and Isotopic Plutonium Concentrations by Gamma-
    Ray Spectrometry," Lawrence Livermore Laboratory,
    Report UCRL-52220 (1977).

2. K. J. Hofstetter, G. A. Huff, R. Gunnink, J. E. Evans,
    and A. L. Prindle, "On-Line Measurement of Total and
    Isotopic Plutonium Concentration by Gamma-Ray Spec-
    trometry," Analytical Chemistry in Nuclear Fuel Repro-
    cessing, ed. W. S. Lyon, Science Press, Princeton,
    pp 266-274 (1978).

3. R. Gunnink, J. E. Evans, and A. L. Prindle, "A Reeval-
    uation of the Gamma-Ray Energies and Absolute Branch-
    ing Intensities of U-237, Pu-238, Pu-239, Pu-240,
    Pu-241 and Am-241," Lawrence Livermore Laboratory,
    Report UCRL-52139 (1976).

4. R. Gunnink and J. B. Niday, "Precise Plutonium Analy-
    sis by Gamma Spectroscopy," Lawrence Livermore
    Laboratory, Report UCRL-76699 (1975).

5. R. Gunnink, A. L. Prindle, J. B. Niday, and
    D. W. O'Brien, "Operations Manual for the Tastex Gamma
    Spectrometer," Lawrence Livermore Laboratory,
    Report-M-106(ISPO-67) (1979).

6. K. J. Hofstetter, G. M. Tucker, R. P. Kemmerlin,
    J. H. Gray, and G. A. Huff, "Application of On-Line
    Alpha Monitors to Process Streams in a Nuclear Fuel
    Reprocessing Plant," Nuclear Safeguards Analysis -
    Nondestructive and Analytical Chemical Techniques, ed.
    E. A. Hakkila, Washington, DC, American Chemical
    Society, pp 124-143 (1978).

7. E. A. Hakkila, D. D. Cobb, R. J. Dietz, J. P. Shipley,
    and D. B. Smith, "Requirements for Near Real-Time
    Accounting of Strategic Nuclear Materials in Nuclear
    Fuel Reprocessing," Analytical Methods for Safeguards
    and Accountability Measurements of Special Nuclear
    Materials, ed. H. T. Yolken and J. E. Bullard,
    National Bureau of Standards Special Publication 583,
    pp 208-220 (1978).

A NONDESTRUCTIVE ASSAY INSTRUMENT FOR MEASUREMENT OF
PLUTONIUM IN SOLUTIONS

D. G. Shirk, F. Hsue, T. K. Li, and T. R. Canada.
University of California, Los Alamos Scientific Laboratory,
Los Alamos, NM 87545, USA

ABSTRACT

A nondestructive assay (NDA) instrument that measures the
^{239}Pu content in solutions, using a passive gamma-ray spectroscopy technique, has been developed and installed in the
Plutonium Processing Facility at Los Alamos Scientific Laboratory (LASL). A detailed evaluation of this instrument has been
performed. The results show that the instrument can routinely
determine ^{239}Pu concentrations of 1 to 500 g/ℓ with accuracies of 1 to 5% and assay times of 1 to 2 x 10^3 s.

INTRODUCTION

Plutonium solutions are generated by a variety of chemical
processes in the Los Alamos Scientific Laboratory (LASL) Plutonium Processing Facility. These solutions include: products
of impure oxide metal scrap dissolutions, ion-exchange column
eluates and effluents, and precipitation filtrates. The plutonium concentrations vary over a wide range, from a few mg/ℓ
to over 500 g/ℓ. The solutions also contain fission products
and daughter isotopes, in particular, ^{241}Am and ^{237}U. The
density of ^{241}Am can vary from a few mg/ℓ to several g/ℓ.

Safeguards interests and process monitoring and control
considerations require a plutonium assay method that is accurate and timely. Transmission-corrected gamma-ray counting
is a nondestructive assay (NDA) method that satisfies these
criteria. Briefly, this method requires the measurement of a
characteristic isotope gamma-ray rate, R, from a sample and

the transmission coefficient, T, of an external gamma-ray
source through the sample at the same energy. The isotope
mass, M, present in the sample is given by

$$(1) \qquad M = \left(\frac{R}{K}\right) CF \quad ,$$

where K is a calibration constant and CF is a sample self-
absorption correction factor that is a function of T.[1-3]

An instrument using this method has been developed and
installed in the Plutonium Processing Facility at LASL. This
paper discusses the hardware configuration, the operational
method, and the instrument evaluation plan. The discussion of
the evaluation plan includes calibration and instrument reliability results and a detailed comparison of the instrument
assay results with the results obtained from the analytical
chemistry laboratory.

INSTRUMENT DESCRIPTION AND OPERATION

The solution assay instrument (SAI) is a minicomputer-based
system with several automated peripherals. The hardware and
peripherals have been designed and applied in such a manner
that operator intervention is minimized; when intervention is
required, it is convenient. There are six major hardware components to the instrument: (1) a germanium detector and sample
holder, (2) NIM electronics, (3) a minicomputer, (4) an operator console, (5) a mobile maintenance-and-graphics cart, and
(6) a digital electronic balance. The system assays the
^{239}Pu sample concentration by a transmission corrected count
of the 414-keV gamma ray from ^{239}Pu.

The sample chamber and germanium detector are shown in
Fig. 1. The sample chamber holds the sample vial, which is a
right-circular cylinder with a volume of 25 ml. With the sample chamber closed, the vial is surrounded by 5 cm of lead
shielding. The transmission source, a plutonium metal disk, is
fixed onto a rotating tungsten shutter located above the sample
vial. The shutter is operated pneumatically under computer
control. Gamma rays from the transmission source and the
sample vial are viewed by the detector through a .75-cm tungsten filter and the glovebox floor. The detector has a resolution of 1.7 keV at 414 keV.

Pulses from the detector are processed by standard high
resolution gamma-ray spectroscopy NIM electronics, including
(1) an amplifier with an internal pulse pile-up rejector, and
(2) a two-point energy stabilized analog-to-digital converter.

A Data General-compatible, 16-bit minicomputer is the
computational and control tool used in the instrument. The

computer chassis contains the CPU, 32K words of core memory,
and the serial communications board. Typical instruction times
are 1200 ns; communication data rates are at 300 baud. Control
functions and measurements carried out by the computer are
actuated from the operator console. A 16-key push-pad is the
interface between the instrument, the peripherals, and the
operator.

The mobile maintenance-and-graphics cart has a paper-tape
reader to load the instrument code or diagnostic programs. The
graphics display, similar to a standard multichannel analyzer
(MCA) display, allows the user to view the pulse height distribution and to enter specified regions of interest around the
photopeaks required by the analysis software. The cart is
attached to the system only during the set-up and maintenance
periods.

To give the SAI the capability of reporting assay results
as g ^{239}Pu/g sample, an electronic balance is located inside
the glovebox. Because the glovebox atmosphere is corrosive,
the balance digital and analog electronics are external to the
glovebox. The balance reading is automatically transmitted to
the computer memory by depressing a key on the operator
console.

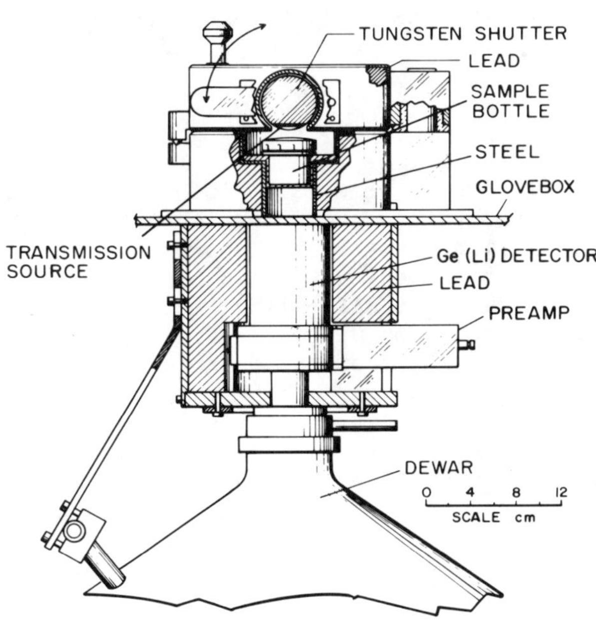

Fig. 1.
Sample chamber assembly
and germanium detector.

To determine the ^{239}Pu mass in a sample aliquot, measurements are required of: (1) the transmission source, (2) the background, (3) the sample, and (4) the transmission source plus the sample. The first measurement determines the unattenuated 414-keV gamma-ray rate from the transmission source. The background measurement records the level of ambient background and is used to correct the rate of the assay peak. These two measurements are performed once or twice a day, depending on glovebox conditions. The last two measurements are performed for each sample and are combined by the system so that no intermediate operator intervention is required.

The third measurement determines the 414-keV gamma-ray rate from the sample; i.e., R in Eq. 1. The fourth measurement sums this rate and the attenuated transmission source signal. The difference between this rate and that measured in the third step, when divided by the 414-keV gamma-ray rate measured in step one, determines the transmission coefficient, T, which in turn allows the sample self-attenuation correction factor, CF, to be calculated (see Eq. 1).

All measurement sequences are initiated by pressing an appropriately labeled key on the operator console. The operator is then led through the required assay steps by a series of computer programmed prompts that appear on the system terminal. When complete, the assay results are printed on the terminal, expressed as g ^{239}Pu/g sample and the associated uncertainty.

RESULTS

The SAI was calibrated using six plutonium standards prepared by the LASL analytical chemistry laboratory. The plutonium concentrations ranged from approximately 0.5 g/ℓ to 300 g/ℓ. The absolute error on the plutonium concentrations was estimated at 0.2%. The SAI calibration results are summarized in Fig. 2, where the percent deviation of the individual calibration constants from the weighted average is plotted versus plutonium concentration. These data show that the calibration constant is concentration independent over a wide range of sample self-attenuation; i.e., the functional dependence of the self-attenuation correction factor upon the measured transmission is correct.

The calculated statistical precision of a single sample assay as a function of plutonium concentration was consistent with the measured precision obtained from repeated assays of the standards, and was found to be consistent. Figure 3 shows the precision obtained for concentrated solutions with a routine 1000-s assay (a 500-s count for the sample and a 500-s count for the sample plus the transmission source). The decrease in precision at higher concentrations is due to an increased uncertainty in the measured transmission. At lower concentrations the precision varies from 4% at 0.5 g Pu/ℓ to 1% at 40 g Pu/ℓ for a routine 2000-s assay (a 1900-s count for the sample and a 100-s count for the sample plus the transmission source). This decreasing precision at lower concentrations is due to an increased uncertainty in the sample count.

As part of the routine measurement control program, one or more standards are assayed per day as unknowns. The assay value must fall within one standard deviation of the standard value before the SAI may be used for plant sample assay. Figure 4 summarizes these measurement control data for a time period of approximately one month. The eight standards used varied in concentration from 5 to 260 g Pu/ℓ. The average assayed value showed a slight positive bias relative to the standard value, 0.3%, with a standard deviation of 0.7%. The latter is consistent with the measurement precision and indicates reasonably long-term instrument stability.

The SAI accuracy for plant samples was evaluated by comparison with three analytical chemistry methods: isotopic dilution mass spectrometry (IDMS), radiochemical alpha particle counting, and coulometric titration. Samples with plutonium concentrations greater than 40 g/ℓ were chemically assayed by coulometric titration. Those with plutonium concentrations less than 40 g/ℓ were assayed by either IDMS or by alpha particle counting. The results of this comparison are summarized in Fig. 5 where the assay differences in percent are plotted versus plutonium concentration. The sample error bars reflect only the SAI uncertainty and include contributions due to statistical precision, sample volume, calibration, and isotopic abundance uncertainties.

The IDMS method has a routine accuracy of 0.2 to 1%, the major fraction of which is due to the complex sample preparation procedures. The SAI-IDMS comparison is shown in Fig. 5a. A small positive bias is observed, 0.24%, with a standard deviation of 1.16%. This standard deviation is consistent with the calculated SAI uncertainty for this plutonium concentration range. The bias is consistent with that observed for the measurement control data discussed above and may indicate a small error in the SAI calibration constant.

Due to the expense and time required for the IDMS method, it is rarely used to assay plant samples. Plant samples were routinely assayed by the alpha particle counting method before installation of the SAI. The accuracy for this method is limited to approximately 5 to 10%, due to sample preparation difficulties.

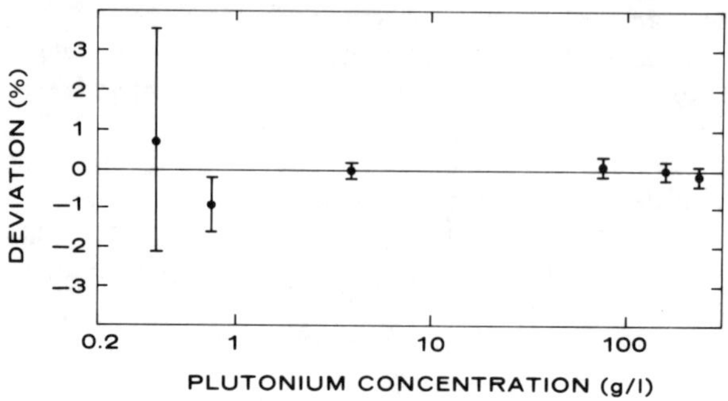

Fig. 2.
Calibration constant deviation as a function
of plutonium concentration.

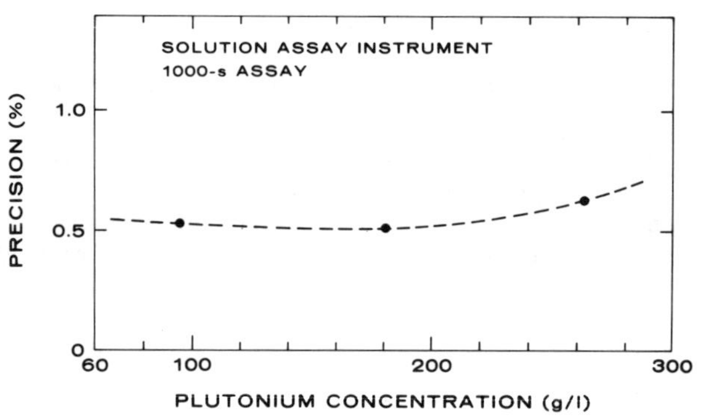

Fig. 3.
Assay precision as a function of plutonium
concentration for a 1000-s count.

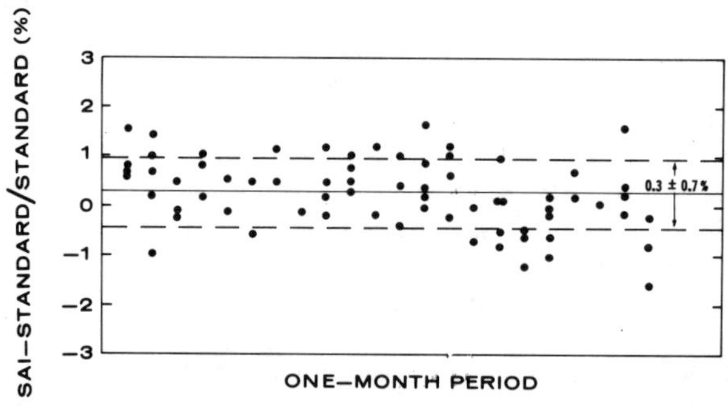

Fig. 4.
Instrument stability as a function of time.

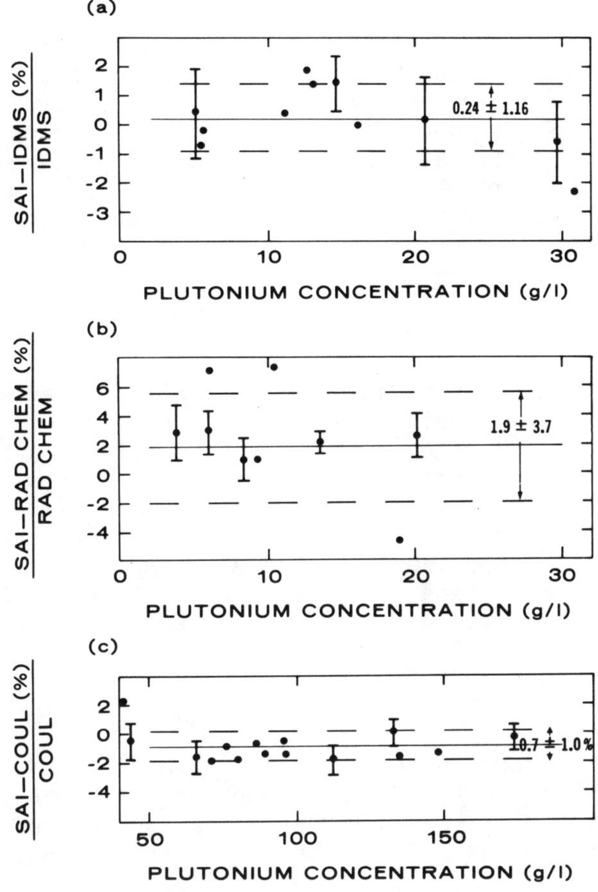

Fig. 5.
Comparison of SAI assay results
with analytical chemistry results.

The SAI-radiochemical comparison is shown in Fig. 5b. The
bias observed here is approximately 2% with a standard deviation of 3.7%. A comparison of these results with those
presented in Fig. 5a shows that the SAI provided increased
accuracy for routine plant sample assay.

The results of the SAI-coulometric titration comparison for
concentrated plutonium solutions are summarized in Fig. 5c.
These data indicate a bias of -0.7% and a standard deviation of
1.0%, which is consistent with the SAI accuracy. The SAI measurement control data show no bias for the standards assayed
during the same period that the plant samples were assayed.
The titration technique has a routine accuracy of 0.2 to 0.5%.
This accuracy depends on proper correction for iron content.
An inadequate correction results in a negative bias.

DISCUSSION

The SAI has proven to be a reliable and accurate instrument
for providing timely assays of plutonium solutions over a wide
concentration range. This instrument is being routinely used
by process personnel who have no detailed knowledge of the instrument. The accuracy and precision of the SAI assay results
compare favorably with analytical chemistry results. Furthermore, use of the instrument does not require complex sample
preparation. These qualities suggest that the SAI can be a
valuable tool for a materials management program and process
control in a plutonium processing facility.

ACKNOWLEDGEMENTS

The authors thank J. L. Parker and K. H. Johnson for their
cooperation and for their support of the prototype SAI. We
also thank M. Stephens for technical support of the SAI units
installed in the LASL Plutonium Processing Facility.

REFERENCES

1. J. L. Parker, "A Plutonium Solution Assay System Based on
    High Resolution Gamma Ray Spectroscopy," ASM/ASTM/ASNT/ANS
    International Conference on Nondestructive Evaluation in
    the Nuclear Industry (February 1978).

2. T. R. Canada, J. L. Parker, and P. A. Russo,
    "Computer-Based Inplant Nondestructive Assay
    Instrumentation for the Measurement of Special Nuclear
    Materials," ANS Topical Conference "Computers in Activation
    Analysis and Gamma-Ray Spectroscopy" (May 1978).

3. R. H. Augustson and T. D. Reilly, "Fundamentals of Passive
    Nondestructive Assay of Fissionable Material," Los Alamos
    Scientific Laboratory Report LA-5651-M (September 1974).

THE MEASUREMENT OF PLUTONIUM OXALATE IN
THERMAL NEUTRON COINCIDENCE COUNTERS

R. S. Marshall and B. H. Erkkila
University of California, Los Alamos Scientific Laboratory,
Los Alamos, NM 87545

ABSTRACT

A coincidence neutron counting method has been developed for
assaying batches of plutonium oxalate. Using counting data from
two concentric rings of ^{3}He detectors, corrections are made
for the effects that water has on the coincidence neutron count
rate. Batches of plutonium oxalate varying from 750 to 1000 g
of plutonium and from 34 to 54% water are assayed with an average accuracy of $\pm3\%$.

INTRODUCTION

Purification of plutonium by the selective precipitation of
plutonium as plutonium oxalate is used extensively at the Los
Alamos Scientific Laboratory (LASL) and Savannah River plutonium
recovery facilities. Plutonium oxalate precipitation not only
serves as a purification step, but also produces a product that
is easily calcined to form low-fired PuO_2.

The plutonium oxalate precipitation process used at LASL
precipitates plutonium as plutonium-III oxalate from a nitric
acid solution. The slurry is filtered in a vacuum filter boat
and air dried for several hours. Water content in the filter
cake varies over a range of 34 to 54%, while the plutonium content varies from 30 to 42%. A typical batch size is about 800 g
of plutonium and about 2200 g of total material.

Both the precipitation process and subsequent calcining
steps are lengthy processes. To maintain real-time accountability of the plutonium moving through these processes, it is
important that the plutonium oxalate be assayed by an

in-line nondestructive assay (NDA) method between the precipitation process and calcining steps. The highly variable water
content of the plutonium oxalate makes accurate NDA measurements
difficult to attain. Methods based on weighing or gamma counting provide only approximate information. Coincidence neutron
counting provides accurate measurements only if corrections are
made for water effects.

Coincidence neutron counting methods for plutonium analysis
are based on counting coincidence neutrons coming from spontaneous fissioning of the isotopes ^{238}Pu, ^{240}Pu, and ^{242}Pu.
The isotope ^{239}Pu can contribute to the coincident neutron
signal if spontaneous fissioning neutrons or (α,n) neutrons
interact with the ^{239}Pu nuclei that cause the fissions. Spontaneous fission neutrons and (α,n) neutrons typically are born
with energies in the 1- to 2-MeV range; the ^{239}Pu fission
cross section for neutrons in this energy range is about 1
barn. With this small neutron cross section, one would expect
the coincidence neutron signal from ^{239}Pu fissions to be a
trivial part of the overall coincident neutron signal in
weapons-grade or reactor-grade plutonium. However, the ^{239}Pu
fission cross section rises rapidly for neutrons of lower
energy.

The water component in the plutonium oxalate studied effectively lowers the spontaneous-fission and (α,n) neutron energies
by proton-neutron elastic scattering. Consequently, ^{239}Pu
fissioning becomes a major contributor to the overall coincidence neutron signal and, more significantly, becomes a function
of the amount of water in the plutonium oxalate.

This paper describes measurements made on a series of plutonium oxalate batches, and the measurement method and calculations we derived to successfully correct for the water effect on
the plutonium oxalate.

INSTRUMENTATION

The neutron counter used for this study is a double-ring
thermal neutron coincidence counter (TNC). The term "double-
ring" refers to two concentric rings of ^{3}He detectors that
surround a central counting well. The ^{3}He detectors are
embedded in polyethylene, which serves as a moderator to decrease the energy of neutrons arriving at the detectors, thereby
increasing the counting efficiency of the TNC. Figure 1 shows
the location and number of ^{3}He detectors in the TNC; the mechanical components are described in detail elsewhere.[1]

The purpose of the double-ring design is to give a signal
that is sensitive to the average energy of neutrons counted by

the [3]He detectors.[2] The thickness of the polyethylene between the inner ring of the [3]He detectors and the counting
well is insufficient to reduce 1- to 2-MeV neutrons to the
optimum energy range for detection. Conversely, the polyethylene thickness between the outer ring of detectors and the
counting well is too thick for optimum moderation. Consequently, if the average neutron energy from material being measured is in the 1- to 2-MeV range, the ratio of count rates
between the inner and outer rings of detectors will be relatively low. If, on the other hand, the average neutron energy
from material being counted is low because of moderation of
neutrons within the material, then the ratio of inner- to
outer-ring count rates will be relatively high.

The 16-cm-diam counting well shown in Fig. 1 is bolted to
the bottom of a glovebox and opens into the glovebox. Containers of plutonium oxalate to be measured are loaded into a sample
chamber in the glovebox and lowered into the counting well by a
motor-driven cable. The outer shielding, consisting of 10-cm-
thick polyethylene lined with a 0.4-mm-thick cadmium sheet,
shields the detector array from background neutrons.

The TNC electronics consists of a high-voltage power supply,
preamplifiers, amplifiers, and shift register counting circuits.[3] Using the shift registers, we are able to count
either single neutron events or coincidence neutron events.
Furthermore, by using two shift registers, we can count single

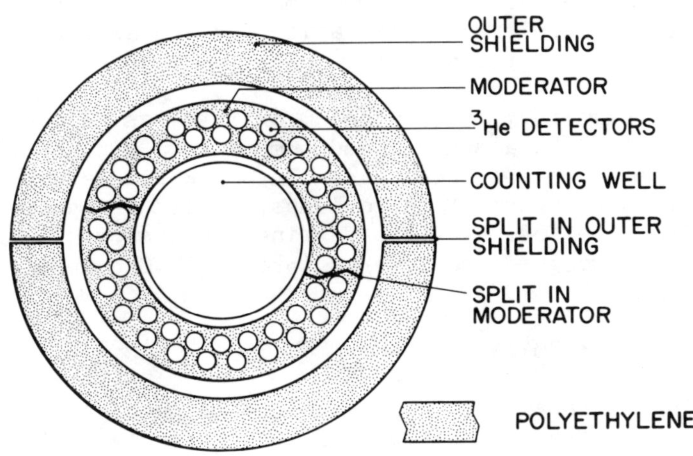

Fig. 1
Cross section of the dual-ring TNC, showing the [3]He
detectors and the counting well.

and coincidence neutron events separately for the inner and outer concentric rings of ^{3}He detectors. The TNC has a single neutron counting efficiency of about 20%. We are able to count over 100 000 neutrons per second without large deadtime losses since the TNC electronics deadtime constant is only about 3.5 μs. A microprocessor, called the THENCS unit,[4] is built into the instrument to store calibration information, to make deadtime and background corrections, and to calculate grams plutonium for measured samples.

CALIBRATION

Five PuO_2 standards, ranging from 5 to 660 g of plutonium were used to calibrate the TNC. The THENCS microprocessor corrected the coincidence counting data for background and deadtime losses, and fitted the corrected count rates to a best-fit curve using regression analysis. The coincidence counting data used for calibration purposes was taken from the inner ring of ^{3}He detectors.

SAMPLING AND PREPARATION FOR COUNTING

Each batch of plutonium oxalate was sampled before counting. Because the plutonium oxalate was not likely to be uniformly dry across the filter boats, a number of core samples were taken from different locations across the filter boats and combined. These samples were weighed, dried, and low-temperature calcined to PuO_2. The dried samples were weighed and sent to the LASL Analytical Chemistry Laboratory for plutonium assay by coulometric titration and for isotopic determination by mass spectroscopy.

After sampling, the plutonium oxalate was removed from the filter boats with a spatula and dumped into a tared 12-cm-diam, 30-cm-high stainless steel can. The dark blue-green plutonium oxalate, in the form of large lumps, filled the cans to within 10 cm from the top. A close-fitting lid was put on the can, and the can was weighed and lowered into the TNC counting well for measurement.

CALCULATIONS FROM ANALYTICAL RESULTS

Using the analytical lab data and weight data, we calculated the percent water and percent plutonium for each plutonium oxalate batch. The plutonium weight fraction for each batch was calculated as follows:

$$(1) \qquad \frac{\text{g Pu}}{\text{g Pu oxalate}} = \frac{\text{g Pu}}{\text{g dried sample}} \times \frac{\text{g dried sample}}{\text{g Pu oxalate}} \quad .$$

Grams of plutonium for each batch were then calculated:

$$\text{(2)} \qquad \text{g Pu} = \frac{\text{g Pu}}{\text{g Pu oxalate}} \times \text{batch weight} \quad .$$

The assay and weighing errors are quite small; more significant errors are associated with cross-contamination and other sampling problems. We estimate that the mass plutonium values listed in Table I have associated uncertainties of about $\pm 1\%$.

TABLE I

MEASUREMENT DATA FOR PLUTONIUM OXALATE BATCHES

Batch Number	Actual Mass Pu (g)	Measured Mass Pu (g)	Ratio of Measured Mass to Actual Mass	Water (wt%)	Ring Ratio
1	887	1291	1.46	44.3	1.070
2	797	1025	1.29	42.7	1.053
3	756	1088	1.44	43.0	1.068
4	781	1260	1.61	48.4	1.095
5	851	1318	1.55	46.5	1.075
6	855	1332	1.56	47.1	1.091
7	917	1164	1.27	37.4	1.054
8	880	1564	1.78	52.7	1.097
9	799	1238	1.54	47.5	1.073
10	925	1450	1.57	47.7	1.083
11	861	1631	1.90	53.9	1.110
12	864	1560	1.81	53.7	1.101
13	912	1019	1.12	34.4	1.036
14	803	991	1.24	36.2	1.029
15	770	1037	1.35	41.9	1.045
16	787	960	1.22	40.5	1.036

The weight fraction of water for each batch was calculated as follows:

$$(3) \quad \frac{g\ H_2O}{g\ Pu\ oxalate} = 1 - \left[\frac{g\ Pu}{g\ Pu\ oxalate} \left(1 + 0.552\ \frac{g\ oxalate}{g\ Pu} \right) \right],$$

where the constant 0.552 is based on the assumption that plutonium oxalate is $Pu_2(C_2O_4)_3 \cdot x\ H_2O$. This assumption is not entirely valid, because surface oxidation to $Pu(C_2O_4)_2 \cdot x\ H_2O$ was observed.

Combining the plutonium oxidation state uncertainty and sampling uncertainties, the percent water values shown in Table I are probably accurate to only ±1% absolute.

MEASUREMENT RESULTS AND CALCULATIONS

Each can of plutonium oxalate was counted in the TNC for 500 s. The count rates were very high (singles, approximately 40 000 counts/s; coincidence, approximately 3 300 counts/s). Hence the precision of the counting data is good. The THENCS microprocessor used the inner ring coincidence count rate for each plutonium oxalate batch to calculate plutonium mass versus the PuO_2 calibration curve.

Table I lists the actual mass plutonium, the measured mass plutonium, the measured-mass to actual-mass ratio, and the weight percent water for each batch. Notice that the measured mass was always biased high and that there is a correlation between the amount of bias and the percent water characteristic of each batch. A correction to the measured mass based on percent water would reduce the measured mass bias, but percent water is not measured directly by the TNC.

As discussed earlier, the ratio of neutron counts between the inner ring of detectors and the outer ring should be sensitive to the average neutron energy and, therefore, to the water content of the plutonium oxalate. The ring ratios and measured-mass to actual-mass ratios from Table I are plotted in Fig. 2; a good correlation is apparent. To evaluate the correlation, we fitted the data to an exponential curve of the following form:

$$(4) \qquad y = ae^{bx},$$

where y = measured mass divided by actual mass,
      a = 0.0015372,
      b = 6.38986, and
      x = ring ratio.

The coefficient of determination value for this fit is 0.9625, indicating that the exponential form fits the data quite well.

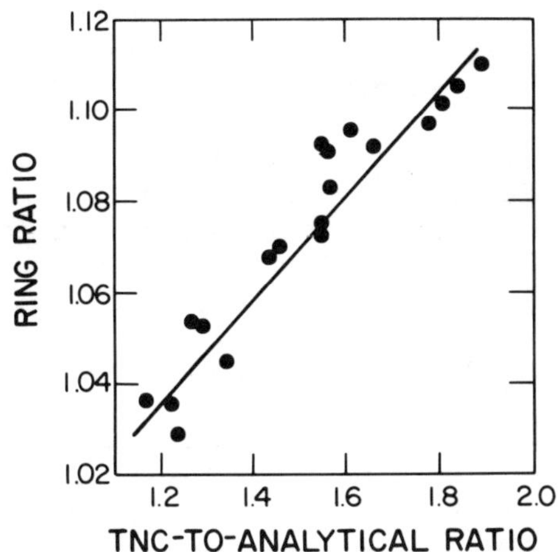

Fig. 2.
Correlation of the ring ratios and the
measured-mass to actual-mass ratios.

Using Eq. 4 and the ring ratios, we then calculated the y
values for each plutonium oxalate batch. We next corrected the
measured plutonium mass values in Table I with their corresponding y values as follows:

$$(5) \qquad \text{corrected mass} = \frac{1}{y} \, (\text{measured mass}) \qquad .$$

Table II shows the results of applying this correction to
the 16 batches of plutonium oxalate. Notice that the average
difference between the corrected mass and actual mass is 2.7%,
and the three worst cases had errors of 5.4%, 4.9%, and 4.5%.

DISCUSSION

The ring ratio correction worked well for the 16 batches of
plutonium oxalate investigated. It is important to note that
the batches ranged over a limited mass range of 756 to 925 g of
plutonium. We have done some preliminary measurements of smaller masses of plutonium oxalate; the ring ratio correction
generated a significant bias for plutonium masses below 500 g.
The 16 batches investigated represented a wide spectrum of water
content. We suspect that the correction equation constants are

incorrect for the very dry plutonium oxalate that might be generated by vacuum drying.

The ring ratios were calculated from inner- and outer-ring
single neutron counts. Using ring ratios of coincidence counts
yields similar, but less precise results. The presence of high
cross-section (α,n) nuclei such as beryllium, lithium, or
fluorine could shift the average neutron energy spectrum and
bias the ring ratio correction. Generally, the oxalate
precipitation process provides good purification of these

TABLE II

RING RATIO CORRECTED MASS PLUTONIUM

Batch ID	Actual Mass Pu (g)	Measured Mass[a] Pu (g)	Error[b] (%)
FOX 1177	887	886	-0.2
FOX 1180	797	787	-1.3
FOX 1178	756	756	0.0
FOX 1181	781	739	-5.4
FOX 1185	851	877	3.1
FOX 1183	855	803	-6.1
FOX 1182	917	886	-3.4
FOX 115C3	880	907	3.1
FOX 115C6	799	835	4.5
FOX 115C7	925	918	-0.8
FOX 116C3	861	871	1.2
FOX 116C4	864	883	2.1
FOX 116C5	912	872	4.5
FOX 116C6	803	816	1.7
FOX 11CC	770	771	0.1
FOX JRC-1	787	826	4.9

[a] Ring ratio corrected.

[b] Average error = 2.7%.

elements, but should these elements be present in 0.1% or greater concentrations, the correction may be biased. None of these limitations pose problems for the particular oxalate precipitation process investigated at LASL because the process is operated within the limiting parameters.

CONCLUSION

Using a double-ring TNC, we have developed a ring ratio correction method that successfully compensates for the effect that water has on the coincidence count rate of plutonium oxalate. Batches of plutonium oxalate can now be assayed with an average measurement error of about 3%. The method provides the rapid in-line measurement capability required to maintain real-time SNM accountability around a plutonium oxalate purification process. At LASL, plutonium oxalate measurements are performed by production technicians, and require a 500-s counting time and approximately 10 min for calculating and recording the data.

ACKNOWLEDGEMENTS

The authors wish to thank Eldon Christensen and Milton Haas for providing the batches of plutonium oxalate used in this study.

REFERENCES

1. N. Baron, "A Correction for Variable Moderation and Multiplication Effects Associated with Thermal Neutron Coincidence Counting," IAEA Symposium on Nuclear Material Safeguards, Vienna (1978).

2. R. S. Marshall, Los Alamos Scientific Laboratory Report LA-7616-PR, 70-73 (1978).

3. M. S. Krick and H. O. Menlove, Los Alamos Scientific Laboratory Report LA-7779-M (1979).

4. B. H. Erkkila and R. S. Marshall, "A Thermal Neutron Coincidence Counting System." To be presented at the American Nuclear Society 1979 Winter Meeting, San Francisco, Calif.

AMERICIUM ASSAY INSTRUMENT

R. S. Marshall, University of California, Los Alamos
Scientific Laboratory, Los Alamos, NM 87545

ABSTRACT

A simple in-line americium assay instrument (AAI) was
installed at an americium recovery process area at Los Alamos
Scientific Laboratory (LASL) for use in process development and
for providing process control information. The AAI counts
59.5-keV ^{241}Am gamma rays, using a NaI(Tl) detector and Eberline SAM II electronics. It has a useful range of 3×10^{-5} to
10 g Am/ℓ and does not suffer from plutonium interference. Com-
parative analyses of samples assayed in the AAI and samples
assayed by the LASL Analytical Laboratory show a combined relative standard deviation of 14%.

Nondestructive assay (NDA) methods have been applied to a
wide range of special nuclear material (SNM) control problems.
Many of the NDA instruments have proven useful as process control monitors, providing more timely determination of material
flow than that available from analytical chemistry laboratories. The safeguards application of these instruments generally
requires a reasonably sophisticated design and a high degree of
measurement accuracy. For process control, however, these requirements can often be less restrictive, especially when quick
installation is required.

An americium recovery process, based on ion exchange and
selective precipitation, recently began operation at the Los
Alamos Scientific Laboratory (LASL) Plutonium Processing
Facility. An in-line NDA instrument, capable of measuring the
americium concentration of process solutions, was designed at
LASL to meet the following criteria for process development:

(1) an ^{241}Am analysis range of 1 x 10^{-3} to 5 g Am/ℓ,
(2) an analysis time on the order of minutes,
(3) analysis accuracy of ±25% relative,
(4) analysis of americium in solutions having plutonium-to-
americium ratios as high as 100:1,
(5) operation in a glovebox with kilogram quantities of plutonium and gram quantities of americium within a 3-m radius,
(6) installation that does not require cutting the glovebox
structure, and
(7) simple operation and calibration.

This paper describes the LASL americium assay instrument (AAI),
which applies a straightforward NDA method to satisfy these
criteria.

The AAI was designed to measure the very intense, relatively
interference-free, ^{241}Am gamma ray at 59.5 keV. This gamma
ray is sufficiently energetic to penetrate the glovebox floor
with only 40% attenuation. Because the solutions measured were
of relatively low concentration (plutonium + americium less than
15 g/ℓ), no correction was necessary for sample self-absorption.

The AAI hardware is shown in Fig. 1. The gamma rays are
detected by a 5- x 5-cm sodium iodide detector. The detector,
photomultiplier tube, and preamplifier are encased in a 12-mm-
thick lead collimator and mounted under the glovebox. Signals
from the detector are processed by an Eberline SAM II electronics package, which contains the required high-voltage power
supply, amplifier, single-channel analyzer, scaler, and timer.
The detector high voltage is stabilized using a reference signal
that is generated by an alpha source in the sodium iodide
detector.

The sample holder, shown above the collimated sodium iodide
detector in Fig. 1, was designed for analyzing low-concentration
americium solutions. After alignment with the sodium iodide
detector, the sample holder was bonded to the glovebox floor
with a silicone adhesive. The holder, machined from one-half of
a lead brick, shields the detector from background radiation
while precisely positioning the sample bottle close to the
detector.

For analyzing higher concentration solutions, a collimator
was placed inside the sample holder. The collimator lengthens
the distance between the sample and the detector, and restricts
the gamma rays counted to those passing through a 3.8-mm-diam
opening. This adapter reduced the americium count rate by a
factor of 180. The lower curve in Fig. 2 shows the pulse-height
distribution measured for an americium sample with a very low
plutonium-to-americium concentration ratio of 1:50, or 0.02. On
the basis of these data, the single-channel analyzer window was

set to accept pulses whose amplitudes corresponded to gamma rays
with energies of 35 to 75 keV.

Plutonium contamination in the samples affects the accuracy
of the americium assay by three different mechanisms:
(1) introduction of plutonium gamma rays into the assay energy
region, (2) introduction of a Compton continuum under the assay
region, and (3) increase in the pulse pile-up and deadtime rate.

Plutonium gamma-ray activity in the 35- to 75-keV region was
calculated assuming a plutonium isotopic composition characteristic of FFTF fuel. These plutonium gamma peaks add about 0.2%
to the ^{241}Am peak area for a 100:1 plutonium-to-americium mixture, and about 2% for a 1000:1 plutonium-to-americium mixture.

Figure 2 shows the influence of the Compton effect upon the
pulse-height distribution for a sample with a plutonium-to-
americium ratio of 3300:1. For a solution with a plutonium-
to-americium ratio of 1000:1, the Compton continuum biases the
americium assay about 2% high.

The increased pulse pile-up and deadtime rates biased the
assay low. Deadtime losses were less than 1% for americium
count rates up to 100 000 counts/min. However, plutonium gamma
rays outside of the counting window contributed to pulse
pile-up. Measurements on standards having varying plutonium-to-
americium ratios indicate that pile-up from plutonium gamma rays
becomes significant when the plutonium concentration exceeds 10
g/ℓ.

The AAI was calibrated with a set of standards whose americium concentrations varied from 1 x 10^{-5} to 10 g/ℓ. The
results for the uncollimated and collimated counting geometries
are summarized in Figs. 3 and 4, respectively, where the count
rates of the standards are plotted versus americium concentration. In the uncollimated geometry, the calibration curve slope
is constant for concentrations less than 5 x 10^{-3} g Am/ℓ.
Similarly, the calibration curve slope is constant in the collimated geometry for concentrations less than 1.5 g Am/ℓ. For
higher concentrations, the curve slope decreases, due to pile-up
and deadtime losses.

To determine the AAI's accuracy on routine plant samples,
the AAI assay was compared with an analytical laboratory assay
based on gamma-ray counting. The routine accuracy of the latter
is approximately ±1%. The results are summarized in Table I.

The AAI analysis consisted of a 60-s background count and a
60-s sample count. The uncertainty of the counting statistics
associated with the measurements is less than 3%. Comparison
between the two assay methods shows that the AAI results are

biased by 1.4%, with an average deviation of $\pm$13.5%. These results are well within the design criteria.

In summary, the AAI is a simple instrument to assemble, install, calibrate, and operate. Using a 1-min background and 1-min sample counting time, solutions varying from 1 x 10^{-3} to 10 g Am/ℓ can be analyzed with an accuracy of better than 14%. Plutonium interference is insignificant if the plutonium concentration in the sample remains below 10 g/ℓ and the plutonium-to-americium concentration ratio remains below 100:1. At LASL the AAI met or exceeded all the initial design requirements. Process personnel used the AAI to obtain the rapid americium analyses needed during development of the americium recovery process.

TABLE I

AAI AND ANALYTICAL LAB ANALYSES OF ^{241}Am SAMPLES

Sample Number	AAI (g Am/ℓ)	Lab (g Am/ℓ)	Percent Difference[a]
LAF 25	0.18	0.26	-31
LAF 26	0.25	0.23	9
LAF 27	0.27	0.44	-39
LAF 28	0.51	0.47	9
LAF 29	0.92	0.81	14
LAF 30	0.66	0.61	8
LAF 31	0.36	0.33	9
LAF 32	0.35	0.27	30
LAF 33	0.56	0.59	-5
LAF 34	0.39	0.40	-3
LAF 35	0.24	0.24	0
LAF 36	0.24	0.25	-4
LAF 37	0.048	0.045	7
LAF 38	0.085	0.085	0
LAF 39	0.079	0.075	5
LAF 40	0.035	0.036	-3
LAF 41	0.041	0.038	8
LAF 42	0.031	0.029	7
LAF 43	0.081	0.076	7
LAF 44	0.057	0.055	4
LAF 45	0.084	0.082	2
LAF 46	0.61	0.61	0
LAF 47	0.27	0.29	-7
LAF 48	0.60	0.61	-2
LAF 49	0.74	0.68	9

[a] $\left[\dfrac{AAI - LAB}{LAB}\right] \times 100$

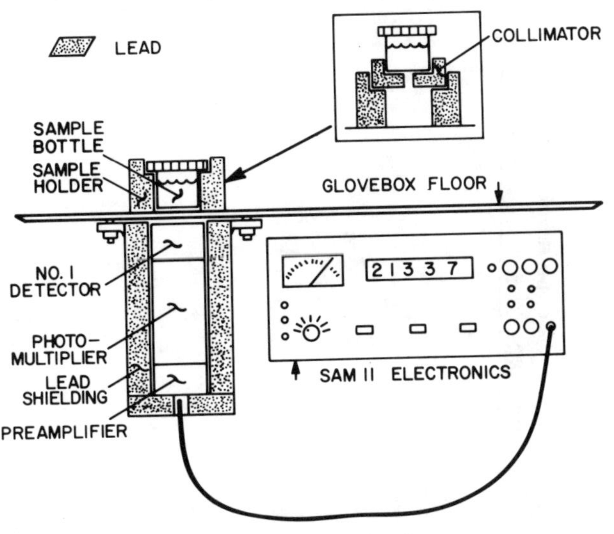

Fig. 1.
AAI components.

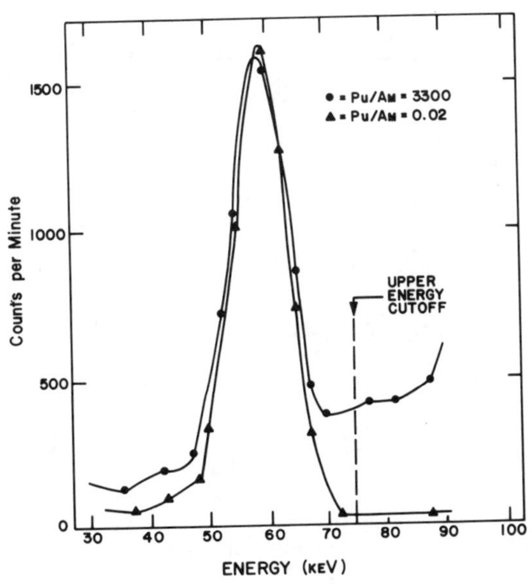

Fig. 2.
Pulse height distribution.

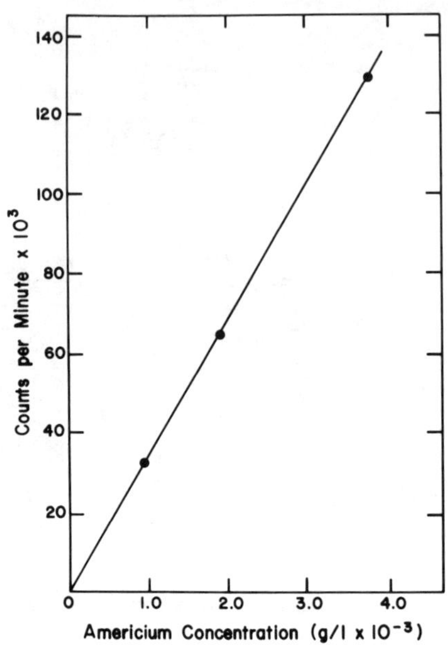

Fig. 3.
Uncollimated calibration curve.

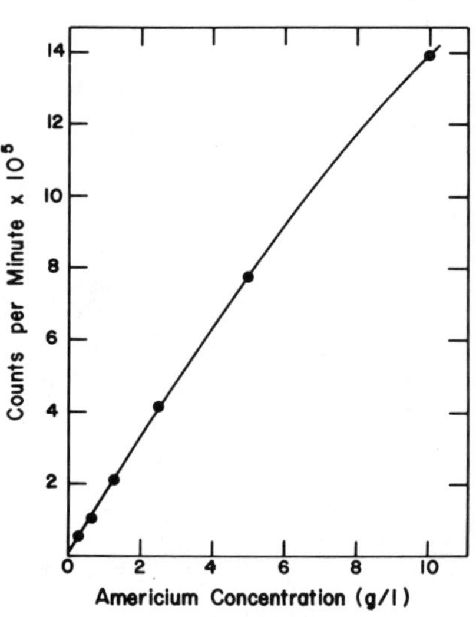

Fig. 4.
Collimated calibration curve.

AN IN-LINE FREE ACID MONITOR FOR
NUCLEAR FUEL REPROCESSING

J. E. Strain, Analytical Chemistry Division, Oak Ridge
National Laboratory, Oak Ridge, Tennessee

ABSTRACT

A free acid monitor has been designed to continuously
measure the excess acid present in nuclear fuel reprocessing
streams. The monitor is based on the relationship of the
partial pressure ratio of water and nitric acid in equilibrium with a nitric acid solution at constant temperature.
The acidity of the condensed equilibrium vapor is proportional to the partial pressure ratio and is determined by
conductivity measurement or direct titration. The operating
range of the monitor has been demonstrated from 0.03M to 15M
feed acid. The accuracy of free acid measurement is only
slightly affected by solute concentration.

INTRODUCTION

Aqueous reprocessing of spent nuclear reactor fuel
involves the initial dissolution of the chopped fuel bundles
in strong acid, usually nitric. This solution is then clari-
fied by filtration or centrifugation and made up to a desired acidity prior to extraction of the uranium and plutonium. The excess acid concentration is crucial since too
much acid results in poor decontamination from fission
products and too little results in a loss of the recoverable
fuel. The conventional methods of excess acid determination
are difficult and time consuming due to the high radiation
levels present in the early stages of fuel reprocessing.

This report details the design and evaluation of a
device intended to continuously monitor the excess acid
(HNO_3) in the early stages of fuel recovery. The monitor
is based upon the partial pressure of water and nitric acid
at constant temperature as a function of acid concentration.

Figure 1 shows the partial pressure of water and HNO_3 at
$70^\circ C$ and $100^\circ C$. The data used to construct the curves of
Figure 1 was taken from Gmelins[1] where data from several
sources is tabulated. The graph shows that if the ratio
of the partial pressures of water (pH_2O) and HNO_3 ($pHNO_3$)
can be measured above an acid solution at constant temperature, the ratio $pH_2O/pHNO_3$ could be related to the acidity
of the original solution. This ratio can be determined
directly by measurement of the acidity of the condensate
formed by condensing the equilibrium vapor phase that exists
above an acid solution whose temperature is held constant.
Measurement of the acidity of the condensate can be accomplished either by direct titration, conductivity or refrac-
tive index measurement. All three readout modes were investigated and each will be discussed later in this report.
Conductivity measurements were used in most of the evaluation due to its simplicity and continuous response.

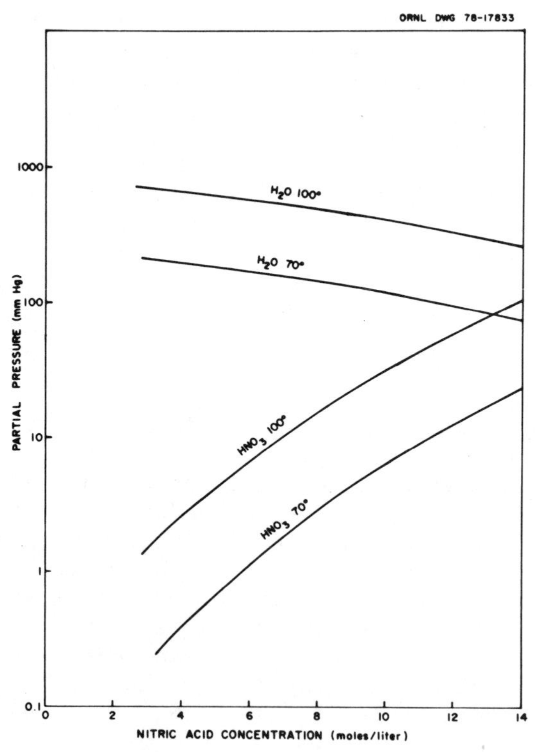

Figure 1. Partial Pressure of H_2O and HNO_3
as a Function of Nitric Acid Concentration

Monitor Design

    The final prototype configuration of the excess acid
monitor, shown in Figure 2, is the end result of a long
series of modifications to correct problems that arose
during the reduction of a principle to a practical device.
The entire assembly, with the exception of the conductivity
cell, was fabricated from 300 series stainless steel. The
drawing of Figure 2 is not drawn to scale but illustrates
the operation of the monitor. The vessel (a) is a solution
reservoir that represents the process vessel to be sampled.
The vertical tube (b) represents an air lift that raises
the sample so that it can flow under the influence of
gravity through the monitor and back to the vessel sampled.
The air lift also provides additional sparging to eliminate
dissolved oxides of nitrogen. The air and liquid phases are
separated in vessel (c) with the air vented to the off-gas
system via (d) and the liquid draining through a course
screen and preheater coil into the evaporation chamber (f).
The heater jacket (e) on the evaporator serves to maintain
the set temperature as the liquid cascades down through the
baffels. These baffels overlap to provide the maximum
air-liquid surface and minimize aerosol formation. The
sample leaves the evaporator and passes through a heat
exchanger (g) that cools it to its original temperature
before returning to vessel (a). Dry air is swept through

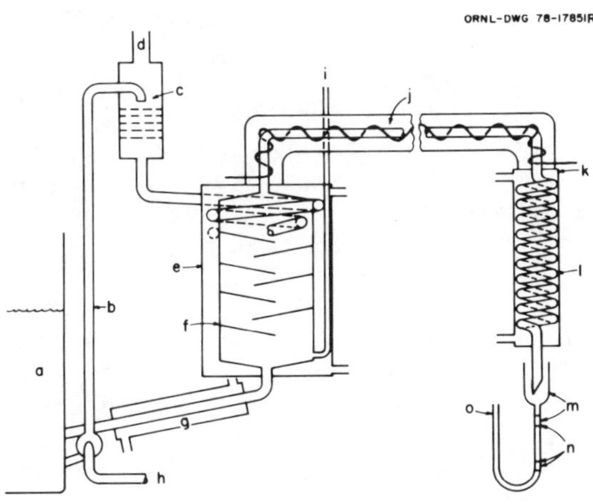

Figure 2. Sectional Drawing of the Free Acid Monitor

the evaporator at the rate of 0.5 liters/min through the
3 mm ID tube (i) to carry the acid vapor into the transport
tube (j). The transport tube is heated electrically to
maintain a temperature of 125°C. The vapor enters the
condenser (k) and is converted to a liquid as it transits a
close wound spiral of 5 mm ID, thin walled, stainless steel,
tubing. The condensate enters the enlarged portion of the
conductivity cell (m) and flows through the annular platinum
electrodes (n). The conductivity cell remains full by vir-
tue of the overflow level (0)

EVALUATION AND RESULTS

A typical response curve of the acid monitor is shown
in Figure 3. In this case, the system was filled with
nitric acid and the effluent was monitored by means of a
conductivity cell. The decrease in slope of the curve at
a feed acid concentration of 10 molar is due to the decreasing conductivity of the condensate. This decrease in conductivity is caused by the decreased ionization of HNO_3 in
strong acid solutions. The maximum conductivity of a nitric
acid solution is observed in a 5.8 molar solution. When a
direct titration of the condensate is used to monitor the
condensate acidity, the operating range of the monitor is
extended to above 14 molar but this reduces the sensitivity
at low acid levels and increases the response time because
a larger volume of condensate is required. The use of a
flow through refractometer to monitor the ratio of HNO_3 to
water in the condensate limits the upper range to <14M. The
response below 4 molar is poor but similar to the direct
titration response.

The response time of the monitor, that is the time
necessary to correctly indicate the new acid level after
an alteration of the feed solution, was found to be about
8 minutes at a condensate flow rate of 10 ml/hr.

Sample Circulation

The prototype monitor used an air lift to raise the
acid solution from a reservoir to a point from which it
could flow through the evaporator and back to the sample
reservoir. An air flow of 12 liters/min provided a 500
ml/hr flow through the monitor at a lift height of 30 cm.

A separate function of the air lift is to aerate the
sample and thus remove dissolved oxides of nitrogen, such
as NO_2. It was found that NO_2 causes erroneously high
results by being vaporized in the evaporator and re-dissolved
in the condensate. A flow rate of 12 liters/min was found

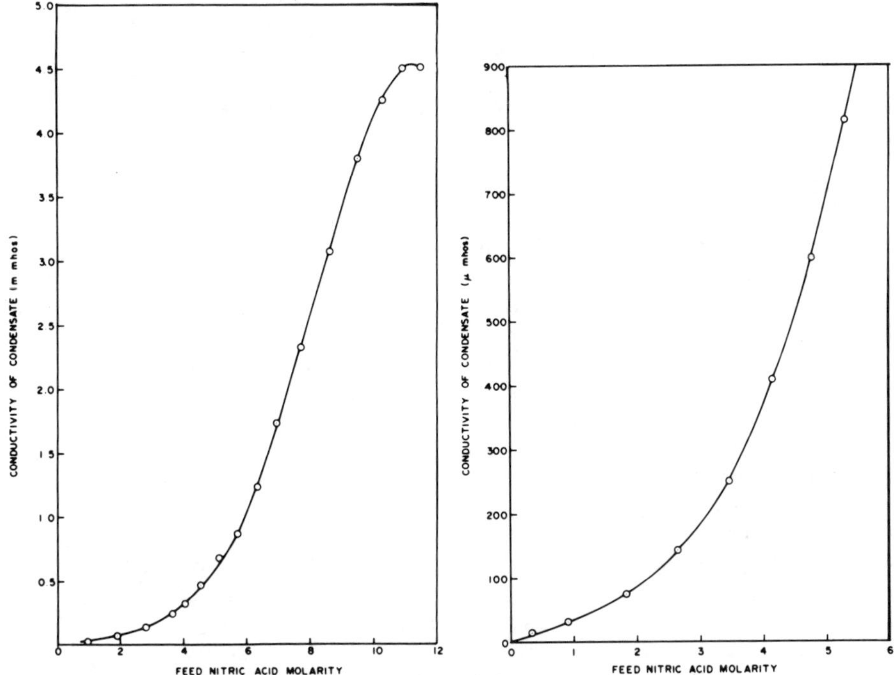

Figure 3. Response of the Free Acid Monitor with a Conductivity Readout. Left--Total Range. Right--Expanded Low Acid Region

to reduce the concentration of NO_2 entering the evaporator to an acceptable level. Dissolution of copper in 4 molar nitric acid circulating through the monitor produced only a 10% increase in indicated acidity; i.e., 4.4 molar during active dissolution. When dissolution was complete, the indicated acidity rapidly returned to the true value.

Evaporator Temperature

Because the partial pressures of water and nitric acid are not equally affected by a change in temperature, the condensate acidity of a given feed acid will increase with increasing evaporator temperature. It was found that a 5^OC change in evaporator temperature from 73^OC will produce a change in the indicated acidity of approximately 5%. At higher evaporator temperatures there is a larger error for the same variation in temperature.

The design temperature of 74° was selected for several reasons: (a) it is above the normal vessel operating temperature: (b) it produces a condensate flow rate

that is high enough to provide rapid monitor response; (c)
it is not so high as to materially change the concentration
of the sample circulating through the evaporator; and (d)
it is low enough that the heating and cooling requirements
are not excessive at the flow rate of 500 ml/hr.

Solute Concentration

The presence of dissolved solids in the acid solution
cause a positive bias in the monitor response. This bias
is due to a common ion effect acting to suppress the ionization of nitric acid and thus increasing the $pHNO_3$. Figure 4 illustrates the correction that must be supplied to
convert the apparent excess acid to the true excess acidity
for three solutes. The curves were generated by adding
solute or concentrated (15.7M) nitric acid to the circulating
acid. After equilibrium was obtained the excess acid was
determined either by direct titration ($NaNO_3$) or complexo-
metric titration ($Al(NO_3)_3$ and $UO_2(NO_3)_2$). The solute concentration was determined gravometrically. The excess acid
ranged from 2 molar to 7 molar. The true excess acid was
then divided by the indicated excess acid to obtain the correction factor. The estimated uncertainties in the data
are indicated by the size of the individual data points.

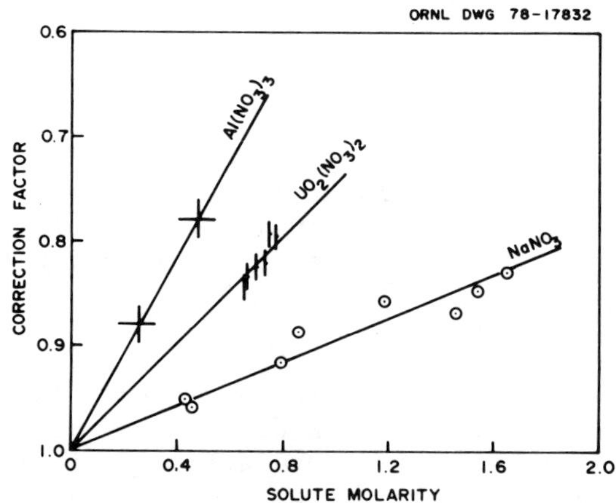

Figure 4.   Effect of Solute Concentration on
Indicated Free Acid

From the figure, assuming that the major solute will be
$UO_2(NO_3)_2$ at about 0.6 molar, it should be possible to estimate the concentration of uranium from feed information and
apply a correction factor that will produce a monitor result
that is within 5% of the true excess acid value.

Insoluble precipitates or residues that do not alter
the partial pressure of either water or HNO_3 have no effect
on monitor response.

Transfer Tube Temperature

The temperature of the transfer tube was found to have
a profound effect on the condensate conductivity. If the
wall temperature of the transfer tube falls below $122^{\circ}C$, the
boiling point of 15.7 molar nitric acid, the HNO_3 vapor con-
denses to form points of secondary evaporation. These drop-
lets, in addition to reducing the concentration of HNO_3 in
the vapor phase, are swept into the condenser at random
intervals to produce serious fluctuations of the measured
conductivity. By using an independently controlled electrical heater and a thermistor to monitor the temperature, the
transfer tube could be maintained above $125^{\circ}C$. Temperatures
above $125^{\circ}C$ have no effect on the monitor's response and the
transfer tube can be of any convenient length. This will
make possible the location of the conductivity cell or even
a direct titration cell in a shielded area remote from the
sampled vessel or pipeline.

Dissolved Oxides of Nitrogen

The oxides of nitrogen that are formed when metals are
dissolved in nitric acid are the most serious impediment to
the performance of the monitor. When an acid sample containing dissolved NO_2 enters the heated evaporation chamber
the NO_2 is vaporized and re-dissolves in the weak acid condensate. There it hydrolyzes to produce HNO_2 and HNO_3.
This increases the conductivity and invalidates the monitor's
response. The airlift and separation chamber are designed
to eliminate dissolved oxides of nitrogen but during dissolution of metals its operation is only partially successful. It is anticipated that the monitor will be inoperative
due to NO_2 until dissolution is complete and the contents
of the digestion tanks have been air sparged for at least
30 minutes. It has also been reported that the use of an
oxygen rich sparge during dissolution of spent reactor fuel
will eliminate the formation of NO_2.[2]

If the monitor is to be installed in a process line or
vessel where there are no oxides of nitrogen present, the
airlift and separation chamber could be eliminated. The
sample circulation can be controlled by adjustment of inlet
tubing diameter. Sample flow can be verified by monitoring
the temperature of the inlet tube to the return heat exchanger. At steady state operation the temperature should remain
constant and just slightly less than the evaporator
temperature.

Conductivity Cell

The conductivity cell was fabricated from 3 mm O.D.
pyrex tubing bent in a "U" shape with two electrodes im-
planted in the inlet, verticle leg. The electrodes consisted of two 5 mm long sections of 2 mm O.D., bright, platinum
tubing separated by 3 cm of flow cell. Electrical connection was made by welding a 0.5 mm diameter platinum wire to
the platinum tubing before sealing it in place. A short
section of 10 mm I.D. pyrex tubing was reduced on one end
and sealed to the inlet leg of the conductivity cell to form
a funnel to admit condensate and permit the escape of the
sweep gas. The length of the vertical, cell-exit-tube is
adjusted so that a liquid level is maintained in the inlet
funnel just high enough to prevent air bubbles from being
drawn into the 3 mm section of the conductivity cell. The
cell, when operated in this orientation, is free from gas
bubble formation.

SUMMARY

The concept of using the partial pressures of components
in a binary mixture to determine concentration has been
demonstrated in this development program. The main objective
has been to develop a free acid monitor to operate in the
range of 2M to 12M nitric acid with an accuracy of ±5%.
This capability has been demonstrated. There is, however,
experimental evidence to indicate that the same analytical
principle can be extended to much lower free acid levels.
For example, a starting solution for one of the calibration
runs containing 180g of U per liter and a free acid level of
0.015N produced a condensate with a conductivity of 16 μ
mhos. If a sensitive gas phase detector can be developed
to simultaneously monitor water vapor and nitric acid vapor
at 125°C, it should be possible to rapidly determine the
free acid in relatively small samples. The minimum sample
volume that could be analyzed at present using a conductivity
cell (vol $\stackrel{\sim}{=}$ 0.5 ml) is 50 ml due to the volume of condensate
that must be generated.

For any planned use, one must consider the relatively
slow response time of the monitor using a condensed vapor
monitor. It will only be of value where a lag time of 8
minutes is relatively short compared to the residence time
of the process solution being monitored; i.e., digestor
tanks, make-up feed tanks, etc. Flow monitors will also be
required to establish air and liquid flow rates. Temperature
sensors will be needed to monitor and perhaps control heating
or cooling. The output signal for the monitor will have

to be modified by a total solute concentration signal that
is applied either automatically or manually. Calibration
of the monitor will have to be verified by off-line analysis.
Proper operation of the conductivity cell can be verified
by injecting 1 ml of standard nitric acid into the cell at
8-hour intervals and noting the indicated conductivity.

ACKNOWLEDGEMENTS

Research sponsored by the Nuclear Power Development
Division, U.S. Department of Energy under contract W-7405-
eng-26 with the Union Carbide Corporation.

REFERENCES

1. Gmelins Handbuch der Amorgan. Chemie, 8 Auflage, N.,
    p. 977-980.

2. "The Fumeless Dissolving of Uranium" G. L. Miles,
    Progress in Nuclear Energy III, Process Chemistry,
    McGraw-Hill Book Co., 1956, pp. 97-101.

A COMPLETELY CONTAINED AND REMOTELY
OPERATED DIGITAL DENSITY METER

C. R. Goergen, Savannah River Plant, E. I. du Pont
de Nemours and Company, Aiken, South Carolina

ABSTRACT

A completely contained and remotely operated density
determination system having unique features was designed,
fabricated, and installed at the Savannah River Plant. The
system, based on a Mettler calculating digital density
meter, provides more precise and accurate results than the
falling drop technique for measuring densities. The system
is fast, simple, easy to operate, and has demonstrated both
reliability and durability.

INTRODUCTION

The Savannah River Plant (SRP) Multipurpose Processing
Facility (MPPF) began operation in July 1978. The MPPF was
designed to separate americium, curium, and californium. A
section of the canyon building for processing high level
radiation material was modified to contain six production and
two analytical cells for the MPPF.[1] A pair of model 8 master-
slave manipulators is used for remote operations in each
cell. The cell walls and shield windows are 6 ft thick.

Density measurements are required for all liquid process
samples before proceeding with sample analyses. These density
measurements are compared with tank instrument indicators to
confirm that samples are representative. Density measurements
are also used in some accountability calculations.

Previously, all in-cell densities at SRP were determined
by the falling drop method. This technique involves timing a
50-μl sample drop as it falls a known distance through an

immiscible oil maintained at a constant temperature. Drawbacks of this method are imprecision, amount of maintenance
required, and the inability to analyze organic samples.
Several oils are needed to cover the density range from 1.0
to 1.6 g/cc, and each requires daily calibration. The falling
drop method was impractical for use in the MPPF due to poor
visibility into the cells and the amount of delicate remote
manipulations required by the method.

Fortsch and Wade first evaluated and installed a Mettler
density meter for remote application.[2] A special system had
to be designed for the MPPF at SRP to permit remote operation
with total containment. Radiation and contamination control
requirements allowed only sealed-electrical and positive-
pressure supplies across the cell walls.

There were several design criteria that had to be met.
Ease of operation was of primary importance. This encompassed
speed for a high throughput rate, and single manipulator
operation to reduce technician dexterity requirements and
fatigue. Since the sample volume from which a variety of
analyses are subsequently performed is less than 10 ml,
minimal sample consumption and prevention of cross-contamination were necessary. The system also had to be durable to
withstand three-shift continuous duty with minimum remote
servicing.

DENSITY INSTRUMENT DESCRIPTION

The instrument selected was the Mettler DMA 45 calculating digital density meter with the DMA 401 external cell.[a]
The DMA 401 has no exposed moving parts, and minimum electronic requirements makes it well suited for remote applications. The principle of operation is based on measuring the
resonance frequency of an oscillating hollow glass U-tube
which is filled with approximately 1 ml of sample.[3] The
environment of the tube is maintained thermostatically at
25°C. The sample tube is electromagnetically excited to
constant amplitude with a solenoid surrounding a small magnet
glued to the lower midsection of the tube. When the oscillation of the tube is resonant, The squared time period (T^2)
of oscillation is linearly proportional to the sample mass
within the constant volume U-tube. The resonant frequency
signal is detected by a solenoid pickup coil and is resolved
by a quartz clock. A microprocessor, using the slope and
intercept calibration constants (A) and (B), calculates the
sample density according to equation (1).

---

[a]The DMA 45 and 401 can be obtained from Mettler
Instrument Corp., Hightstown, NJ.

(1)
$$Density = \frac{T^2 - B}{A}$$

The result is then displayed digitally, with a resolution of 0.0001 g/cc. The calibration is linear over a range of 0 to 3 g/cc.

SYSTEM DESCRIPTION AND OPERATION

The density meter system consists of a remote equipment rack and a controlled temperature bath located inside the MPPF cell, and an external equipment cabinet with control console located outside the cell.

The remote equipment rack that was installed in the MPPF cell is shown in Figure 1. To introduce the sample into the density meter, the decapped sample vial is raised to the sample introduction tube, which is fitted with a disposal 100-μl tip to prevent cross-contamination. A sample aliquot is then drawn into the measurement tube by a motor-driven[b] peristaltic pump.[c] Once the measurement tube is filled, a level sensor[d] disables the pump to prevent inadvertant depletion of the sample. The sensor detects the dielectric field change when liquid enters the tube and produces a signal which is used to stop the pump. By connecting the sensor to a metal sheath surrounding the tubing, the sensitivity can be increased to detect organic streams.

After allowing 1 minute for thermal equilibrium to be obtained, the density of the sample is read from the DMA 45 LED display. The disposable sampling tip is then removed and a funnel swung under the sample introduction tube to receive waste. The rinse cycle is activated and a motor-driven, "Teflon" four-way valve[e] introduces, in sequence, 2N nitric acid, methanol, and nitrogen, to backflush, rinse, and dry the measurement tube, respectively. These rinse supplies are located outside the cell and the liquids are gravity fed from a height of 8 meters.

---

[b] The servomotor is available from Honeywell, Inc., Minneapolis, MN.

[c] The "Masterflex" peristaltic pump head can be obtained from Cole-Parmer Instrument Co., Chicago, IL.

[d] The level sensor is a Lab Monitor II manufactured by Pope Scientific, Inc., Menomonee Falls, WI.

[e] The motorized valve driver with four-way "Teflon" valve is available from Hamilton Co., Reno, NV.

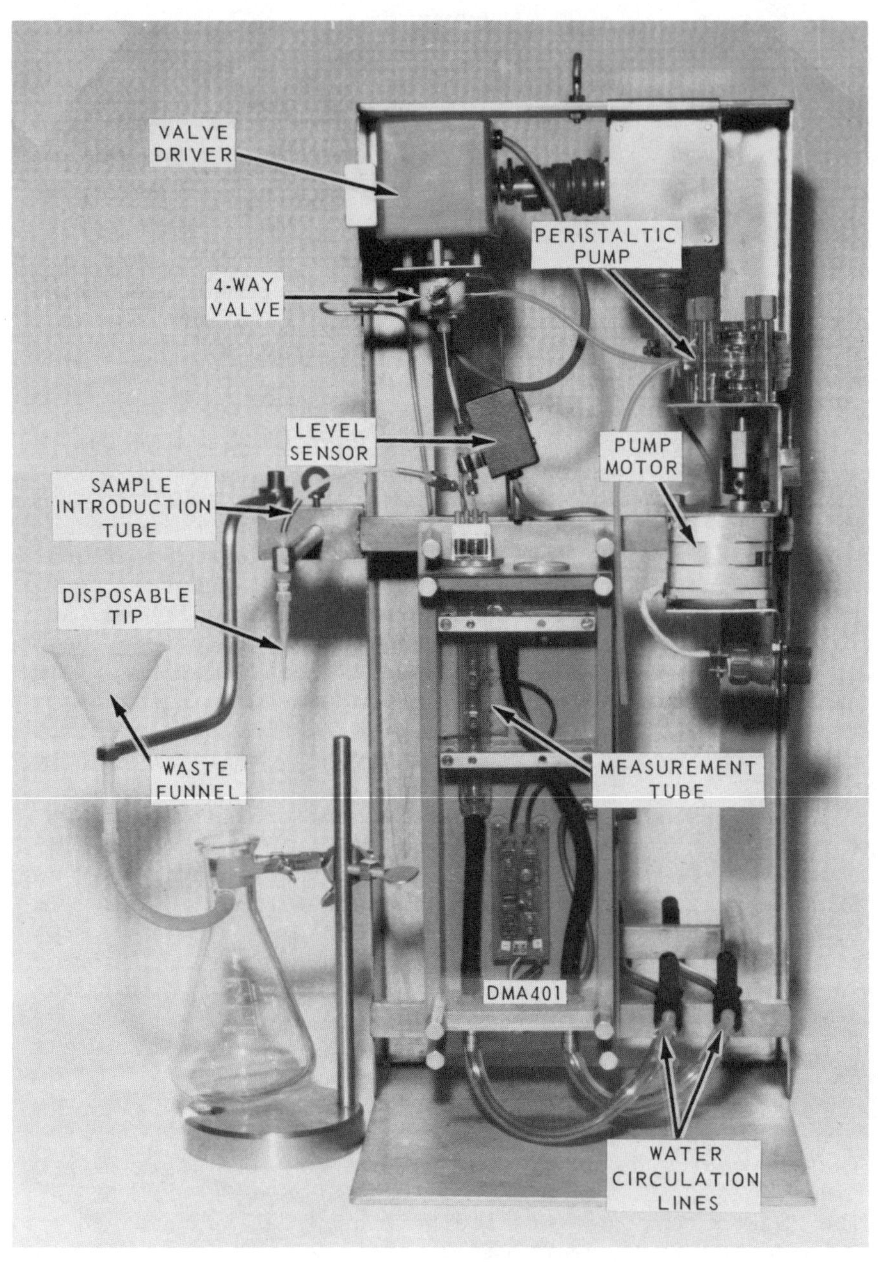

Figure 1. Remote equipment rack

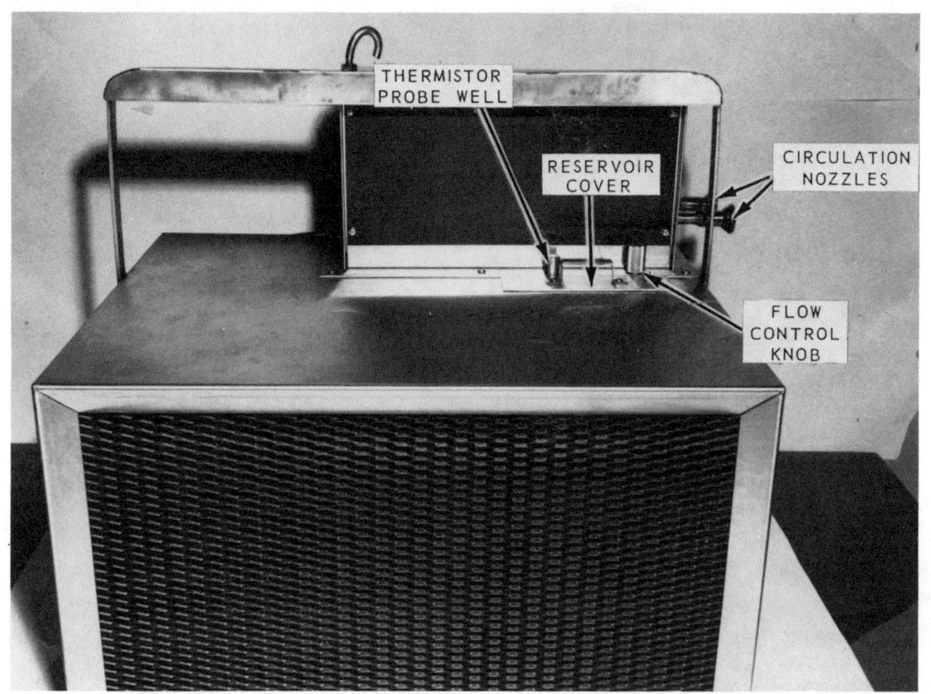

Figure 2. Remote water bath

The controlled temperature bath that was installed in the MPPF cell is a commercially modified, refrigerated recirculating bath[f] and is shown in Figure 2. It circulates distilled water through the measurement tube jacket to maintain the temperature within ± 0.02°C. A thermistor probe senses the remote bath temperature.

The external equipment cabinet with the control console is shown in Figure 3. Electrical connections to an in-cell junction box are made through curved service tubes at the sides of the cell window. The DMA 45 density meter is mounted in the top section of the panel. Only three pushbuttons are required for routine operation: FILL, EXPEL, and RINSE. The FILL and EXPEL buttons control the direction of the sampling pump. Depressing the RINSE pushbutton initiates an automated cycle which prepares the measurement tube for the next sample. The rinse cycle sequence is controlled by a digital

---

[f] The constant temperature bath is a modified RTE-4 obtained from Neslab Instrument Co., Portsmouth, NH.

331

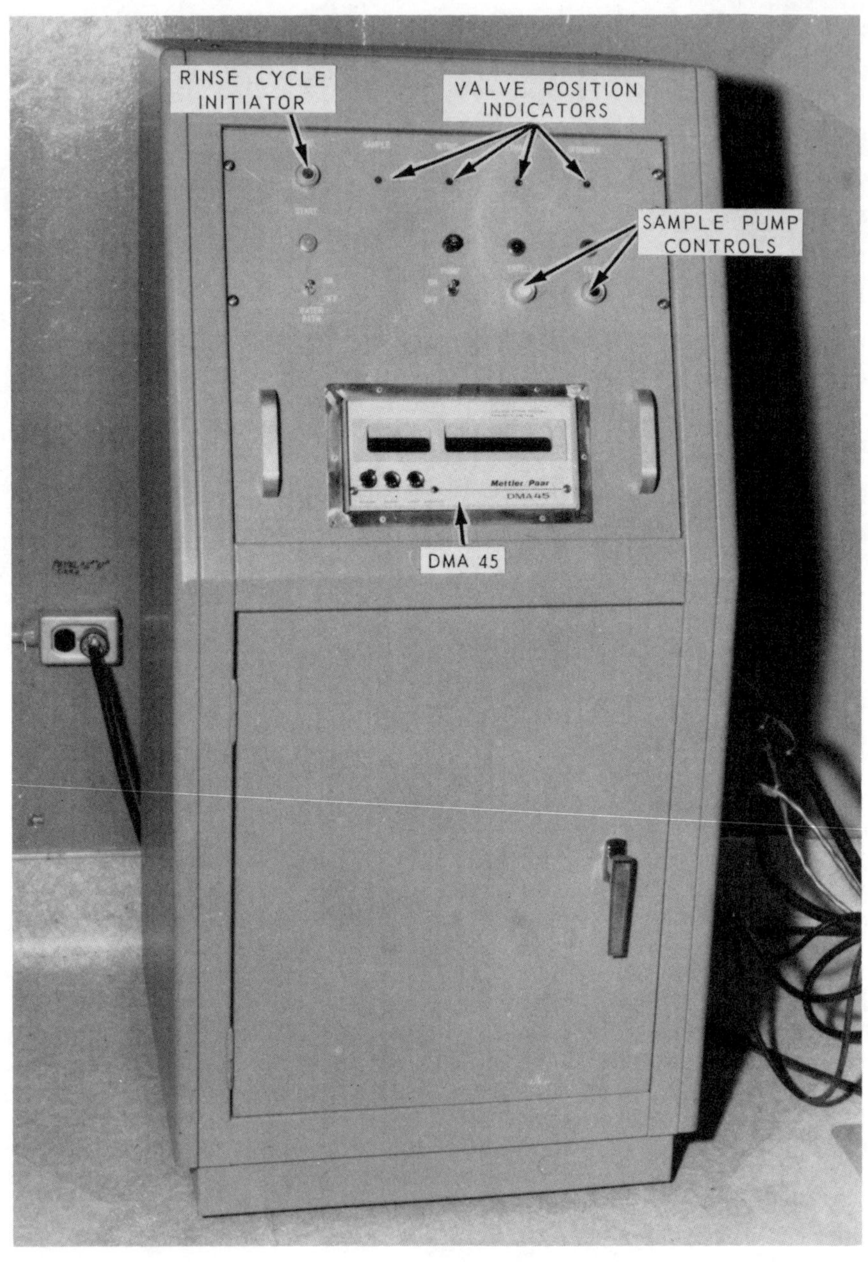

Figure 3.   External equipment cabinet

timer[g] that is also connected to LED indicators to show the
position of the four-way valve. The timer in the MPPF meter
is preset for an 8-second nitric acid wash, a 10-second
methanol rinse, and a 55-second drying step with nitrogen.
All the peripheral controller modules for the density meter
system are located in the lower section of the equipment
cabinet. These include the digital timer, liquid level
sensor, water bath controller, and the digital thermistor
thermometer[h].

After the rinse cycle is completed, the correct density
of air should be displayed by the meter before the next
sample is started. The analysis time involving a complete
cycle takes less than 5 minutes.

MOCKUP AND OPERATING EXPERIENCES

Prior to installation, all components were tested for
performance and serviceability in a mockup facility. Remote
adjustments and equipment changes with the use of a single
manipulator were demonstrated during mock up. In the design
of the system, optimum use was made of pin-guided drop-in
and single-bolt clamping for component mountings. To further
facilitate maintenance, flanged "Teflon" tubing connections
were used, which require only fingertight pressure to
achieve a proper seal. In addition, rinse supplies are
connected by ball joints to a service strip at the back of
the remote equipment rack; clamp release, quick disconnects[i]
are used at the four-way valve.

The downtime of the installed density meter system in
the MPPF has been less than 1% during its first year of
operation. The only equipment failures have been external
and where related to electronic problems in the peripheral
modules. These modules are commercially available and can be
easily repaired or replaced.

PERFORMANCE

With this new system, the precision of density measurements has improved 15 fold over the falling drop technique
and the speed of analysis has increased. The precision for a

---

[g]The digital timer valve controller is available from
Hamilton Co., Reno, NV.
[h]The thermistor thermometer is a model 5810 from Digitec,
Dayton, OH.
[i]The quick-disconnect fittings are SS model LF3000 available
from Legris, Inc., Rochester, NY.

single determination is ± 0.0002 g/cc at the 95% confidence level. This high degree of measurement precision enables the quality of replicate samples to be monitored.

Quality control synthetic samples are analyzed daily by the density meter, and the results have demonstrated excellent accuracy. Over a 1-year period the instrument remained within the $\pm 0.1\%$ control limits without recalibration or bath temperature adjustments, which demonstrates excellent long range stability. The only routine maintenance required during this period was refilling the water bath and replenishing the rinse supplies.

OTHER INSTALLATIONS

Successful operation of the remote density meter in the MPPF led to the installation of two similar instrument systems in operating intermediate analytical cells. A DMA 45 meter was modified to enable it to operate the two remote DMA 401 measurement cells and provide a significant cost saving. Their operating experiences after 9 months of continual use are similar to those for the MPPF installation.

Density meters have also been installed in four radio-bench locations; however, these have a different system configuration and are neither automated nor electrically operated.

SUMMARY

Remotely operated density meters have replaced the falling drop method for measuring the density of highly radioactive samples at SRP. These meters have significantly improved the precision and accuracy of the density results. The system for containment and remote operation was designed and fabricated at SRP. After 1 year of routine operation, this system has proven to be fast, reliable, easy to operate, and easy to maintain.

REFERENCES

1. R. D. Kelsch, A. J. Lethco, and J. B. Mellen, _Proceedings of the 20th Conference on Remote Systems Technology_, p. 235 (1972).

2. E. M. Fortsch and M. A. Wade, _Proceedings of the 22nd Conference on Remote Systems Technology_, p. 30 (1974).

3. J. P. Elder, _American Laboratory_, p. 75 (April 1978).

A FACILITY FOR THE ANALYSIS OF RADIOACTIVE SAMPLES
BY INDUCTIVELY COUPLED PLASMA SPECTROSCOPY

C. S. Homi, and R. M. Manabe, Chemical Sciences Group,
Research Department, Rockwell Hanford Operations,
Richland, Washington, USA.

ABSTRACT

A 29-channel polychromator and a scanning monochromator
are used to determine over 60 elements by inductively coupled
plasma (ICP) spectroscopy. The addition of a sample hood,
special torch containment box, and an off-gas scrubber system
will allow the analysis of low- to intermediate-level radioactive samples. The simultaneous multi-element analysis
capabilities of the ICP reduces sample handling and total
analysis time. Consequently, the potential for radiation
exposure to personnel will be reduced.

INTRODUCTION

The identification and measurement of nonradioactive
elements in nuclear waste are essential to Rockwell Hanford
Operations (Rockwell) waste management programs. Processes
which isolate and concentrate radionuclides are dependent on
the nonradioactive constituents of the chemical matrix.
Measurement of toxic nonradioactive elements are also important for treatment and disposal of cold wastes.

Major to trace-level cation measurements in radioactive
samples are currently done by an atomic absorption spectrometry (AAS) system. This system has provided analyses of
radioactive samples since 1972 but will not meet future pro-
grammatic and radiological safety criteria.[1] These
requirements include more accurate and precise analysis, more
safety and contamination control redundancy plus reduction in
personnel radiation exposure limits.

This report discusses how the implementation of an ICP
system will satisfy future requirements.

DISCUSSION

Simultaneous, multi-element analysis by optical emission
spectroscopy has been applied to a large number of sample
types for many years. The analysis of liquid samples by
previous optical emission sources has been limited by poor
precision and low sensitivity. The development of the ICP
source changed this by providing the emission spectroscopist
with a more stable and efficient excitation source for solution analysis.

The ICP excitation source is contained in a specially
designed quartz tube. A spark from a Tesla coil starts the
ionization process. The plasma is sustained by absorbing
electromagnetic energy supplied by a radio frequency generator. Temperatures in excess of 10,000 K are produced in
certain regions. Sample aerosol is injected into the plasma
via the innermost tube and passes through the center of the
intense toroidal-shaped plasma torch (Figure 1). The high
temperature and inert environment of the ICP efficiently
decomposes the sample aerosol into free atoms and ions. Minimal chemical recombination and interaction of ions or atoms
occur in the ICP. These reactions are a major source of
potential error in flame atomic absorption spectroscopy (AAS).

Spectral emission from the region 15-18 mm above the
induction coil is observed by the spectrometers (Figure 2).
Diffraction gratings in both spectrometers resolve the emission into elemental line spectra. The polychromator has
29 fixed exit slits in the diffracted beam plane located at a
characteristic emission line position for each element. The
scanning spectrometer or monochromator can be adjusted to any
characteristic line in the ultraviolet or visible spectral
regions. Over 30 additional elements not accessible by the
polychromator can be detected on a single element basis. The
monochromator can also be used in the scanning mode to record
emission spectra and to detect possible spectral interferences from complex sample types. The analytical channels
programmed into the polychromator and their approximate
detection limits are shown in Table 1. Additional elements
that can be measured with the scanning monochromator are
shown in Table 2.

Raw signal data from both spectrometers are transmitted
to the dedicated instrument computer where the appropriate
calibration is applied to obtain final results and real time
quality control and data validation are provided.

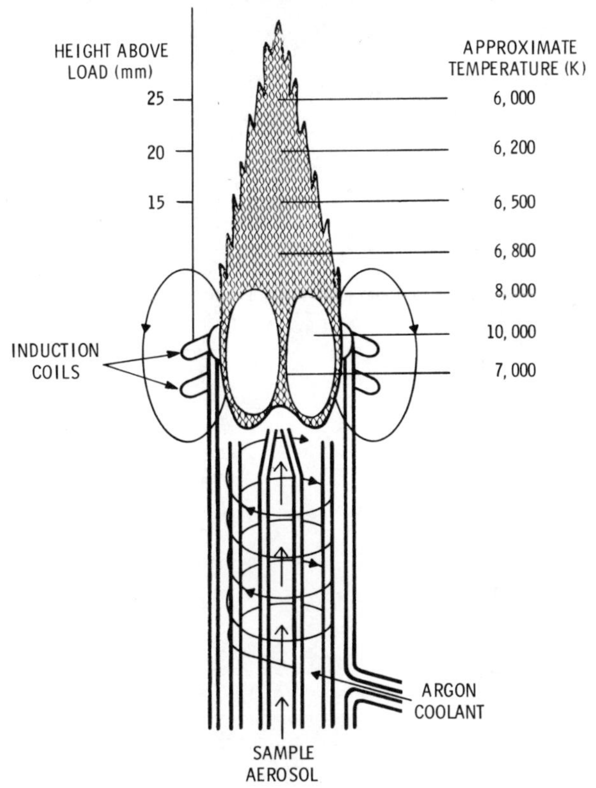

FIGURE 1. Plasma Torch.

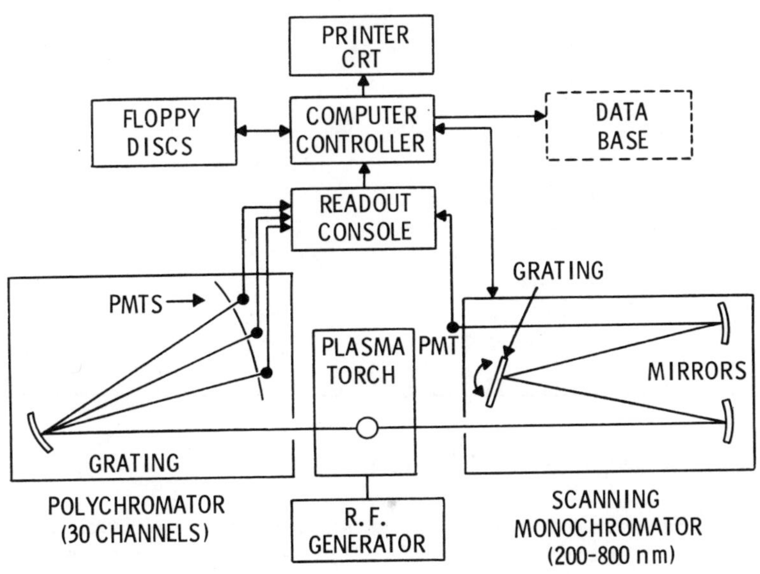

FIGURE 2. Plasma Analytical System.

	TABLE 1 Polychromator Elements		TABLE 2 Single Element Capabilities of Scanning Monochromator
Element	Approx. Detection Limit μg/mℓ	Element	Approx. Detection Limit μg/mℓ
Ag	0.005	As	0.4
Al	0.01	Au	0.1
B	0.005	Be	0.005
Ba	0.005	Ga	0.05
Bi	0.10	Ge	0.2
Ca	0.10	Hf	0.1
Cd	0.01	In	0.1
Ce	0.03	Nb	0.1
Co	0.01	Pt	0.1
Cr	0.003	Rh	0.02
Cu	0.002	Sb	0.2
Fe	0.002	Sc	0.03
K	0.10	Se	0.1
La	0.01	Ta	0.1
Mg	0.001	Te	0.1
Mn	0.02	Th	0.03
Mo	0.03	Tl	0.2
Na	0.02	U	0.03
Nd	0.05	V	0.05
Ni	0.02	W	0.02
P	0.2	Y	0.002
Pb	0.10	Lanthanides	0.05
Pd	0.05		
Si	0.01		
Sn	0.2		
Sr	0.0002		
Ti	0.005		
Zn	0.1		
Zr	0.005		

ANALYSIS OF RADIOACTIVE SAMPLES

The analysis of radioactive samples by both AAS and ICP techniques requires special facilities. Radioactive gases and solutions must be totally contained and disposed of in a safe manner.

Atomic Absorption Scrubber

In the present AAS facility, off-gases from the flame are
collected in a chimney above the burner and drawn through two
perforated scrubber plates (Figure 3). The plates are
sprayed with water to moisten and wash the off-gases. Excess
water from the spray nozzles returns to the reservoir and the
moistened gases are pulled down and up through the section
filled with berl saddles. The moist gases are then drawn up
through a condenser and exhausted through HEPA filters into
the building air filter system.

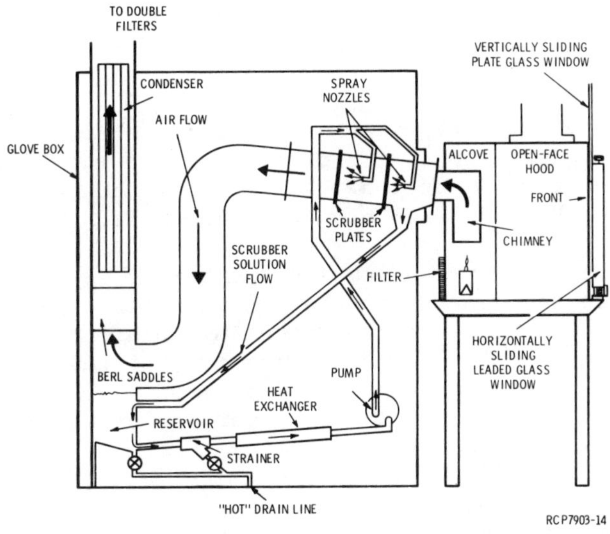

FIGURE 3. Existing Atomic Absorption Scrubber System

Access to the horizontal section of the burner chimney
was provided because preliminary tests showed the greatest
deposition of radioactive materials occurs here. Removable
aluminum foil liners protect this section and are changed
once a week. The vertical part of the chimney is removable
and replaced when necessary.

The scrubber is housed in a glovebox and all valves are
remote controlled from the outside. As a result the system
is cumbersome to maintain and operate. On several occasions
carbon particles from the flame source have deposited in the
scrubber solution and plugged the spray nozzles. This allows
passage of radioactive materials out of the system and onto
the first effluent air filter. Extra precautions must be

made to minimize radioactive buildup on this filter because
changing it is not only costly but increases the potential
for personnel radiation exposure.

Inductively Coupled Plasma Scrubber

Off-gases from the ICP torch will be directed into the
bottom of a vertical scrubber (Figure 4). These gases will
immediately come in contact with water from the lower spray
nozzle and pass up through the section filled with pall
rings. Water from the upper spray nozzle will flow counter-
current to the off-gases in the pall ring section. Excess
water returns to the reservoir where it can be recirculated
to the spray nozzles or released to the contaminated drain.
The system can be flushed by adding water or dilute acid to
the reservoir and releasing the solution after circulation
through the scrubber to the contaminated drain.

Experience with the existing AAS scrubber system indicates the greatest deposition of radioactive material will
occur in the chimney section. As a result the entire chimney
elbow inside the ICP torch box is replaceable.

After the system is assembled and before any radioactive
solutions are aspirated into the plasma torch, cold tests
will be made. Nonradioactive strontium solutions will be
aspirated into the torch and the interior of the torch box
and chimney will be analyzed for deposited strontium. A
removable filter will be placed at the top of the scrubber
during these tests to check for break-through. Overall efficiency of the total system will be calculated from these
tests. The detection limit of 0.2 μg/l for strontium will
provide excellent sensitivity for these studies.

Portable shielding will be used to protect personnel from
radioactive samples in the open-faced hood. Mirrors and long
handle tongs will be used to insert the aspirator tube into
the samples. It is planned to maintain exposure at the front
of the hood to near background levels. Intensely radioactive
samples will be diluted to reduce exposure. Radiation detectors will be positioned at locations based on the results of
the cold tests. Final testing will have to be made with
low-level radioactive solutions.

CONCLUSIONS

Advantages of the ICP System

● <u>Simultaneous multi-element analysis</u>. All the elements
listed in Table 1 and any single element from Table 2 can be

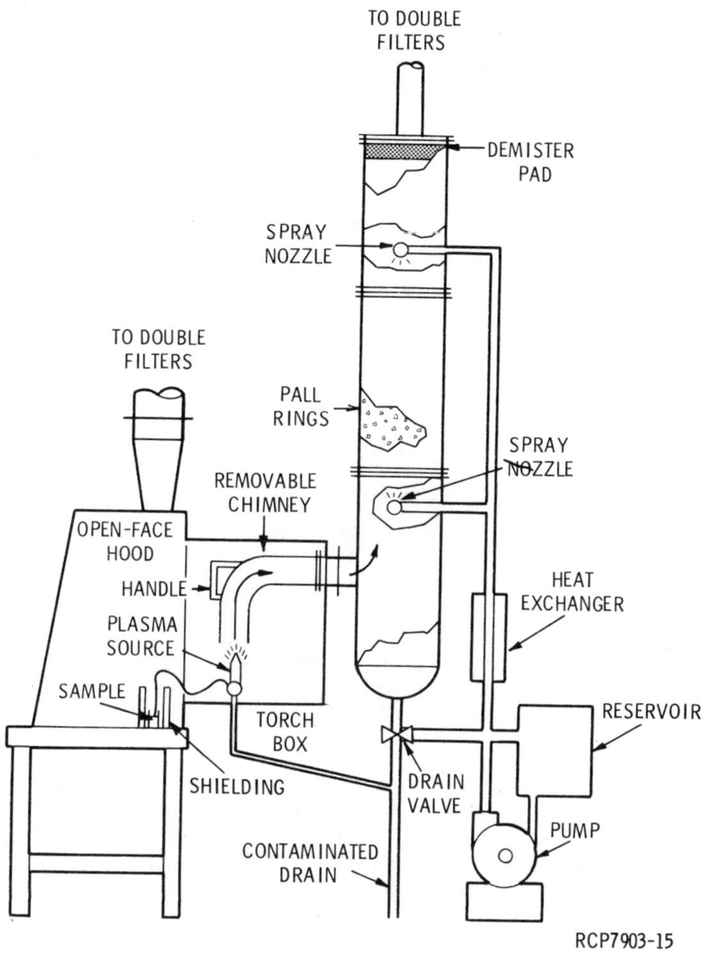

FIGURE 4. Inductively Coupled Plasma Scrubber System

measured simultaneously. As a result, more information is
provided to the customer and sample handling time is greatly
reduced. Based on past experience for a typical sample, a
10-fold reduction in handling time can be realized. Also,
much less radioactive material per sample is introduced into
the off-gases and liquid containment systems.

● <u>Large dynamic analytical range</u>. Changes in the analyte
concentration between samples can range beyond four orders
(10^4) of magnitude and still remain within the analytical
calibration. This reduces the need for dilution, resulting
in fewer sample handling operations.

• <u>Fewer analytical interferences</u>. The high energy of the
plasma eliminates most chemical effects encountered in flame
atomic absorption. Therefore, higher reliability can be
attached to the measurements.

• <u>Sensitivity comparable to flame atomic absorption</u>. In
general, sensitivity is as good as, if not better than, flame
atomic absorption.

• <u>Elimination of combustible and carbon producing gases</u>.
Flame atomic absorption requires combustible gas mixtures
such as air-acetylene or nitrous oxide-acetylene. Inert,
noncombustible argon gas is used for the ICP torch.

• <u>Better scrubber system</u>. The ICP scrubber will be more
efficient and easier to maintain than the AAS system. Since
no organic fuels are used for the torch plugging of the spray
nozzle with carbon particles will not occur.

Disadvantages of the ICP System

• <u>Stray light interferences</u>. Stray light scatter inside the
spectrometer is caused by the high intensity of the plasma
torch. The scatter increases background in certain regions
of the spectrum. Bandpass filters on each phototube reduce
this problem to tolerable levels.

• <u>Nebulizer effects</u>. The argon flow rate through the
nebulizer section has to be maintained at approximately
1.0 l/min to prevent extinguishing the torch. As a result
small orifice nebulizers, which are prone to plugging, must
be used. Nebulizer design development is expected to reduce
this disadvantage.

• <u>Stability of standards</u>. Due to the multi-element capability of the ICP, the most efficient method for standard
preparation is by mixing several elements (up to 10) together
in a single calibration series. Selection of element groups
for these standards may be difficult because of conflicting
chemistries. Also, the stability of these standards is less
than that of a single element.

REFERENCE

1. J. M. Harnly, "A Facility for the Analysis of Intensely
    Radioactive Samples by Atomic Absorption", <u>AEC Report</u>,
    Atlantic Richfield Hanford Company, Richland, WA,
    <u>ARH-SA-146</u>, February 1973.

A HIGH-CAPACITY NEUTRON ACTIVATION ANALYSIS FACILITY*

R. C. Hochel
E. I. du Pont de Nemours and Company
Savannah River Laboratory
Aiken, South Carolina 29801

ABSTRACT

A high-capacity neutron activation analysis facility,
the Reactor Activation Facility, was designed and built and
has been in operation for about a year at one of the
Savannah River Plant's production reactors. The facility
determines uranium and about 19 other trace elements in
hydrogeochemical samples collected in the National Uranium
Resource Evaluation program, which is sponsored and funded
by the United States Department of Energy, Grand Junction
Office. The facility has a demonstrated average analysis
rate of over 10,000 samples per month, and a peak rate of
over 16,000 samples per month.

Uranium is determined by cyclic activation and delayed
neutron counting of the U-235 fission products; other
elements are determined from gamma-ray spectra recorded in
subsequent irradiation, decay, and counting steps. The
method relies on the absolute activation technique and is
highly automated for round-the-clock unattended operation.

INTRODUCTION

The Reactor Activation Facility (RAF) is a highly automated neutron activation analysis (NAA) system. The facility
is installed at the Savannah River Plant's C-Area production
reactor and has been in operation since September 1978. The

---

*The information contained in this article was developed
 during the course of work under Contract No. AT(07-2)-1
 with the U.S. Department of Energy.

RAF was built to provide NAA for the large number of hydrogeochemical samples collected in the National Uranium Resource Evaluation (NURE) program, which is sponsored and funded by the United States Department of Energy, Grand Junction Office. The Savannah River Laboratory (SRL) is one of three laboratories participating in the hydrogeochemical and stream sediment reconnaissance portion of the NURE program, and is responsible for sampling, analyzing, and reporting on an area of about 1,500,000 square miles in 30 eastern and 7 western states.

The RAF is intended to provide analyses for an estimated 400,000 samples to be collected in SRL's portion of the NURE program, and replaces a smaller pilot scale facility which was operated from September 1975 to November 1977. The facility has a demonstrated average analysis rate of over 10,000 samples per month, and a peak rate of over 16,000 samples per month. Uranium and up to 19 other elements are determined in sediment samples, and in ground and stream water samples after concentration on ion-exchange resin. Sediment and resin samples are prepackaged into special polyethylene irradiation capsules. Up to 2400 capsules can be loaded into the system for 72 hours of round-the-clock unattended operation.

Uranium is determined by cyclic activation and delayed neutron counting of the U-235 fission products, other elements are determined from gamma-ray spectra recorded in subsequent irradiation, decay, and counting steps. The method relies on the absolute activation technique. Details of the absolute activation technique are given by MacMurdo and Bowman,[1] and additional information about installation, checkout, and calibration of both the pilot facility and the RAF are documented in SRL-NURE quarterly and semi-annual reports.[2]

RAF LAYOUT AND PNEUMATIC TRANSPORT SYSTEM

A schematic of the RAF is shown in Figure 1. The three modules shown in the schematic are located about 20 feet above and 60 feet laterally from an irradiation assembly which butts up to the reactor containment wall. A very well moderated flux of about 5×10^{12} n/cm -sec is available.

The Control Module contains a computer and assorted process control and data acquisition instrument components. As its name implies, it is the control center of the entire facility. Similarly, the Counting Module contains the various neutron and gamma detectors required for sample analyses. The Counting Module also contains the bulk of

344

the RAF's pneumatic transport system including connections
to the six identical irradiation positions within the
irradiation assembly. The Storage Module contains two
identical sample storage devices called stacks. The stacks
provide computerized control of all sample loading and un-
loading, and storage of irradiated samples for intermediate
decay.

The pneumatic transport system is a network of one-inch
I.D. polyethylene transport tubing and four-way (4W) diverter units. Each terminus point of the transport system is
connected by pneumatic valves to a propulsion air manifold
and an exhaust air manifold. Transport of a sample capsule
is a simple matter of opening a propulsion valve at the
sending end, and an exhaust valve at the receiving end.
Capsules are tracked for position and time by photodetectors
located at all terminus points.

Two ports on each of the two stacks (S1 and S2) are
connected through Diverter 4W1 to Diverter 4W2. Diverter
4W2 is used to select either gamma-neutron Detector Station
D1 or D2. Diverters 4W3 and 4W4 connect D1 and D2 to
Irradiation Positions R1-R3 and R4-R6, respectively. One
port on each stack is connected to a turnaround station (T)
by Diverter 4W5. Once a capsule is at the turnaround, any
of the Gamma Counting Stations G1-G4 or G5-G8 can be
accessed through Diverters 4W6 and 4W7, respectively.

Control Module

Hardware of the Control Module is shown schematically
in Figure 2. The heart of the RAF is the Systems Engineering Laboratories (SEL) 32/55 computer. It is a full 32-bit
machine with 128K words of memory. Besides the usual I/O
devices, are two 80-megabyte moving-head disks. The disks
are used to store programs and data, and increase effective
system memory considerably.

The computer performs all process control and data
reduction tasks.[3] The real-time peripheral unit provides
bidirectional interfacing between the computer and all of
the pneumatic components of the Counting and Storage
Modules. Action commands by the computer are answered by
responses such as tripped photodetectors or set limit
switches. Twelve scales and ten pulse-height analyzers
(all of SRL design) support the various neutron and gamma
detectors.

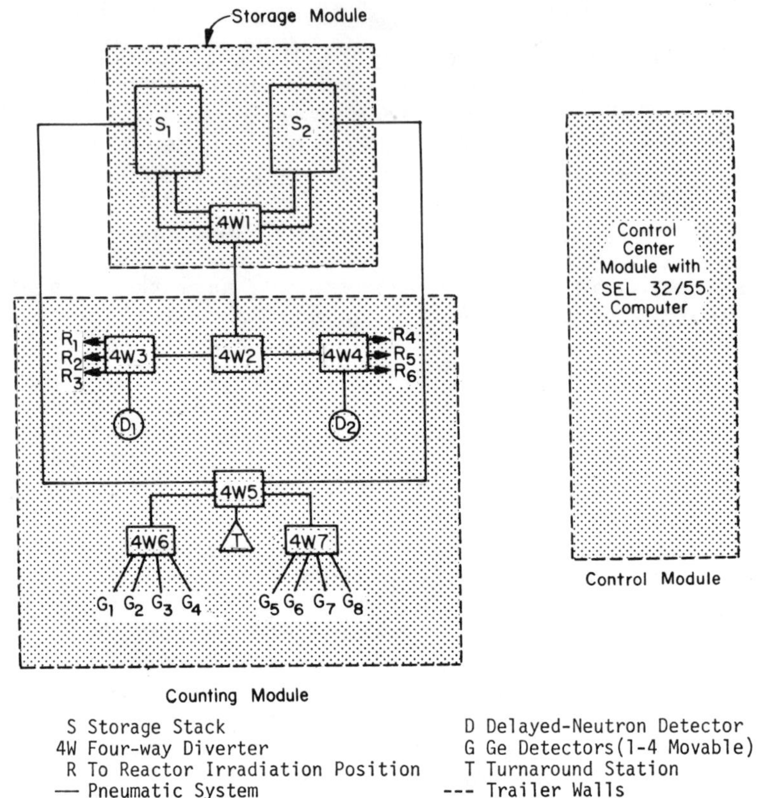

Figure 1. Hardware configuration for operating system
(stack loading/unloading system not shown)

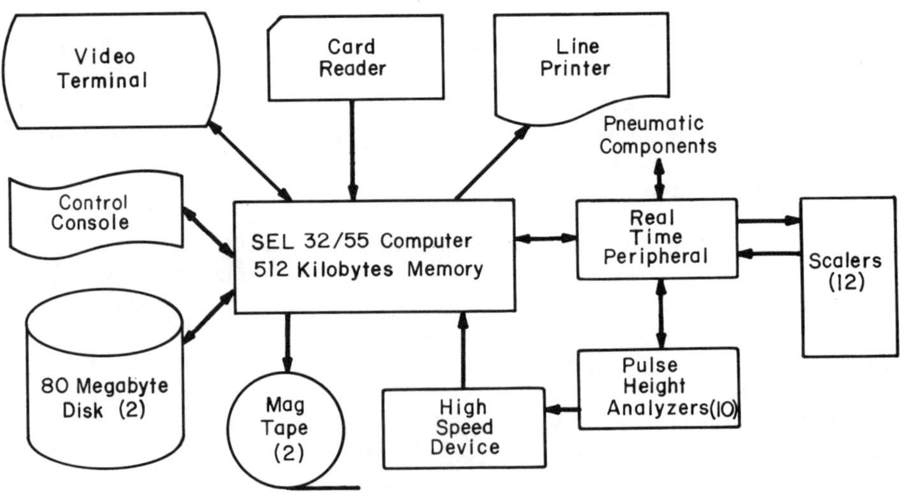

Figure 2. Configuration for SEL 32/55 computer

Counting Module

The Counting Module (Figure 1) contains the neutron and gamma detectors which provide for the analysis of uranium and other elements, respectively.

Positions D1 and D2 are identical combined neutron and gamma counting stations. The neutron detectors each consist of an annulus of 10 stainless steel tubes containing boron trifluoride. The gamma detectors are intrinsic or high-purity germanium devices each of about 10% efficiency and 2.0 keV resolution. A sample at detector station D1 or D2 is counted simultaneously for delayed neutrons (uranium) and gamma-rays (short-lived activation products).

The gamma detectors at Gamma Counting Stations G1-G8 are also intrinsic devices all of about 8% efficiency and 1.8 keV resolution. Stations G1-G4 are variable-geometry counting stations. Atop the detectors are vertical drop tubes. Each drop tube can provide up to five different counting geometries by extending or retracting stops in the tube which vary the sample-to-detector distance. Stations G1-G4 are used primarily for measuring intermediate-lived activities present in samples 10 minutes after irradiation.

Stations G5-G8 each have a single counting geometry which is identical to those of Stations G1-G4 with all their stops retracted (i.e., closest counting geometry). Stations G5-G8 are used primarily for measuring long-lived activities present in samples seven days after irradiation.

Storage Module

Each of the two stacks in the storage module can hold about 6000 samples; both are mechanically and operationally identical. A stack is a 16 x 16 array of vertical storage tubes held on 1½-inch centers by a precision lattice plate. Each tube can hold 25 samples. In the bottom of each tube is a grabber, a circular device of finger-like springs. Spreading the fingers allows a sample to fall from the bottom of the tube. Conversely, a sample can be pushed up into the tube (by a piston) to a position where it is held by the fingers.

A store/retrieve or Z-motion device under the lattice plate can either store or retrieve the bottom sample in any of the stack's 256 tubes. Once a sample is in the Z-motion, it can be moved to any tube, or pneumatically fired from any of the stack's three exit/entry port tubes to the

Counting Module. The Z-motion rides on the ways of a Y-axis
track, which in turn rides the ways of an X-axis track.
Motion along the X- and Y-axes is controlled by four
binary-incremented air cylinders with strokes of 1½, 3, 6,
and 12 inches. By extending or retracting the appropriate
air cylinders, the computer can position the Z-motion under
any tube. Feedback from the air pressure that operates the
cylinders signals their states to the computer through pressure switches. A photodetector senses the presence or
absence of a sample in the Z-motion device.

REFERENCES

1. K. W. MacMurdo, W. R. Bowman, Nuclear Instruments and
    Methods, 141 (2), 299-306 (1977).

2. Savannah River Laboratory Quarterly Report, Hydrogeo-
    chemical and Stream Sediment Reconnaissance - Eastern
    United States, National Uranium Resource Evaluation
    Program, E. I. du Pont de Nemours and Company, Savannah
    River Laboratory, Aiken, South Carolina. Published
    in quarterly issues January-March 1975 through January-
    March 1978; then changed to semi-annual report. Reports
    available on microfiche from DOE-GJO.

3. W. W. Bowman, "Rapid Analysis of Germanium Spectra,"
    Nucl. Instr. Meth. 96, 135 (1971).

MASS SPECTROMETRY

THE MASS SPECTROMETRIC ANALYSIS OF NANOGRAM LEVELS
OF RUTHENIUM

J. E. Delmore, Exxon Nuclear Idaho Company, Inc.,
Idaho Falls, Idaho 83401

ABSTRACT

Previous investigations have demonstrated the possibility
of using ruthenium originating from the spontaneous fission of
^{238}U for geochronological purposes. Because the abundance of
ruthenium is low in geologic specimens and isotopic data are
required, an indepth study of the production of steady mass
spectrometer ion beams from nanogram amounts of ruthenium was
undertaken. The results of this continuing mass spectrometric
investigation are presented with some illustrative data.

INTRODUCTION

This laboratory has been analyzing microgram level samples
of various fission products, including ruthenium (Ru), by isotope dilution mass spectrometry for approximately fifteen
years.[1] Based on this experience, a program was initiated in
1974 to analyze ore from the ancient natural reactors of Oklo.[2]
Nine fission product elements were analyzed; three (Rb, Sr, Ba)
showed natural isotopics, five (Nd, Sm, Mo, Zr, Ce) were mixed
fission product-natural isotopics, and one (Ru) had an isotopic
composition corresponding to that of essentially pure, stable
fission product material.

This led to the speculation that if Ru is that rare and
immobile in nature, perhaps ^{238}U spontaneous fission produced
Ru could be measured in uranium (U) ores. This in turn could
be used to study the stability of Ru in nature and possibly
as an age dating tool. Subsequent measurements[3] of non-Oklo
ore samples indeed showed that fission product Ru was present,
and that the predominant source of Ru was from the spontaneous
fission of ^{238}U.

The analytical method had to be improved before this type
of measurement could be utilized for some practical purpose.
The goal, therefore, was to improve the method such that data
good to at least 1% could be obtained on samples no larger
than five nanograms of total Ru. This level of performance
is routinely exceeded for many elements, Pu being an example.
However, the physical and chemical characteristics of Ru make
these objectives more difficult to achieve. While the early
data was good to 5-10%, easily demonstrating the existence of
^{238}U spontaneous fission origin Ru in a wide variety of U
ores, better accuracy was desired to more fully evaluate the
use of Ru as a geochronology tool.[3,4,5]

With this potential application in mind, a program was
initiated to improve the Ru analytical method. The basic
phases of the method are: a) sample dissolution, b) separation chemistry and c) mass spectrometry. After sample dissolution, the Ru is distilled as RuO_4^3, which gives a clean
separation for Ru at the 1 PPB level for all samples except
those with high organic content. The major difficulty is the
generation of ion beams of sufficient intensity and stability
to allow the collection of the necessary mass spectrometric
data from nanogram sized samples. A 90° sector tandem magnet
mass spectrometer equipped with pulse counting and a computerized data system was used. The criteria established for acceptable data were 100-200 counts/sec on the smallest peak of
interest, which is about 0.5% of the total sample, and an
intensity stable to within 5% per minute.

THE THERMAL IONIZATION PROCESS

Ruthenium is ionized in the mass spectrometer by the
positive thermal ionization process. This involves evaporating Ru from a metallic surface that has a high affinity for
electrons. This affinity for electrons is measured by the
work function (W). The material which is used almost universally in the thermal generation of positive ions is rhenium (Re), because it is malleable, has good strength at high
temperatures, and a desirable work function (4.7-5.1 volts).
The work function is a surface property, which varies with the
crystalline face or the polycrystalline character of the surface. It is also strongly affected by surface cleanness. The
ability of an element to give up an electron to such a surface
is measured by its ionization potential (IP), which for Ru is
7.36 volts.

When the assumption is made that the only species leaving
the filament are the positively charged and neutral species of
the element, the Saha-Langmuir equation may be applied:

$$n+/n° \propto \exp (W-IP)/kT = \exp \delta/kT$$

Where n+ and n° are the number of positively charged and neutral species, respectively, W and IP the work function and ionization potential, respectively, δ is the difference between W and IP, k is the Boltzmann constant, and T is the absolute temperature. A more complex form of this equation exists, but efforts to verify it experimentally[6] have not been successful. This simplified form of the equation[7] has been verified experimentally,[6] including the change in temperature dependence with δ. The sign and magnitude of δ strongly effects the temperature dependence of n+/n°. The δ for the Ru-Re system is -2.36 volts, which indicates a large increase in ion production with increased temperature. A positive δ would give the opposite temperature dependence, as is the case for cesium (IP = 3.89 volts, δ = +1.11 volts).

There are two broad categories of thermal ion sources in widespread use; the single filament and the multiple filament arrangements. When using the single filament the sample is loaded directly on the ionizing filament. The major advantage of this source is that all of the sample contacts the ionizing filament. The drawbacks are: a) the possibility of the sample poisoning the work function of the ionizing surface and, b) the inability to independently vary the temperature of the ionizing surface and the vapor pressure of the element.

When using the multiple filament arrangement, the sample is loaded on one or two of the side filaments, which are heated to vaporize the sample in the direction of the ionizing filament. The ionizing filament is lined up with the optics axis of the mass spectrometer. This source has the advantage of permitting independent adjustment of the ionizing filament temperature and the vapor pressure of the element to be ionized. The major disadvantages are: 1) not all the material vaporizing from the side filament strikes the ionizing filament, and 2) a compound containing the element of interest may be too volatile to permit a sufficient residence time on the hot filament for ionization or for both dissociation and ionization of that element. Uranium is an excellent element to analyze by the multiple filament technique, because the Re filament at 2100°C can both dissociate and ionize the uranium oxide vapor species to yield U$^+$ ions with good efficiency.

There are, however, two major problems when Ru is ionized using the multiple filament technique. The first problem is that the multiple filament source requires a very hot center filament to ionize a high work function element efficiently.

This requirement presents another problem which is, that at
high ($\sim$2100°C) temperature, a measureable Mo spectra appears
which causes a serious spectral interference with Ru. This
in itself precludes the use of the multiple filament technique.

The second major difficulty with the multiple filament
source is that it normally does not produce Ru ions with an
efficiency that is analogous to other elements. An understanding of this phenomenon helps to explain some other an-
omolies as well. The compounds of the higher oxidation states
of Ru are volatile, and if a volatile compound strikes a very
hot surface, it will either dissociate or re-evaporate. With
uranium oxides, dissociation and ionization occur because the
oxides have a much lower vapor pressure which allows a longer
residence time on the filament; this results in a highly
efficient ion source. In contrast, the various Ru species we
have worked with apparently re-evaporate, or possibly dissociate and then evaporate without ionizing. It is probable
that Ru would ionize more efficiently in the multiple filament
source if it were present as the metal on the side filament;
this method would be worth exploring were it not for the spectral interference problem of Mo from the hot center filament.

To date the single filament ion source has been far more
successful for Ru than the multiple filament ion source, provided that certain exacting conditions are met. These conditions are: a) using Re filaments that are low in Mo contamination and intensely prebaked; and b) the adjustment of
sample handling and mounting conditions to prevent the formation of the higher oxidation states of Ru. The close control
of these parameters has led to the most significant improvements in sensitivity, and also offers the greatest potential
for future refinements.

The presently recommended procedure entails loading Ru
from a strong HCl or HBr solution directly onto a single
filament. The solution must never go dry during any phase
of the sample handling until the last drop evaporates on the
filament. Ruthenium will be lost to the container walls if
the solution goes dry during sample handling, and if repeated
drops are evaporated to dryness on the filament a significant
loss in Ru ion intensity results. The solution must be evaporated very slowly on the filament to obtain the best sensitivity. A 10 μL drop must be dried at a temperature that takes
about forty-five minutes. Faster drying rates with higher temp-
atures causes a significant loss in sensitivity.

These results suggest that different chemical forms of
Ru are being produced on the filament as a function of drying
time or temperature. One possible explanation is that the Ru,

which is predominantly in the +3 valence state in concentrated
HCl or HBr, is converting to a higher oxidation state. The
very gentle drying process may be preserving the +3 oxidation
state. Higher oxidation state Ru compounds are more covalent,
and hence more volatile. The greater volatility results in
Ru vaporizing in the ion source at lower temperatures, giving
reduced ion production. The increase in covalency means a
decrease in the ionic character of the chemical bonds, and
hence a lower probability of ionization.

The logical extension of this work is the reduction of
Ru to the metal on the filament. Reduction was attempted in
two different ways: a) with hydrogen reduction, and b) with
the resin bead method. Both techniques produced sensitivity
increases of several fold, but both also had drawbacks that
prevent application unless further refinements are made. The
problem with the hydrogen reduction was a migration of Mo impurities in the Re filament to the surface, negating the reduction in Mo accruing from the extensive pre-baking procedure.
A major problem with the resin bead method is the requirement
for the chemistry and techniques for loading 10 to 50 ng
quantities of Ru nearly quantitatively onto a single resin
bead.

DATA AND CONCLUSIONS

Based on the work performed to date, the preferred method
is the conventional single filament method, providing the necessary precautions are taken during the sample handling and
drying steps. The results given in Table 1 are representative
of the analysis of a uranium ore sample which contained $\sim$40 ng
of Ru prior to separation chemistry. Mass 100 is used to
correct for natural Ru (^{100}Ru is not produced in fission)
after subtraction of ^{100}Mo based on the analysis of ^{95}Mo.
This correction must be small for the data to be acceptable.
The data in Table 1 easily exceeds the criteria of 1% data
for the fission product nuclides.

Developments that would provide further significant improvements in the Ru analysis are: a) Re filaments with sig-
gnificantly lower Mo content, and b) a technique for reducing
Ru to the metal on the filament without drawbacks of the
methods tested to date. In spite of these development needs,
the current method, in its state of development, is adequate
to allow continued evaluation of the use of Ru from ^{238}U
spontaneous fission for geochronology studies.

TABLE 1.   FISSION ORIGIN Ru FROM U ORE DEPOSIT

Scan	99	100	101	102	104
1	25.42	1.14	27.84	29.64	15.96
2	25.50	1.19	27.68	29.70	15.92
3	25.31	1.09	27.72	30.13	15.75
4	25.29	1.16	27.83	29.86	15.85
5	25.26	1.16	27.82	29.93	15.83
6	25.37	1.02	27.61	29.95	16.05
7	25.36	1.02	27.67	30.01	15.95
Average	25.36	1.11	27.74	29.89	15.90
St. Dev.	.03	.03	.04	.06	.04
Net Ru F.P.	26.42	0.00	28.58	29.50	15.50

REFERENCES

1.   F. L. Lisman, R. M. Abernathy, W. J. Maeck, J. E. Rein;
Nuclear Science Engineering: 42, 191-214 (1970).

2.   W. J. Maeck, F. W. Spraktes, R. L. Tromp, J. H. Keller;
The Oklo Phenomenon, IAEA Pub. 405 (1975) pg 319.

3.   W. J. Maeck, J. E. Delmore, R. L. Eggleston,
F. W. Spraktes; Natural Fission Reactors, IAEA Pub. 475
(1978) pg 521.

4.   W. J. Maeck, K. E. Apt, G. A. Cowan; Natural Fission
Reactors, IAEA Pub. 475 (1978) pg 541.

5.   W. J. Maeck; Proc. Conf. Mass Spec & Allied Topics,
pg 434 (1977).

6.   A. Persky, E. F. Green, A. Kuppermann; J. Chem. Phys.;
49, 5, 2345-2357 (1968).

MASS SPECTROMETRIC ISOTOPE METROLOGY OF URANIUM
ON RESIN BEADS

J. D. Fassett, W. R. Kelly, L. A. Machlan, and
L. J. Moore, Inorganic Analytical Research Division,
National Bureau of Standards, Washington, D. C., U.S.A.

ABSTRACT

The potential for making high-precision, high-accuracy
isotopic ratio measurements on submicrogram amounts of
uranium absorbed on anion-exchange resin beads has been
investigated utilizing a single stage thermal ionization
mass spectrometer with pulse counting detection. The errors
inherent in this measurement process have been evaluated.
These errors include isobaric interferences, contamination,
ion scattering and baseline correction, mass discrimination,
and isotopic fractionation. A mass spectrometric procedure
for loading and analyzing beads has been developed. A
single bead is loaded onto a rhenium V-filament wetted with
cyclohexanone. At 1660 °C a stable U^+ signal of greater
than 100,000 counts per second is produced from a nanomole
of uranium. Analyses of the NBS SRM's U-100 to U-900
indicate that the mass spectrometric errors can be minimized
and fractionation pattern reproduced such that the $^{235}U/^{238}U$
ratio can be measured to 0.1 percent (95% confidence limit).
For minor isotopes the precision is counting statistics
limited.

INTRODUCTION

The resin bead sample loading technique in thermal
ionization mass spectrometry was first described in 1970.[1]
The technique as has been more recently developed by Oak
Ridge National Laboratory (ORNL)[2-4] has been singled out by
the international safeguards community for its promising
advantages of facilitating shipping and minimizing both
sample size and sample handling in the analysis of uranium

and plutonium. The potential danger that could arise from
accidental or purposeful diversion of these substances
demands increased accuracy in the procedures used for their
accountability. The burden is now placed on the analytical
chemist to provide the routine, accurate, and precise analyses of these materials. The advantages of the resin bead
technique result from (1) the selective absorption by the
beads of U and Pu from the fission and actinide products in
dissolver solutions, (2) the convenience of handling the
nanogram quantities of material on the beads, (3) the
ability to load and analyze beads directly in the mass
spectrometer, and (4) the minimization of radioactivity
involved. The large reduction in sample size with the
concomitant reduction in radioactivity makes this analytical
procedure far safer than previous methods.

The disadvantages and concerns associated with the
resin bead technique are the problems common to the handling
and analysis of small samples. These problems are (1) the
non-representative sampling of the original, bulk material,
(2) the increased danger of contamination, (3) the increased
difficulty of controlling isotopic fractionation, and
(4) the limitation imposed by counting statistics in pulse
counting detection.

The role of the National Bureau of Standards (NBS) in
resin bead research is to provide measurement assurance for
the accurate isotopic analysis of uranium and plutonium. To
fulfill this role, NBS will develop the chemical and mass
spectrometric methodology for calibration and standardization. Furthermore, NBS has the capabilities to provide
resin bead Standard Reference Materials (SRM's) when the
need exists.

The extent of the research described in this work is
the identification and evaluation of the potential sources
of error involved in the technique and an indication of the
precision and accuracy to be expected. The analysis of
nanogram quantities of uranium and plutonium requires the
high sensitivity that is achieved by pulse counting detection.[4-7] The parameters involved in the use of pulse
counting detection have been studied. Although only uranium
loaded beads have been studied to date, the considerations
should also be applicable to the analysis of plutonium.

EXPERIMENTAL

Instrument

The mass-spectrometer is of the basic NBS design.[8,9]
It is a single focussing instrument with a 30 cm radius of
curvature and 90° deflection magnetic sector. The ion
source is a thin lens source with "Z"-focussing. The detector
consists of an electron multiplier and a movable Faraday
cage in the plane of the focal point. The electron multiplier is 17-stages with the "Rajchman" structure and a Cu-Be
conversion dynode. The operational gain of the multiplier
is 10^8. Output pulses from the multiplier are fed into a
high-speed amplifier-discriminator specifically designed to
be sensitive and to be free from the errors associated with
signal overload.[10] The pulses from the amplifier-discriminator are fed into a commercial counter operating in its
high frequency mode. The 1-second integrated signals from
the counter are transferred to a calculator where deadtime
and baseline corrections are applied and the isotopic ratios
calculated.

Chemistry

The strongly basic anion resin Bio-Rad AG 1x8 100-200
mesh used in this study. The beads are characteristically
spherical with a diameter of 0.100-0.125 mm. The beads in
their original chloride form were loaded by equilibration
with 8M HCl solutions of uranium. The nitrate form of the
resin was purposely avoided due to the possibility of bead
degradation during long-term storage. The beads used in
this study were then washed in acetone and air-dried. The
approximate uranium content of a single bead was calculated
to be 500 ng.

Bead Handling

The handling of individual resin beads requires
patience, a steady hand, a tungsten needle, and a high
quality stereomicroscope. All manipulation of beads and
filament loading was done in a Class 100 clean air hood.
As observed by ORNL,[3] manipulation of beads appears more
formidable than it proves to be. A bead will readily attach
itself to a tungsten needle by electrostatic attraction and
detach when touched to the filament. Success with maintaining the dry bead on a dry filament was poor. Crimping
increased success but was not 100 percent reliable. Wetting
the filament with a few microliters of cyclohexanone and
then dropping the bead into the wetted filament has been 100
percent successful at maintaining the bead on the filament,

a technique developed in earlier pioneering work with resin beads.[1]

Filaments

The filaments are constructed from zone-refined rhenium ribbon, 0.025 x 0.762 mm, folded lengthwise in half and spot welded to the filament legs. The filament legs are made of tungsten wire, 15 x 0.5 mm diameter, spot welded to stainless steel rods.[11] The filaments are degassed in a vacuum and potential field for 30 minutes at approximately 2000 °C prior to use.

DISCUSSION

The primary sources of error in a mass spectrometric analysis are isobaric interferences, uncontrollable contamination (blank), baseline determination, measurement system calibration, and isotopic fractionation control. In the mass spectrometric analyses in this laboratory using Faraday cage detection and vibrating reed electrometer measurement, the control of isotopic fractionation is most often the limiting error. The goal of NBS resin bead research is to elucidate and minimize sources of systematic error in the overall measurement process. Major sources of residual error are expected to accrue from counting statistics and isotopic fractionation.

The selective absorption characteristics of the anion resin bead removes the elements in the reactor fuels whose ions could interfere in the uranium and plutonium mass regions.[2] The potential exists for molecular ion isobaric interferences from the organic matrix of the bead and the cyclohexanone used in the loading procedure. This potential is not realized when the ion source is precisely focussed on the bead in the V-filament. It is observed that a very precise focus is possible due to the "point source" that the bead approximates and the focussing characteristics of the V-filament.[11] It has been observed that the source when defocussed can produce an ion spectrum with peaks at every mass position in the uranium mass region with intensities up to 150 counts per second (CPS). These ions are presumably organic molecular ions being emitted at some distance from the filament. In all analyses analog display of the mass spectrum is observed before data acquisition. The absence of this spectrum is easily monitored.

The handling and analysis of nanogram quantities of materials demands that contamination from the environment be scrupulously controlled. Air concentrations of U have been measured at 0.03–0.10 ng/m^3 in rural air but up to 90 ng/m^3 near a major nuclear facility.[12,13] Only by providing clean areas for sample handling and analysis can confidence in contamination control be sustained. In this regard sample loading and manipulation at NBS is done in Class 100 air.

Controlling the blank contribution from the bead, filament material, and source memory are also critical. The loading blank is estimated at 10^{-14} moles of U. The source is cleaned after 25 analyses to prevent the possible appearance of memory. No memory has been observed for this number of analyses.

The limit to the accurate measurement of minor isotopes are the quiescent background or dark current of the electron multiplier and ion scattering in the mass spectrometer. The average quiescent background is 0.14 CPS. The abundance resolution due to ion scattering in the single stage mass spectrometer is 1 to 2 parts in 10^5. The baseline during the bead analyses of the NBS SRM's U-100 to U-900 ranged from 0.6 – 1.5 CPS at the 236.5 mass position, equivalent to a noise level of 10^{-19} amps.

Pulse Counting

The primary requirement for high-accuracy electron multiplier detection is that the electron multiplier operate at nearly 100 percent efficiency where each ion impinging on the first dynode produces a measureable pulse from the last dynode. If this condition is not met there will exist a discrimination which will be efficiency dependent. Since efficiency of the ion-to-electron conversion process at the first dynode is dependent upon the angle of incidence of the ion beam, this mass discrimination will also be focal dependent. A movable Faraday cage installed in front of the electron multiplier allows both measurement of the efficiency and protects the multiplier from intense ion beams. This efficiency is 100 percent within measurement error.

The amplifier-discriminator can also introduce error in the measurement process if it is not capable of handling large signals. Here the dangers are the introduction of non-linear dead time and double counting of pulses. These effects are not corrected by the standard dead time correction. The NBS amplifier-discriminator is specifically designed to avoid these effects.[10]

For high accuracy the dead time correction of the measurement system must be calibrated. This calibration was accomplished empirically by analyzing U-100 and U-900 and plotting of the theoretical/experimental ^{235}U/^{238}U ratios

for each standard versus dead time correction values. The
method presumes control of the fractionation in the measurement of the experimental ratios.[14] The calibrated dead time
with the NBS measurement system is presently 53.5 nsec.
This dead time will be decreased when the operation of the
counter is optimized.

RESULTS

The standard procedure for determining the limit of
error in the isotopic measurement process at NBS is the
execution of a systems calibration which consists of the
determination of the $^{235}U/^{238}U$ ratios in the well characterized series U-050 to U-930.[9]

The results of a systems calibration using beads loaded
with U-100, U-350, U-500, U-750, and U-900 are presented in
Table I. The mass spectrometric heating pattern consisted
of the stepwise increase in temperature from 1000 °C to
1500 °C over a 15 minute period. A final adjustment was
then made until a total U$^+$ count rate of 200,000 CPS was
reached, typically at a temperature of 1660 °C. The instrument focus was optimized and baseline and peak shape checked

---

Table I

SYSTEMS CALIBRATION

Sample	Temp. °C	Observed Ratio 235-U/238-U	RSD %	Theoretical/ Observed
U-100	1695	0.11340	0.30	1.00148
U-350	1670	0.54599	0.12	1.00093
U-500	1660	0.99852	0.08	1.00118
U-750	1670	3.1634	0.08	1.00086
U-900	1655	10.372	0.14	1.00027
			Average	1.00094
			S.D.	0.00045
			95% C.L.	0.00125

362

before data were taken. Data were recorded 35 minutes into
the run for 20 minutes. Measurements were made by magnetic
stepping between the uranium peaks and integrating each peak
for 10 seconds. The observed change in fractionation was
typically 0.2 percent during data acquisition. The average
theoretical/experimental value in the systems calibration
was 1.00094 for the $^{235}U/^{238}U$ ratio. The standard deviation
of 0.00045 produces a 95 percent confidence limit of 0.12
percent for a single determination.

As an indication of the accuracy and precision of the
resin bead technique for measuring minor isotopes, the
$^{234}U/^{238}U$ ratios were measured in the 20 minute period
following acquisition of the major isotope data. The
results are presented in Table II. The accuracy and precision are counting statistics limited which is reflected in
the RSD between the five ratio sets measured in the 20
minute period.

Table II

234-U/238-U RATIO DETERMINATION

Sample	234-U Atom Fract.	Observed Ratio 234-U/238-U	RSD %	Theoretical/ Observed
U-100	0.000676	0.000757	1.89	1.0052
U-350	0.002498	0.003862	0.71	1.0048
U-500	0.005181	0.010392	0.40	1.0029
U-750	0.005923	0.02474	0.36	1.0058
U-900	0.007777	0.08952	0.98	0.9994
			Average	1.0036
			S.D.	0.0025
			95% C.L.	0.0072

REFERENCES

1.   I. L. Barnes, K. M. Sappenfield, and W. R. Shields, in
     Recent Developments in Mass Spectroscopy, K. Ogala and
     T. Hayakawa, Ed., University of Tokyo Press, Tokyo,
     Japan, 1970, p. 692.
2.   R. L. Walker, R. E. Eby, C. A. Pritchard, and
     J. A. Carter, Anal. Lett., $\underline{7}$, 563 (1974).
3.   R. L. Walker, C. A. Pritchard, J. A. Carter, and
     D. H. Smith, USDOE Report, ORNL/TM-5505, July 1976.
4.   D. H. Smith, R. L. Walker, L. K. Bertram, and
     J. A. Walker, USDOE Report ORNL/TM-6563, October 1978.
5.   J. W. Arden and H. H. Gale, Anal. Chem., $\underline{46}$, 687
     (1974).
6.   J. R. Rec, W. B. Myers, and F. A. White, Anal. Chem.,
     $\underline{46}$, 1243 (1974).
7.   R. S. Strebin, Jr. and D. M. Robinson, Anal. Chim.
     Acta, $\underline{91}$, 267 (1977).
8.   W. R. Shields, Ed., NBS Tech. Note 426, September 1967.
9.   E. L. Garner, L. A. Machlan, and W. R. Shields, NBS
     Spec. Publ. 260-27, April 1971.
10. R. W. Shideler, Int. J. of Mass Spectrom. and Ion
     Phys., $\underline{21}$, 213 (1976).
11. E. S. Gladney and H. L. Rook, Anal. Chem., $\underline{47}$, 1554
     (1975).
12. L. A. Dietz, Rev. of Sci. Inst., $\underline{30}$, 235 (1959).
13. J. E. Martin, E. D. Harwook, and D. T. Oakley, "Com-
     parison of Radioactivity from Fossil Fuel and Nuclear
     Power Plants", HEW, Bureau of Radiological Health,
     November 1969.
14. W. R. Shields, personal communication.

DISCLAIMER

DEVELOPMENT OF CHEMICAL ISOLATION AND CONCENTRATION
TECHNIQUES FOR TECHNETIUM-99 ANALYSIS BY RESIN-BEAD
MASS SPECTROMETRY

T. J. Anderson. E. I. du Pont de Nemours & Co.,
Savannah River Laboratory, Aiken, South Carolina, USA.

ABSTRACT

A novel, highly sensitive, isotope-dilution analytical
technique for the determination of technetium-99 has been
developed around single ion-exchange bead mass spectrometry
in collaboration with R. L. Walker of the Oak Ridge National
Laboratory. Mass spectrometry is much more sensitive than
direct counting for the low-energy, low-specific activity,
Tc-99 isotope. Further, the point source provided by a
single ion-exchange bead leads to a greater signal-to-noise
ratio in the mass spectrometric measurement than does conventional application of a solution to the source filament.
Recent results indicate a sensitivity greater than 0.1 picogram. Isolation of technetium from the samples occurs after
addition of Tc-97 as a yield tracer. A combination of ion-
exchange chromatography and ion-association solvent extraction provides decontamination from the potential interferences, Mo-97 and Ru-99. Subsequently, the technetium is
loaded onto a pair of anion-exchange beads (diameter
$\sim$0.3 mm). The noncritical isolation and bead-loading scheme
typically concentrates the technetium in the sample by a
factor of about a million with overall recoveries exceeding
50%. A variety of environmental samples from the Savannah
River Plant (SRP) has been analyzed by this method.

INTRODUCTION

Technetium-99 is one of several long-lived fission products whose direct radiometric detection at environmental levels is not normally feasible because of low specific activity
and low energy emissions. Its high mobility in aqueous

systems, including water-saturated soils, and its several
volatile forms provide transport opportunities from processes
and disposal operations.[1,2] Furthermore, technetium-99 is
produced in high yield from fission (6% from U-235).

Numerous studies have been conducted on the uptake of
Tc-99 from various soil and plant systems at relatively high
Tc-99 spike concentrations.[3,4,5] (Some controversy
exists, however, in the transfer coefficients in such systems.)[6] Likewise, much is known about the acute behavior of
large amounts of Tc-99m in the human body as a result of the
widespread medical use of this isotope.[7,8,9,10] Unfortunately, few data exist describing the behavior of technetium
in the body from chronic intake of environmental forms and
levels.[1] In both areas, the lack of a sufficiently sensitive technetium analysis has hampered the definition of these
systems under more realistic conditions.

Most recent analytical research aiming for a Tc-99
analysis with the highest possible sensitivity has centered
around the use of mass spectrometry. Its principal attrac-
tions are high sensitivity and selectivity, as well as versatility for other applications. J. H. Kaye et al. of Pacific
Northwest Laboratories have made notable contributions with
this and other techniques.[12,13] The Chemistry Group of
Savannah River Laboratory's Environmental Transport Division
in collaboration with R. L. Walker of Oak Ridge National Laboratory's Mass Spectrometry Group has also concentrated on
mass spectrometric techniques for Tc-99 analysis. This paper
describes the resultant isolation chemistry of a novel technique which enhances mass spectrometry's already high sensitivity by concentrating the isolated technetium into a single
anion exchange resin bead for filament loading. The resultant
"point source" provides increased ion throughput for greater
sensitivity than conventional solution loading of the filament. R. L. Walker will describe the mass spectrometry
techniques in the following paper entitled "Development of
Resin Bead Isotope Dilution Mass Spectrometric Techniques for
Tc-99 Analysis."

ANALYSIS GOALS

Several goals were of primary importance in the design
of this Tc-99 analysis: Foremost was the selection of mass
spectrometry as the measurement system because of its ability
to detect as little as 1 x 10^{-14} g of Tc-99. An isotope
dilution technique using Tc-97 as a yield tracer was chosen
to correct for isolation and detection inefficiencies. From
these choices came the need for chemistry with high discrimination against the isobars of the measurement isotopes
(Tc-97 and Tc-99), namely Mo-97 and Ru-99.

The decision to use ion-exchange resin bead mass spectrometry sample loading was based on the success of the technique for zirconium, uranium, and plutonium analysis at
ORNL.[14,15] While it affords higher signal-to-noise mass
spectrometric measurements than direct solution loading of
mass spectrometer filaments, it dictates only the final phase
of the chemical isolation scheme in a noncritical way.

Further goals included high recovery of technetium
through the chemical isolation procedure, the elimination of
technique-critical steps (e.g., distillation from concentrated H_2SO_4, electroplating, special reagent purification,
etc.), elimination of reagents with potential for molybdenum
and/or ruthenium contamination (e.g., CuS precipitation), and
the use of a minimum volume of reagents, again to lessen contamination potential. Of lesser importance and suitable for
later consideration are speed and cost of the analysis.

ANALYSIS METHODOLOGY

In its current state, the core of the Tc-99 analysis is
designed to accept aqueous technetium solutions of low to
moderate ionic strength (Figure 1). It is based on two

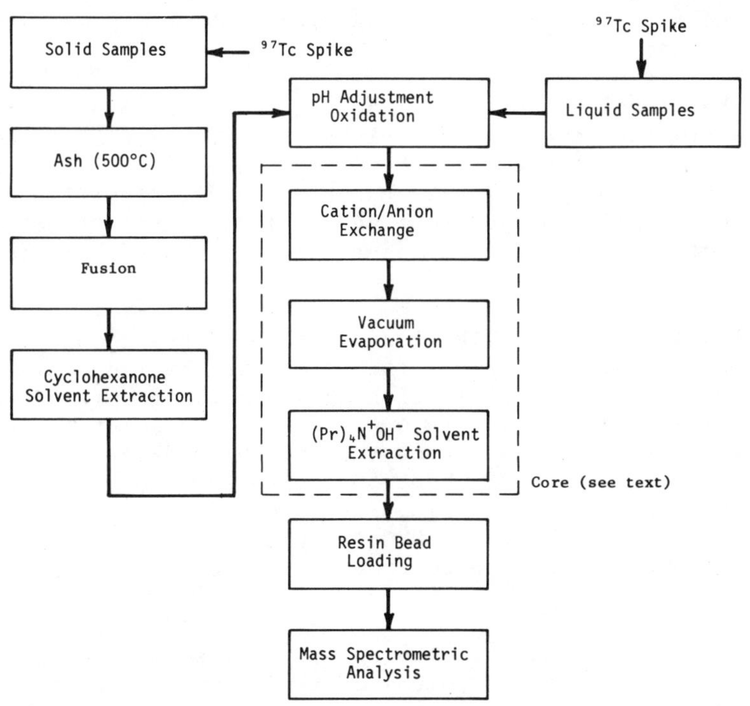

Figure 1. Procedural Diagram

chemical isolation techniques: tandem column ion-exchange and ion-association solvent extraction. The former is a modification of the work of Huffman et al.[16] and the latter is adapted from Maeck et al.[17]

Spiking of samples with Tc-97 precedes any chemical treatment. Aqueous samples are then oxidized with H_2O_2 and adjusted to a pH <5 with HCl, as necessary, to ready them for ion-exchange. With solid samples, the spike is applied in multiple portions to the finely divided sample with as wide a distribution as possible. Uniformity is enhanced by slurrying with a small excess of concentrated NH_4OH. Following drying at 100-120°C, samples with significant organic content are charred and ashed. Ashing temperatures must be kept low (∿500°C) to prevent loss of technetium by volatilization.

The ash or soil is then fused with either a Na_2O_2 flux or a mixed flux of (5.08) K_2CO_3: (3.92) Na_2CO_3: (1.00) $NaNO_3$ (w:w:w). The latter flux requires a somewhat higher temperature than Na_2O_2, but the fusion can be performed in inexpensive Ni crucibles without significant dissolution of the crucible. This avoids the need to use zirconium or platinum crucibles or to deal with large quantities of hydrated nickel oxide when dissolving the fuse. The mixed flux leaves very little residue when applied to the sand and clay soils and pine straw common to the SRP site. About 4 g of flux are used for each gram of soil.

The dissolved fuse is subjected to a single extraction/ back-extraction cycle with cyclohexanone and cyclohexane[18] to transfer the technetium to an aqueous medium of low ionic strength suitable for ion-exchange. Freshly distilled cyclohexanone greatly speeds phase separations. Typical technetium recoveries for the procedure through back-extraction are 87%.

The apparatus for the tandem column ion-exchange is shown in Figure 2. Its purpose is a gross cationic cleanup and some initial discrimination against molybdenum and ruthenium. Tandem operation, as opposed to two discrete column steps, significantly shortens the time needed for ion-exchange. Elution of much of the molybdenum and ruthenium from the anion exchange column by 1M HCl is followed by elution of technetium with 4M HNO_3. Typically, 99% of the input technetium is recovered from the tandem ion-exchange step.

Before proceeding to the solvent extraction step, it is necessary to reduce the volume of the column effluent and its content of NO_3^-. The latter suppresses recovery of technetium in the subsequent solvent extraction step. Both objectives

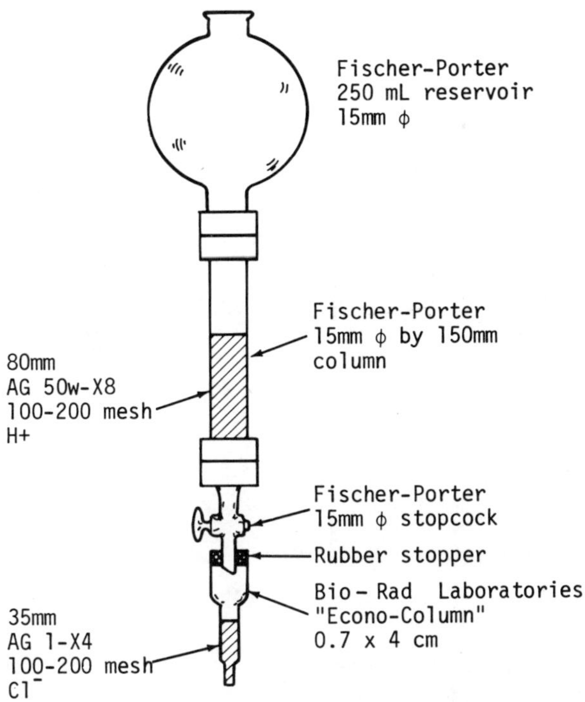

Figure 2.   Tandem Ion-Exchange Column

are accomplished by a room temperature vacuum evaporation
of the column effluent to a final volume of about 0.2 mL.
The apparatus is shown in Figure 3.   Recovery of input technetium is normally 91%.   This technique conforms to an early
suggestion of Anders[19] and solves the problem of low and inconsistent recoveries of technetium experienced by evapora-
tions at higher temperatures.

Solvent extraction of the tetrapropylammonium ion-
association compound of technetium into 4 methyl-2-pentanone
follows the vacuum evaporation.   Back extraction into an
aqueous phase is accomplished by diluting the 4 methyl-2-
pentanone with cyclohexane after the work of Boyd et al.[18]
A double extraction cycle is used to reduce the contamination
potential from incomplete phase separation.   Early experiments with Mo-99 tracer indicated that the distribution ratio
for molybdenum in the forward extraction was less than
$\sim 1.3 \times 10^{-5}$.   Recovery of technetium through the solvent
extraction procedure has been about 95%.

Technetium is then loaded onto pairs of anion exchange
beads following another room temperature evaporation of the
aqueous back-extract to a volume of 0.2 mL.   Beads of 2.7 to
3.3×10^{-4}m diameter are microscopically selected from a

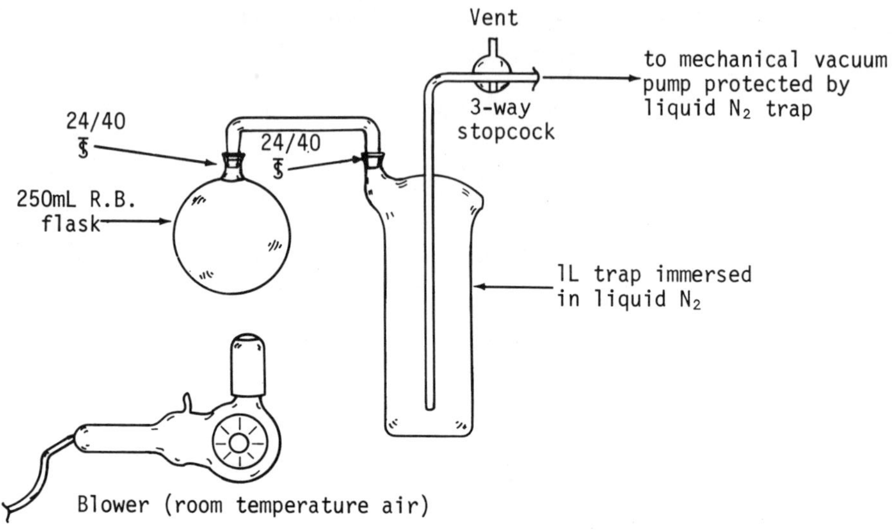

Figure 3.   Vacuum Evaporation Apparatus

supply of AG-1X10, Cl⁻, 100-200 mesh beads. A pair of such
beads is combined with the 0.2 mL of evaporated back-extract
in a 400 μL micro test tube (Brinkman 22-36-390-5) and agi-
tated vigorously for two hours. The second bead is a backup
in case the first is lost during handling. Both a vortex
mixer and a pipet shaker (TechniLab model 31) have been used.
The latter allows facile agitation of at least four micro
test tubes at once. In a series of trials with simulated
evaporated back-extract under the above conditions, the uptake of technetium for seven pairs of beads averaged 86% with
a standard deviation of only 1.7%. Earlier experience with
small beads ($<2.5 \times 10^{-4}$m diameter) and 4% cross-linkage gave
inconsistent uptakes, some as low as 32%.

Following uptake, the beads are isolated from the
solution and mounted on glass microscope slides with 1:8
(vol:vol) flexible collodion:ethanol for storage or shipment.
The handling of the beads is conducted with the aid of an 8X
to 40X stereoscopic zoom microscope. Transfer tools consist
of drawn Pasteur pipets and a tungsten needle probe. The
pipets are drawn to tips larger than the bead diameters for
transferring beads plus solution; and to tips smaller than
the bead diameters for transferring solution alone. After

the solution is removed, beads can be transferred by the
tungsten needle when tipped with a globule of adhesive teased
from cellophane tape. Experience with this handling system
has been excellent; most novices can be trained to handle
beads without loss in a matter of hours. Overall recoveries
from aqueous sample through bead uptake are typically 73%.

In addition to purification, another necessary function
of the isolation scheme is concentration. Considering an
aqueous sample of 100 mL and a 0.3-mm diameter pair of anion-
exchange beads, an overall concentration of $\sim 1 \times 10^6$ is effected by the procedure, ~ 3500 being attributable to the bead
uptake step. The important aspect of this is that the spa-
tial distribution of the technitium on the mass spectrometer
filament remains confined by the bead to a "point source"
during filament loading.

The procedure for mass spectrometric measurement will be
discussed by R. L. Walker in a following paper.

RESULTS

Standards

A number of standards has been prepared spanning the
values from 0.1 pg to 12 ng of Tc-99. All contained 917 pg
of Tc-97. The standards were subjected to the entire analysis procedure for aqueous samples (Table I). Good consist-
ency in the normalized Tc-99/Tc-97 corrected count ratios
is evident throughout the 6-decade range. The method clearly has potential sensitivity beyond 0.1 pg.

Samples

Various environmental samples, both aqueous and solid,
from the vicinity of SRP have been analyzed for Tc-99 by this
method (Table II). It should be noted that these data were
generated with less than normal bead uptake and some results,
namely F-Area Spring and the herb-litter sample, show poor
internal precision for the bead pair. Nonetheless, one con-
cludes that Tc-99 in the F-Area seepage basin easily migrates
with the groundwater to the F-Area Spring (~ 9 years travel
time based on H-3 studies).[20] From there it is transported
to the spring's receiving stream, 4-Mile Creek, and finally
the Savannah River. These data, when combined with Savannah
River flow rates, generate an annualized instantaneous Tc-99
release rate from SRP of 0.05 Ci/yr to the river. In all
cases, the measured Tc-99 aqueous concentrations have been
well below the Federal concentration guides for releases to
uncontrolled areas.[21]

Table I. Analysis of Standard Bead Sets

^{99}Tc Added, ng	$\dfrac{\text{Counts } ^{99}\text{Tc}}{\text{Counts } ^{97}\text{Tc}} \div {}^{99}\text{Tc Added, ng}$
11.6	1.08 ± 0.03[a]
1.16	1.05 ± 0.03
0.116	1.10 ± 0.04
0.0106	1.65 ± 0.2
0.0116	1.09, 1.13[b]
0.00105	1.1, 1.1
0.000105	1.8, 1.1

a. Counting standard deviation shown.
b. Results from both beads of a pair shown.

Table II

^{99}Tc Results from Environmental Samples

Sample	^{99}Tc Concentration (pg/L)[a]
H-Area Seepage Basin	$(1.4, 1.5) \times 10^{5}$
F-Area Seepage Basin	$(3.8, 3.9) \times 10^{3}$
F-Area Spring	$(7.3, 10) \times 10^{3}$
4-Mile Creek	10.4, 11.0
Savannah River (Upstream SRP)	$<8 \times 10^{-2}$
Savannah River (Downstream SRP)	0.4, 0.5
Herb-Litter from F-Area	$(8.1, 16) \times 10^{2}$ (pg/m^{2})

a. Results calculated for each resin bead of a sample pair.

In summary, the current Tc-99 analysis can determine as
little as 0.1 pg of Tc-99. It has been applied successfully
to a variety of environmental samples and thus provides a
powerful tool for realistic determination of the environmental behavior of Tc-99. Through further development,
increased sensitivity can be expected. Productivity gains
might be possible through elimination of the second ion-
association solvent extraction cycle and a shorter, less
tedious solubilization procedure for solid samples.

ACKNOWLEDGMENTS

The information contained in this article was developed
during the course of work under Contract No. DE-AC09-
76SR00001 with the U. S. Department of Energy.

REFERENCES

1. D. J. Brown. "Migration Characteristics of Radionuclides
    Through Sediments Underlying the Hanford Reservation."
    Disposal of Radioactive Wastes into the Ground. IAEA
    Document STI/PUB/156, Vienna; p. 215 (1967).

2. R. C. Routson, G. Jansen, and A. V. Robinson. "^{241}Am,
    ^{237}Np, and ^{99}Tc Sorption on Two United States Subsoils
    from Differing Weathering Intensity Areas." _Health
    Physics_, _33_, 311 (1977).

3. E. R. Landa. _The Behavior of Technetium-99 in Soils and
    Plants_. Thesis, University of Minnesota (1975).

4. R. C. Routson and D. A. Cataldo, "Accumulation of ^{99}Tc
    by Tumbleweek and Cheatgrass Grown on Arid Soils."
    Health Physics, _34_, 685 (1978).

5. R. G. Gast, E. R. Landa, L. J. Thorvig. _The Behavior of
    Technetium-99 in Soils and Plants_. USERDA Report COO-
    2447-5, Department of Soil Science, University of Minne-
    sota, St. Paul, Minnesota (1976).

6. J. E. Till, F. O. Hoffman, and D. E. Dunning, Jr.
    _Assessment of ^{99}Tc Releases to the Atmosphere--A
    Plea for Applied Research_. USDOE Report ORNL/TM-6260,
    Oak Ridge National Laboratory, Oak Ridge, Tennessee
    (1978).

7. T. M. Beasley, H. E. Palmer, and W. B. Nelp. "Distribu-
    tion and Excretion of Technetium in Humans." _Health
    Physics_, _12_, 1425 (1966).

8.   S. K. Shukla, G. B. Manni, and C. Cipriani. "Behaviour
     of Pertechnetate Ion in Humans." J. of Chromatography,
     143, 522 (1977).

9.   E. J. Baumann, N. Z. Searle, A. A. Yalow, E. Siegel, and
     S. M. Seidlin. "Behavior of the Thyroid Toward Ele-
     ments of the Seventh Periodic Group." Am. J. of Physi-
     ology, 185, 71 (1956).

10. R. Herbert, W. Kulke, and R. T. H. Shepherd. "The Use
     of Technetium 99m as a Clinical Tracer Element." Post-
     grad. Med. J., 41, 656 (1965).

11. V. J. Sodd and B. J. Jacobs. "Analysis of Human Thy-
     roids for ^{99}Tc." Health Physics, 14, 593 (1968).

12. J. H. Kaye, M. S. Rapids, and N. E. Ballou, "Determina-
     tion of Picogram Levels of Technetium--99 by Isotope
     Dilution Mass Spectrometry." Proceeding of Third
     International Conference on Nuclear Methods in Environ-
     mental and Energy Research, Columbia, Missouri. In
     press. (1978).

13. J. H. Kaye and N. E. Ballou, "Determination of Tech-
     netium by Graphite Furnace Atomic Absorption Spectrom-
     etry." Anal. Chem., 50, 2076. (1978).

14. R. L. Walker, R. E. Eby, C. A. Pritchard, and
     J. A. Carter, "Simultaneous Plutonium and Uranium Iso-
     topic Analysis from a Single Resin Bead--A Simplified
     Chemical Technique for Assaying Spent Reactor Fuels."
     Anal. Lett., 7, 563 (1974).

15. R. L. Walker, J. L. Botts, J. A. Carter, and
     D. A. Costanzo, "Mass Spectrometric Determination of
     Zirconium from a Resin Bead." Anal. Lett., 10(4), 251
     (1977).

16. E. H. Huffman, R. L. Oswalt, and L. A. Williams.
     "Anion-Exchange Separation of Molybdenum and Technetium
     and of Tungsten and Rhenium." J. Inor. Nucl. Chem., 3,
     49 (1956).

17. W. J. Maeck, G. L. Booman, M. E. Kussy, and J. E. Rein.
     "Extraction of the Elements as Quaternary (Propyl,
     Butyl, and Hexyl) Amine Complexes." Anal. Chem., 33,
     1775 (1961).

18. G. E. Boyd and Q. V. Larson. "Solvent Extraction of
     Heptavalent Technetium." J. Phys. Chem., 64, 988 (1960).

19. E. Anders. "The Radiochemistry of Technetium."
     National Academy of Sciences Nuclear Sciences Series,
     NAS-NS-3021, p 13, USAEC (1960).

20. J. W. Fenimore and H. H. Horton. _Operating History and
     Environmental Effects of Seepage Basins in Chemical Sep-
     arations Areas of the Savannah River Plant._ USAEC Re-
     port DPST-72-548 (Declas.), E. I. du Pont de Nemours &
     Co., Savannah River Laboratory, Aiken, South Carolina.

21. "Standards for Protection Against Radiation." _Title 10,
     Code of Federal Regulations_, Part 20 (10 CFR 20), Ap-
     pendix B, 6343 (1976).

DEVELOPMENT OF RESIN-BEAD ISOTOPE-DILUTION MASS
SPECTROMETRIC TECHNIQUES FOR TC-99 ANALYSIS

R. L. Walker. Analytical Chemistry Division, Oak Ridge
National Laboratory, Oak Ridge, Tennessee, USA

ABSTRACT

An isotope dilution mass spectrometric method has been
developed for the analysis of Tc-99 after isolating it onto
anion exchange resin beads. A single resin bead containing
Tc-99 and Tc-97 spike is loaded onto a rhenium V-shaped filament for thermal emission mass spectrometry. The application
of this technique requires the use of a mass spectrometer of
high abundance sensitivity and pulse counting capability for
the necessary ion detection sensitivity.

The use of reducing agents to improve metal ion production and increase ionization efficiency of Pu and U samples
loaded as solutions onto filaments is an accepted practice in
many mass spectrometry laboratories.[1,2] The mechanism for
the increased ion emission obtained from a resin bead in comparison to loading samples as solutions probably results from
the point source for the ion optics represented by the bead
combined with the reducing action supplied by the carbona-
ceous material of the bead when it is decomposed at about
1000-1200°C during the initial filament heating. This prevents loss of sample as volatile oxides and enhances metal
ion emission.

This paper discusses the development of the technique,
including the mass spectrometer, choice of filament material,
scanning modes, interferences, and present achievable sample
sensitivities.

INTRODUCTION

This method is an extension of the resin-bead sample
loading technique developed at ORNL primarily for small samples of uranium and plutonium.[3] The resin-bead method was

chosen since a very sensitive and specific method is necessary
for measuring low specific activity Tc-99 in environmental
samples. The high fission yield (6.3%) of Tc-99 and very
long half-life (2.1 x 10^5 y) make this radionuclide potentially hazardous in animal food chain systems.

The usual method of Tc-99 analysis is beta counting,
which has a sensitivity limit of about 10 pg (0.2 pCi).
Since this is a non-specific method with relatively poor
sensitivity, an isotope dilution mass spectrometric method
using a spike of Tc-97 has been developed here and at Pacific
Northwest Laboratories by Kaye[4] and co-workers. The main
difference between the methods is the filament loading technique. In Kaye's method, technetium is loaded on the filament and reduced to the metallic state with H_2, whereas in
our technique the technetium is adsorbed onto anion resin,
and a single bead is loaded onto the filament. The ionization efficiency of Tc is approximately 50 times greater
loaded on resin beads when compared to loading it as solutions. The limit of detection is 0.]-0.2 picogram and is
near the expected limit due to uncertainties caused by hydrocarbon interference and that caused by Mo in the Re filament
at the mass 97 position.

Mass Spectrometer

The mass spectrometer used in this work is of the single
focusing, tandem magnet type.[5] The design used at ORNL is
based on that developed by White and Collins,[6] which utilizes
the "C" configuration. The instrument shown in Figure 1, has
30-cm radius and 90° deflection in each magnet, which gives a

Figure 1. Mass spectrometer

clean spectrum with high abundance sensitivity. Ions are
detected by a secondary electron multiplier behind the re-
ceiver slit. The detector used is an RCA type 6810A 14-stage
multiplier and has a gain of 10^6 in the pulse-counting mode.
Data are stored in a PDP-11 computer for later reduction and
processing.

Filaments

Filaments for this work are made from zone-refined
rhenium. They are of the V-shaped design and are prebaked in
an auxiliary vacuum system for at least 30 minutes at 2000°C
to remove impurities and reduce the molybdenum background as
much as possible. Measurements have been made to determine
the best filament material for use in Tc analyses. We found
that of the three metals (Re, W, and Ta) tested, the order of
sensitivity for Tc is Re>W>Ta. The sample changer used has
been described by Christie and Cameron.[7] It is a "ferris"
wheel upon which are mounted six sample filaments. It is
driven from outside the vacuum system by a rotary-motion
feed-through.

Tc-97 Preparation

Approximately 2 μg of Tc-97 was made by irradiating 25
mg of isotopically enriched ^{96}Ru for 20 days in the High Flux
Isotope Reactor at ORNL at a neutron flux of 6 x 10^{14} neutron
cm^{-2} sec^{-1}. Half of the target has been chemically processed
to recover technetium isotopes and the other half retained
for future use. The yield from the processed target was 1.6
μg having the following isotopic composition:

97	98	99
79.642 ± 0.097	20.175 ± 0.100	0.183 ± 0.006

This is sufficient technetium to run several hundred samples
using approximately 1 ng per sample. We were surprised to
find the high ^{98}Tc and ^{99}Tc content of the product, but the
amount of ^{99}Tc in the spike will not affect its usefulness
except perhaps in samples containing very low (near 1 picogram) concentrations. Assuming the contribution of the
sample to be 20% greater than the ^{99}Tc in the spike, the
detection limit would be about 0.2 picogram.

Isotopic Analysis Procedure

Resin beads are loaded onto filaments in a chip of col-
lodion with the aid of two low-powered microscopes: one for
removing the bead from the microscope slide on which it is
stored, and another for putting the bead in the V-shaped

rhenium filament. Loaded filaments are placed in the source
region and evacuated to <5 x 10^{-7} torr before analysis.

The resin bead is decomposed in a slow, step-wise manner
by applying d.c. current. Usually decomposition is complete
at 1000-1200°C and mass scanning is started. The Tc masses
at 97 and 99 are scanned, and it is also necessary to scan
masses 98 and 100 for correcting the Mo contribution to the
^{97}Tc signal. Mass position 100.5 is scanned to make a correction for electronic noise. Analysis is made at the lowest
possible temperature to minimize Mo interference. Data are
usually taken at about 1900°C. Peak hopping is used for data
accumulation to improve statistics and sensitivity, since
this method of data taking eliminates time lost scanning between peaks.

RESULTS AND DISCUSSION

Two significant interferences have been uncovered with
mass measurements at 97 and 99 during the course of this work.
These are Mo at 97 and hydrocarbons primarily affecting low
levels of Tc-99. Ruthenium at mass 99 has been eliminated
as a possible interference since it has not been observed in
the analysis of blank filaments, chemical blanks, and technetium standards.

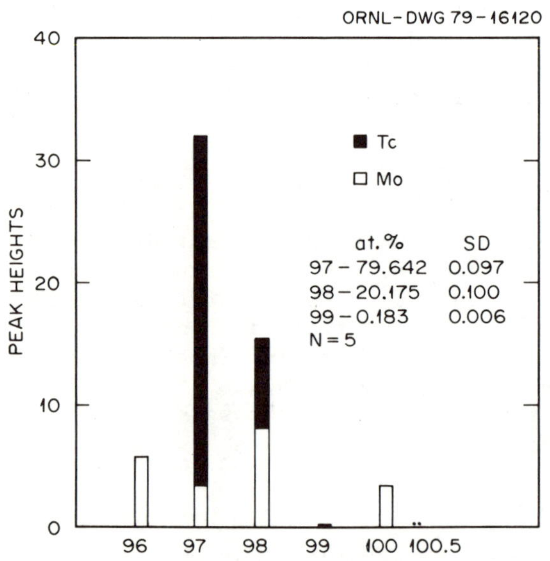

Figure 2. Tc spectrum

The correction for ^{97}Mo can be reliably made if the ^{97}Tc
ion signal equals or exceeds the ^{97}Mo ion signal. Normally
this ratio is exceeded. Reducing filament length and using
initial bake-out before bead loading minimizes the Mo signal.
We are now getting average ratios of about 10 for ^{97}Tc/^{97}Mo
for 0.5 ng ^{97}Tc standards. For estimating the Mo correction,
we use the average obtained from natural ^{96}Mo and ^{100}Mo.
A drawing of a Tc spectrum is shown in Figure 2. This shows
the spectrum obtained from the combination of the ORNL spike
and natural Mo from the rhenium filament.

During the course of this work, hydrocarbon background
has been a real problem, especially when trying to achieve
the highest sensitivity for ^{99}Tc. It is impossible to correct it by monitoring another mass, since it varies from
filament to filament and ratios are not constant. We have
looked for the source of the contamination and have not been
successful. We find it in blank filaments as well as loaded
ones; therefore, we have ruled out the resin bead as the
source. Oil diffusion pumps are used for partial evacuation
of the source region before turning on the ion pumps, and it
was postulated that this could be the source. Liquid nitrogen pumping was added to the source region without any
obvious effect. A promising method that we have recently
developed is to resolve the hydrocarbon from the metal ion
peaks. The hydrocarbon masses are approximately 0.2 amu
above the metal ion masses, and with greatly reduced slit
widths between the magnets, we can resolve the hydrocarbon
and metal peaks as shown in Figure 3. This makes the
analyses more difficult for the operator and reduces the
signal, but is more than worth the sacrifice of intensity.
A comparison of the results obtained on the ORNL-prepared
spike using narrow center slits to resolve out hydrocarbons
is shown in Table I. A tremendous improvement in precision
is noted; also, the lower values for the minor isotopes
would be expected since there is no hydrocarbon contribution
to them.

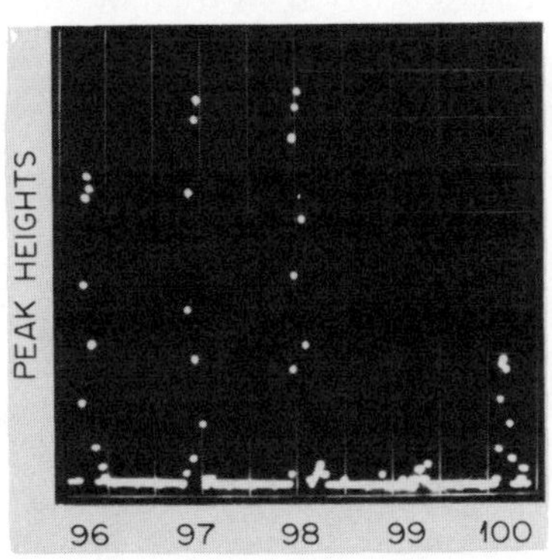

Figure 3. Tc spectrum with
hydrocarbons resolved
(ORNL-DWG 79-16119)

Table I

Isotopic Analysis of ORNL Tc-97 Standard

Mass	Atom Percent	
	a	b
97	79.260 ± 0.574	79.642 ± 0.097
98	20.509 ± 0.523	20.175 ± 0.100
99	0.231 ± 0.061	0.183 ± 0.006

a) normal conditions.
b) narrow center slits.

Mixtures of ^{97}Tc and ^{99}Tc have been analyzed to determine the sensitivity and reliability of the method. For this experiment, a graded set of ^{99}Tc standards from 11.6 to 1 x 10^{-4} ng were mixed with a constant amount of ^{97}Tc ($\sim$0.9 ng). The results are shown in Figure 4. Some problems are noted in

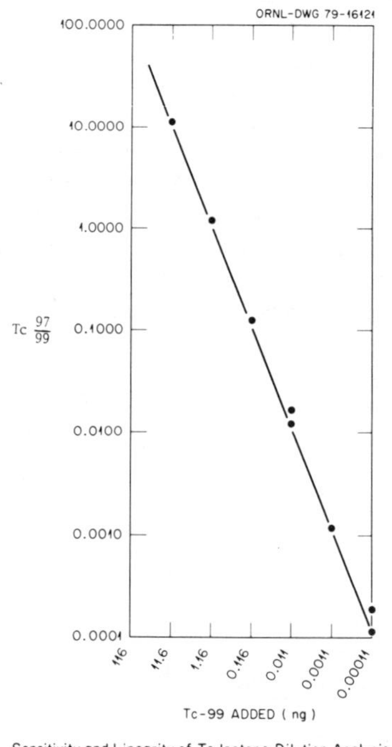

Figure 4. Sensitivity and Linearity of Tc Isotope Dilution

the low concentrations, but the linearity is quite good down to 10^{-4} ng. This work was done in collaboration with T. J. Anderson at the Savannah River Laboratory. The ^{97}Tc spike used was from his laboratory and does not contain measurable amounts of ^{98}Tc and ^{99}Tc; therefore, the sensitivity is much better than for the ORNL spike discussed earlier.

ACKNOWLEDGMENT

Research sponsored by the U. S. Department of Energy, Division of Basic Energy Sciences, under Contract W-7405-eng-26 with the Union Carbide Corporation.

REFERENCES

1. L. A. Dietz and H. C. Hendrickson, "Selective Measurement Methods for Plutonium and Uranium in the Nuclear Fuel Cycle," R. J. Jones, ed., USAEC (1963).

2. J. W. Arden and N. H. Gale, "Separation of Trace Amounts of Uranium and Thorium and Their Determination by Mass Spectrometric Isotope Dilution," Anal. Chem. <u>46</u>, 687 (1974).

3. R. L. Walker, R. E. Eby, C. A. Pritchard, and J. A. Carter, "Simultaneous Plutonium and Uranium Isotopic Analysis from a Single Resin Bead--A Simplified Chemical Technique for Assaying Spent Reactor Fuels," Anal. Lett. <u>7</u>, 563 (1974).

4. J. H. Kaye, M. S. Rapids, and N. E. Ballou, "Determination of Picogram Levels of Technetium-99 by Isotope Dilution Mass Spectrometry," Proceedings of Third International Conference on Nuclear Methods in Environmental and Energy Research, Columbia, Missouri, in press (1978).

5. D. H. Smith, ed., DOE Report, ORNL/TM-6485 (1978).

6. F. A. White and T. L. Collins, Appl. Spectros. <u>8</u>, 169 (1954).

7. W. H. Christie and A. E. Cameron, "Reliable Sample Changer for Mass Spectrometer," Rev. Sci. Instrum. <u>37</u>, 336 (1966).

QUALITY ASSURANCE
AND STANDARDS

TREATMENT AND REPORTING OF UNCERTAINTIES FOR
ENVIRONMENTAL RADIATION MEASUREMENTS

R. Collé. Office of Radiation Measurement, Center for
Radiation Research, National Bureau of Standards,
Washington, D.C., USA.

ABSTRACT

Recommendations for a practical and uniform method for
treating and reporting uncertainties in environmental radiation measurements data are presented. The method requires
that each reported measurement result include the value, a
total propagated random uncertainty expressed as the standard
deviation, and a combined overall uncertainty. The uncertainty assessment should be based on as nearly a complete
assessment as possible and should include every conceivable
or likely source of inaccuracy in the result. Guidelines
are given for estimating random and systematic uncertainty
components, and for propagating and combining them to form
an overall uncertainty.

INTRODUCTION

During the past few years, the need for improvements
in the quality and usability of environmental radiation
data has been emphasized many times. One of the problem
areas frequently identified concerned the need for greater
uniformity in the methods used to report data, particularly
in the treatment and reporting of uncertainties associated
with measurement results. Unfortunately, almost all environmental radiation data are presently reported either without
any statements of uncertainties, or with unspecified uncertainties, or with uncertainties which are based on incomplete assessments. Further, federal practices pertaining
to the reporting of uncertainties not only contribute to
these abuses, but are also as diverse as the number of

agencies and requirements specified within an agency's
regulations.

To address these problems, the National Bureau of
Standards (NBS) coordinated an interagency and multi-orga-
nizational project with broad national participation. The
project was initiated in 1977 at the request of the Task
Force on Radiation Measurements of the Conference of Radiation Control Program Directors (CRCPD). At approximately the
same time, a Health Physics Society (HPS) ad hoc committee
was organized for the purpose of upgrading environmental
radiation data. One of the nine project objectives identified by that HPS committee had a scope nearly equivalent to
the CRCPD Task Force recommendation. As a result, NBS agreed
to coordinate and implement the project under the auspices of
the HPS committee and an appropriate subcommittee was formed
to serve as a working group.[1] The subcommittee prepared a
detailed *Guide and Recommendations* for reporting of environmental radiation measurements data which will be published as an NBS Special Publication.[2] A condensed version
of the *Guide* will be included, along with eight other reports, in a publication of the HPS that will present the
combined output of the ad hoc committee.[3] Some highlights
from that part of the *Guide* concerned with the assessment,
propagation and reporting of uncertainties are presented
here. Of necessity, this summary must be very brief, and
the reader should consult the complete *Guide*[2] for additional
detail.

REQUIREMENTS FOR UNCERTAINTY STATEMENTS AND THE COMPLETE
UNCERTAINTY ASSESSMENT PROCESS

There are three necessary conditions for the reporting
of uncertainties. They are:
   i)   Every reported value must include an assessment of
      its uncertainty.
  ii)   The reported uncertainty must be clearly understood
      and must convey sufficient information so that its
      meaning, using correct terminology, is unambiguous.
 iii)   The reported uncertainty must be based on as
      nearly complete an assessment as possible.
These requirements should be self evident. A reported value
without an accompanying uncertainty statement is for nearly
all purposes worthless. Similarly, an unspecified uncertainty statement forces the user to speculate as to its
meaning. Lastly, if the uncertainty is meant to be an
estimate of the likely accuracy or the likely limits to the
absolute error in the measurement result, then the uncertainty must be based on as nearly complete an assessment as
possible. The uncertainty assessment must consider every
conceivable or likely source of inaccuracy in the result.

A large part of the *Guide* is devoted to this process
of making a complete uncertainty assessment of all of the
conceivable sources of inaccuracy, and the mechanics of
treating the individual components from their initial
assessment through their propagation. The philosophical
basis for this approach has been described[4] and will not be
considered here. Figure 1 attempts to illustrate this
complete assessment process, and the division of the sources
of inaccuracy between random and systematic components.
Every conceivable source of inaccuracy should be classified
into one of the two categories depending on how the uncertainty is estimated. Until combined for the measurement
result, the individual uncertainty components in both
categories should be retailed separately.

The random uncertainty components are statements of
precision (or more correctly, of *im*precision) and can be
considered to be those sources of inaccuracy which *can be*
and *are* assessed and propagated by statistical methods.
Sample statistics, such as the standard deviation s, which
are computed entirely from the measurement data, are used
to estimate the population parameters such as σ. For com-
pleteness, the systematic uncertainty components (δ) can
be considered to be the residual set of conceivable sources
of inaccuracy which are biased and not subject to random
fluctuations, and those which may be due to random causes
or stochastic process but *cannot be* or *are not* assessed by
statistical methods. The systematic uncertainty includes
sources of inaccuracy in addition to biases in order to
obtain a complete accounting of all sources of inaccuracy
in the overall uncertainty. Finally, the overall uncertainty (u) used as an estimate of the inaccuracy is then
just some combination of the random and systematic uncertainties.

A third category of conceivable sources of inaccuracy
is that termed "blunders" (Figure 1). These are the out-
right mistakes in the measurement process. If one or more
blunders are known to exist, then the value should be
discarded, or a correction should be applied to the measurement. One must recognize that blunders do and will occur
and an internal system of independent checks should be
established for each stage of the measurement and data
reduction system to detect and avoid them.

STATISTICAL TREATMENT OF RANDOM UNCERTAINTIES

Each random uncertainty component should be estimated
in terms of its standard deviation, or the standard error

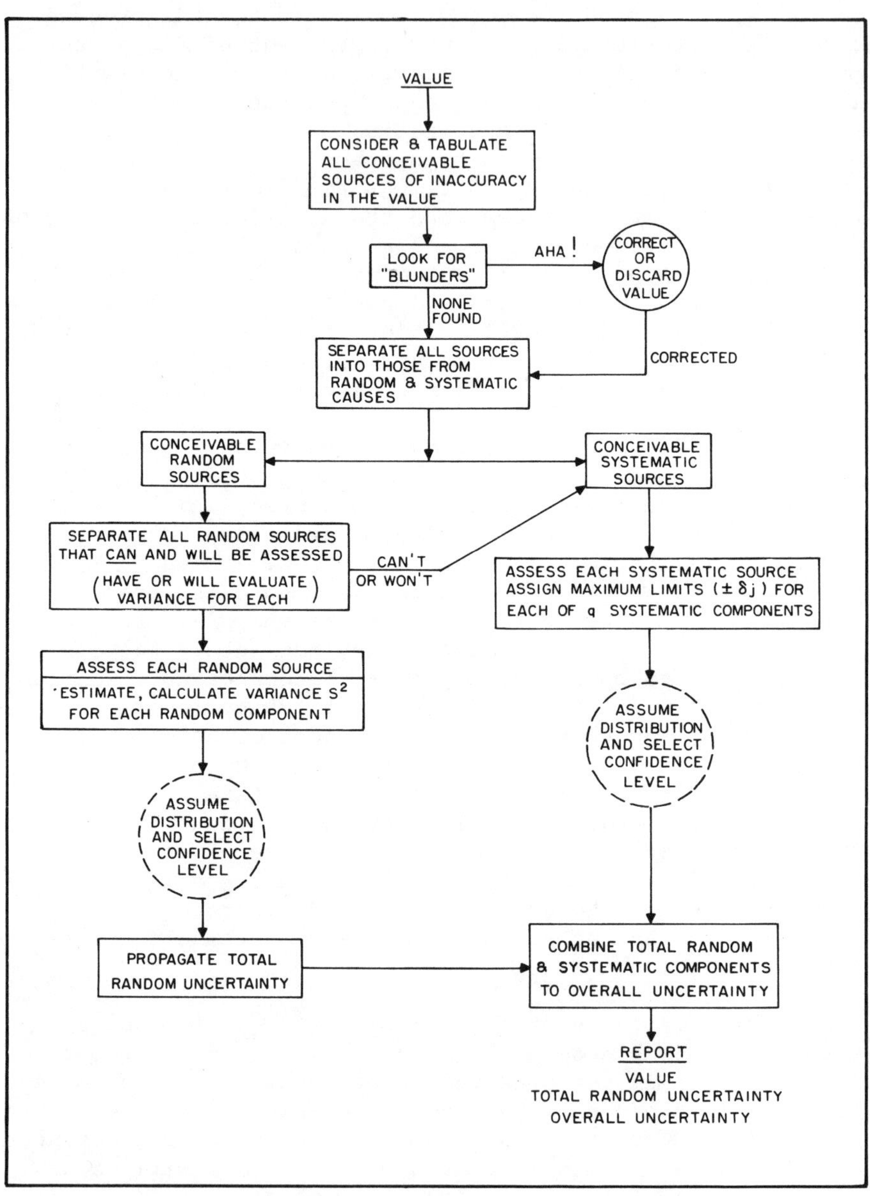

Figure 1.   Complete Uncertainty Assessment Process.

of the mean for independent multiple measurements. These
components are estimated by a statistical analysis of replicate measurements. The random "counting error" contribution may be estimated from single measurements by making
the familiar Poisson assumption. Although the use of the
term "counting error" will probably continue to prevail,
the term is a gross misnomer. It takes into consideration
only the random scatter about the mean from the radioactive
decay process itself. To presume that this is the only
source of random fluctuation in the overall measurement
process is fatuous. Unfortunately, federal guidance, as it
pertains to reporting environmental radiation data, does
not suggest otherwise. Other sources may be random timing
uncertainties, random variations in the sample preparation,
positioning of the sample at the detector, etc. The list
is nearly endless. Many of these are addressed in a detailed appendix to the *Guide*.

The only way to realistically assess the overall
random uncertainty is to make replicate determinations and
calculate the standard deviation of the mean by the usual
statistical methods. For reported values derived from
several component quantities, the individual uncertainty
estimates (obtained from replicate measurements for each
independent variable must be propagated to obtain the overall standard deviation of the final value. It is recognized
that replicate measurements of environmental samples are
uncommon and may not be feasible because of time and cost
constraints. Yet, an uncertainty assessment based on some
procedure beyond calculating the square root of the total
number of counts is clearly needed. The *Guide* recommends
that the total random uncertainty be obtained by propagating the individual variances for each random uncertainty
component. This would include not only the Poisson counting error and random uncertainty derived from the particular
measurement under conconcern, but also the random components
from corrections, constants, calibration factors and any
other measurements that make up the final result. These
contributions may be combined by summing in quadrature or
by the use of other propagation of error formulae given in
the *Guide*. It should be emphasized that one benefit of
making a complete assessment is that the process will require
the evaluation and statistical control of many previously
unevaluated measurement parameters. This will ultimately
improve the quality of measurements and aid quality control
within laboratories.

ASSESSMENT OF SYSTEMATIC UNCERTAINTIES

A clear distinction between random and systematic uncertainties is often difficult and troublesome. In part,

this is because many experimental processes embody both
systematic and stochastic (random) elements. The *Guide*
contains a number of examples which attempt to clarify the
distinction. Although a general guideline for the approach
to the assessment can be formulated, there are, unfortunately, no rules to objectively assign a magnitude to the
systematic uncertainties. This is a subjective process based
mainly on the judgement and experience of the person making
the measurement.

In estimating the magnitude of the systematic uncertainties, the *Guide* recommends that they be considered as
upper bounds, overall or maximum limits. By what method
then should the magnitude of these maximum limits to the
systematic uncertainties be assigned? It may be based on a
comparison to a standard or verification with two or more
independent and reliable measurement methods. Additionally,
it may be based on judgement; based on experience; based on
intuition; based on other measurements and data; or just a
pure guess (which may be better than nothing). Or it may
include combinations of, or all of the above factors. The
Guide also attempts to make this subjective process clearer
through the use of examples of typical measurement situations. The fact that a conceivable source of inaccuracy is
ascribable to random causes is not a sufficient condition
for treating it as a random uncertainty in error reporting.
It must be assessable by statistical methods derived from an
analysis of repeated measurements. In the absence of these
data, the effect of this random cause must be treated as a
systematic uncertainty.

APPROACH TO AN OVERALL UNCERTAINTY

As mentioned, one can obtain a combined total random
uncertainty stated in terms of some dispersion statistic
such as the standard deviation (s). Conceptually, one can
propagate and express the total random uncertainty in terms
of some confidence limits at any chosen confidence level.
This requires a number of assumptions, such as to the under-
lying population distributions, and can very rapidly become
exceedingly complicated requiring computations based on the
number of measurements involved in each random uncertainty
component, the number of degrees of freedom, the magnitude
of the counts in the Poisson counting error contribution,
etc. One is therefore confronted with the choice between
making simplifying assumptions (which ultimately may provide
results that are misleading), or performing complex calculations (whose efficacy, in terms of the time and cost
required, can be seriously questioned).

Additionally, one can obtain a list of each conceivable source of systematic uncertainty expressed as a confidence limit ($\pm\delta_j$) which can be at some high (say 99% or greater) confidence level.

Actually, without making assumptions about the distributions of the systematic uncertainties it is impossible to proceed. One of the simplest set of assumptions that can be made is that 1) the component systematic uncertainties are all independent, and 2) they are distributed such that all values within the estimated limits are equally likely (rectangular distribution). This approach, often termed the PTB approach[5], is gaining in popularity. With the above two assumptions, the rectangular systematic uncertainty distributions can be folded together to obtain a combined probability distribution for which the variance may be computed. This may then be combined in quadrature with that for the random uncertainty. In effect, the assumed Normal distribution of the random uncertainty is convoluted with the combined systematic uncertainty distribution to obtain an overall distribution. With this, the overall uncertainty limits at any given confidence level can be evaluated.

RECOMMENDED STATEMENT OF OVERALL UNCERTAINTY AND FINAL RESULT

There is no one clearly superior method out of the range of approaches to reporting an overall uncertainty. Yet, the diversity of current practices and methods either recommended or required by governmental bodies, or used by those who report data make it difficult if not impossible to combine or compare these data in a manner that would be useful for determining trends or possible necessary actions. There is, therefore, a great need for uniformity in the method used to report the uncertainties of environmental radiation data. In hope of achieving greater uniformity, the PTB method is recommended. In its simplest form, the random and systematic uncertainties should be combined to form an overall uncertainty in the result x by

$$u = \sqrt{s^2 + 1/3 \sum_{j=1}^{q} \delta_j^2}$$

where s is the standard deviation corresponding to the total random uncertainty and δ_j is the magnitude of the estimated systematic uncertainty for each of the q systematic uncertainty components. Additional details and other special cases are given in the *Guide*.

The recommended method requires that each reported
measurement result include the value, the total random uncertainty expressed as the standard deviation, and a combined overall uncertainty. Examples illustrating the method
are given in a detailed appendix to the *Guide*. This recommended approach may not be the best for all purposes. It is
intended to be useful and practical without imposing un-
necessary burdens on the time, money and personnel resources
of laboratories. Adoption of this method may eventually
demonstrate its most serious shortcomings and ultimately
lead to better methods.

REFERENCES

1. Serving on the subcommittee during the preparation of
    the *Guide and Recommendations* were: H.H. Abee (Union
    Carbide), L.K. Cohen (NRC), R. Collé (NBS), D. Ed
    (Illinois Dept. Public Health), E.H. Eisenhower (NBS),
    A.N. Jarvis (EPA), I. Fisenne (EML-DOE), M. Jackson
    (Albama Power), R. H. Johnson, Jr. (EPA), D. Olson
    (RESL-DOE), and J. Peel (DOE).

2. *A Guide and Recommendations for Reporting of Environ-
    mental Radiation Measurements Data,* Ad Hoc Subcommittee
    on Data Reporting, National Bureau of Standards Special
    Publication, to be published.

3. *Upgrading Environmental Radiation Data,* Committee for
    Upgrading Environmental Radiation Data, Health Physics
    Society, to be published. The work of this committee
    was described in a special session of the HPS 24th
    Annual Meeting, Philadelphia, July 12, 1979.

4. R. Collé, Reporting of environmental radiation measure-
    ments data, *11th Annual National Conference on Radia-
    tion Control,* Oklahoma City, May 10, 1979. Proceedings
    to be published by HEW.

5. S. Wagner, PTB Mitteilungen <u>79</u>, 343 (1969). An English
    version, *How to Treat Systematic Errors in Order to
    State the Uncertainty of a Measurement,* is available
    from the Physikalisch-Technische Bundesanstalt,
    Braunschweig, as report number FMRB 31/69. Additional
    details may be found in P.J. Campion, J.E. Burns and A.
    Williams, *A Code of Practice for the Detailed Statement
    of Accuracy,* National Physical Laboratory, London
    (1973); and A. Williams, P.J. Campion and J.E. Burns,
    Statement of results of experiments and their accuracy,
    Nucl. Instr. and Meth. <u>112</u>, 373-6 (1973).

QUALITY ASSURANCE FOR RADIONUCLIDE MEASUREMENTS AT
ENVIRONMENTAL LEVELS

Isabel M. Fisenne and George A. Welford. Environmental
Measurements Laboratory, U. S. Department of Energy,
New York, New York, U.S.A.

ABSTRACT

The Environmental Measurements Laboratory administers
and participates in two quality control programs. The first
is the intralaboratory program for evaluation of staff and
Environmental Measurement Laboratory's contractor results
obtained for environmental samples. The second Quality
Assurance Program is operated for the U. S. Department of
Energy, Division of Operational and Environmental Safety. The
Quality Assurance Program is an interlaboratory program
designed for major DOE contractors reporting environmental
radionuclide data.

This paper contrasts the problems associated with the
data evaluation process under both programs.

INTRODUCTION

The Environmental Measurements Laboratory (EML) has for
many years collected and analyzed environmental samples for
the assessment of global fallout. The sample matrices have
included air, soil, fresh and seawater, rainwater, tissue and
vegetation. The results obtained from the sample measurements
must be of sufficiently high quality to permit their use in
tracing and modeling global transport of fallout debris and

estimating the radiation doses to the general population. To
insure the usefulness of the measurement data, EML has maintained an intralaboratory quality control program and has
participated in national and international intercomparison
exercises.

In recent years, the U. S. Department of Energy (DOE)
has required its contractors handling radioactive materials
to submit annual reports detailing their environmental
monitoring measurements and the radiation dose assessments to
the local population. These contractor reports are submitted
to the Division of Operational and Environmental Safety (OES),
the group charged with enforcing DOE's regulations for
occupational and environmental exposures. In 1976, OES
recognized the need for establishing a formal program to
assist in the evaluation of the accuracy of the contractors'
monitoring data. EML was requested to develop and administer
a quality assurance program (QAP) directly geared to the type
of environmental samples which form the usual monitoring
program.

In this paper, we will describe the EML internal quality
control program and QAP, the data treatment and evaluation of
each program, and the decision making process at EML and for
the contractor laboratories by OES.

THE EML QUALITY CONTROL PROGRAM

The EML quality control program has the advantage that
the majority of the samples are collected, coded, prepared
and aliquoted by one Division and analyzed by another
Division. Typically, a group of samples has contained 10 to
15% standards, blanks and blind duplicates. The "standard"
sample analyses allow us to check the accuracy of an analysis
while the blind duplicate results monitor the precision of
the analytical and measurement processes. Table I shows the
replication in a "standard" human bone ash analyses for ^{90}Sr
during calendar year 1978 compared to the average obtained
for this sample from 1957 to 1977. The mean deviation for
EML analyses of ^{90}Sr in blind duplicate bone ash samples has
been <10% for 13 of the past 17 years. Following each year
in which the deviation was >10%, review of the laboratory
practices and at one point a change of staff reduced the mean
deviation to an acceptable level.[1]

TABLE I

REPLICATE BLIND ANALYSES OF ^{90}Sr IN A
"STANDARD" HUMAN BONE ASH SAMPLE

Analysis	Aliquot (g)	dpm ^{90}Sr per gram decay corrected to Jan. '79	July '57
1	10	0.80	1.36
2	10	0.76	1.30
3	5	0.74	1.27
4	10	0.69	1.18
5	10	0.74	1.26
Mean ±SD		0.75±.04	1.27±.07

dpm ^{90}Sr per gram (based on 149 blind analyses at EML from 1957 to 1978):

1.30±.15 (12%) corrected to July '57
0.76±.09 (12%) corrected to Jan. '79

Similar efforts in quality control are expended on contractor work. Public invitations are extended to commercial laboratories to bid on a particular set of analyses. The laboratories must perform within specified limits on a group of qualification samples. The contractor selected is subject to a quality control program similar to the EML program. Although the method of analysis is not specified, the contractor must follow specifications on reporting data. The requirements are in terms of single Poisson standard deviations geared to the activity level of the sample. Failure to meet the performance criterion can and has resulted in the termination of a contract.[1]

In addition to the intralaboratory program, EML participates in formal intercomparison programs. Our Laboratory's results from International **Atomic** Energy Agency (IAEA)

intercomparisons have been documented elsewhere.[2] On a
national basis, EML regularly receives and analyses water
samples from the Environmental Protection Agency's Las Vegas
Quality Assurance Branch. Some of these intercomparison
results for ^{239}Pu and ^{89,90}Sr in water are listed in Table
II.

The essential point for the internal quality control
program and participation in intercomparison programs is to
provide a basis for action. The action may take a compli-
mentary form or require review of current analytical
practices. In either case, feedback to the analyst is
mandatory to any quality control program.

OES QUALITY ASSURANCE PROGRAM

The Environmental Measurements Laboratory administered
OES Quality Assurance Program consists of a quarterly
distribution of 4 or 5 sample types to 34 laboratories. The
soil, tissue and vegetation samples are natural matrices
while the air filters and water samples are usually spiked
samples. Accompanying each distribution is a description of
the samples, a list of suggested radionuclide analyses and
the approximate activity of these nuclides. The participants
are given 90 days in which to submit their results and within
30 days (120 days after sample distribution) a compilation of
the results are issued as a coded report. Each participating
laboratory is notified of its code and OES receives an uncoded tabulation.

For the quarterly reports, the data reported by each
laboratory, including replicates and percent error, are
listed by sample type, laboratory code number and nuclide.
The uncoded EML value is listed and the ratio and error of
the ratio of the reported value to the EML value is calculated. A simplistic statistical summary based on the number
of laboratories within 20% and 50% of the EML value is
included in the report. The annual statistical summary for
the calendar year is somewhat more elaborate. Outlier
testing is applied to obtain the overall mean value for a
particular nuclide-matrix pair. The ratio of the reported
values (or mean values in the case of replicates) to the EML
value are represented graphically. An example from the
statistical summary report is given in Table III.

TABLE II

EML RESULTS FROM EPA QUALITY ASSURANCE BRANCH -
^{239}Pu and ^{89,90}Sr in Water Cross-Check Program

Date	Nuclide	No. of Labs.	Known Value	pCi per liter EML Values	Grand Mean
Jan. '78	^{239}Pu	17	1.7	1.8	1.5
				1.5	
				1.5	
July '78	^{239}Pu	17	3.6	3.2	3.1
				2.9	
				3.1	
Jan. '79		21	4.8	4.1	4.0
				4.0	
				4.0	
Aug. '77	^{89}Sr	31	14.	11.	15.
				12.	
				9.	
	^{90}Sr	33	10.	7.	9.
				8.	
				8.	
Jan. '78	^{89}Sr	38	25.	23.	24.
				25.	
				20.	
	^{90}Sr	42	31.	27.	30.
				28.	
				27.	
May '78	^{89}Sr	39	16.	14.	19.
				14.	
				12.	
	^{90}Sr	41	27.	23.	24.
				25.	
				27.	
Sept. '78	^{89}Sr	39	19	15.	14.
				16.	
				14.	
	^{90}Sr	41	16.	16.	16.
				16.	
				16.	
Jan. '79	^{89}Sr	44	14.	12.	13.
				12.	
				13.	
	^{90}Sr	46	6.	6.	6.
				5.	
				6.	

TABLE III

STATISTICAL EVALUATION

| DATE | TYPE | ISOTOPE | | EML VALUE | |------ALL LABORATORIES INCLUDED------| | | | | |-----------CUTLIERS EXCLUDED---------| | | | |
					LABS REPT	MEAN	STD ERROR OF MEAN	RATIO	LABS INCLD	MEAN	STD ERRCR OF MEAN	RATIC	CUTLIERS
78 07	WATER	H	3	0.367E 03	17	0.353E 03	0.253E 02	0.96	15	C.371E C3	0.131E 02	1.01	2
78 07	SOIL	K	40	0.197E 02	13	0.192E 02	0.624E 00	0.97	13	0.197E C2	0.624E 00	1.00	0
78 07	TISSUE	K	40	0.222E 02	7	0.172E 02*	0.107E 01	0.78	7	0.172E 02*	0.124E 01	0.78	0
78 07	VEG	K	40	0.272E 02	9	0.246E 02	0.199E 01	0.90	8	0.262E 02	0.150E 01	0.96	1
78 07	TISSUE	CO	60	0.108E 01	10	0.114E 01	0.130E 00	1.06	7	0.104E 01*	0.171E-01	0.96	3
78 07	VEG	CO	60	0.257E 01	9	0.399E 01	0.111E 01	1.55	7	0.246E 01	0.231E 00	0.96	2
78 07	WATER	CO	60	0.240E 00	12	0.356E 01	0.333E 01	14.84	10	0.243E 00	0.617E-02	1.01	2
78 07	AIR	SR	89	0.127E 02	3	0.147E 02	C.104E 01	1.16	3	C.147E 02	0.116E 01	1.16	0
78 07	AIR	SR	90	0.115E 02	6	0.115E 02	0.373E 00	1.00	6	0.115E 02	0.387E 00	1.00	0
78 07	SOIL	SR	90	0.540E 00	12	0.672E 00*	0.646E-01	1.25	11	0.600E 00	0.474E-01	1.11	1
78 07	TISSUE	SR	90	0.406E 01	9	0.345E 01	0.337E 00	0.85	9	0.345E 01	0.386E 00	0.85	0
78 07	VEG	SR	90	0.878E 01	10	0.837E 01	0.833E 00	0.95	8	0.854E 01	0.336E 00	0.97	2
78 07	WATER	SR	90	0.360E 00	10	0.337E 00	0.235E-01	0.94	9	0.356E 00	0.151E-01	0.99	1
78 07	AIR	CS 137		0.257E 03	14	0.263E 03	0.161E 02	1.02	13	0.264E 03	0.120E 02	1.03	1
78 07	SOIL	CS 137		0.720E 00	13	0.704E 00	0.284E-01	0.98	12	0.678E 00*	0.153E-01	0.94	1
78 07	TISSUE	CS 137		0.302E 01	10	0.262E 01*	0.207E 00	0.87	9	0.279E 01*	0.124E 00	0.92	1
78 07	VEG	CS 137		0.662E 01	9	0.633E 01	0.479E 00	0.96	8	0.670E 01	0.335E 00	1.01	1
78 07	WATER	CS 137		0.540E-01	13	0.527E 00	0.433E 00	9.76	10	0.492E-01*	0.946E-03	0.91	3
78 07	SOIL	RA 226		0.850E 00	9	0.966E 00	0.950E-01	1.14	8	C.887E C0	0.562E-01	1.04	1
78 07	AIR	PU 238		0.410E-02	6	0.279E-01	0.215E-01	6.80	4	0.435E-02	0.136E-02	1.06	2
78 07	SOIL	PU 238		0.680E-03	3	0.703E-02	0.648E-02	10.34		SEE NOTE (2)			
78 07	TISSUE	PU 238		0.180E-02	5	0.460E-02	0.182E-02	2.55	5	0.460E-02	0.168E-02	2.55	0
78 07	VEG	PU 238		0.410E-02	6	0.479E-02	C.116E-02	1.17	5	0.375E-02	0.574E-03	0.91	1
78 07	AIR	PU 239		0.540E 00	12	0.553E 00	0.313E-01	1.02	10	0.516E 00	0.256E-01	0.95	2
78 07	SOIL	PU 239		0.630E-02	7	0.572E-02	0.108E-02	0.91	6	0.661E-02	0.785E-03	1.05	1
78 07	TISSUE	PU 239		0.194E-01	6	0.196E-01	0.290E-02	1.01	5	0.172E-01	0.202E-02	0.89	1
78 07	VEG	PU 239		0.441E-01	8	0.466E-01	C.110E-01	1.06	7	C.362E-01	0.417E-02	0.82	1
78 07	WATER	PU 239		0.400E-04	7	0.424E-03	0.254E-03	10.59	6	0.174E-03*	0.581E-04	4.36	1
78 07	AIR	AM 241		0.486E 00	5	0.487E 00	0.498E-01	1.00	5	0.487E 00	0.562E-01	1.00	0
78 07	TISSUE	AM 241		0.450E-02	3	0.817E-02	C.183E-02	1.81		SEE NOTE (2)			
78 07	VEG	AM 241		0.104E-01	3	0.201E-01	0.566E-02	1.94	3	0.201E-01	0.669E-02	1.94	0

* : MEAN DIFFERS FROM EML VALUE AT 10 % LEVEL OF SIGNIFICANCE
(1): FEWER THAN 3 VALUES INITIALLY
(2): FEWER THAN 3 VALUES REMAIN AFTER OUTLIERS REMOVED

The EML intercomparison of the body of the reported
values is stated in the broadest terms. Individual decisions
on the quality of the measurements reported by a laboratory
are the responsibility of the DOE area operations office. To
increase the information base available to the operations
offices, EML has distributed some samples two or three times.
The introduction of these blind duplicate samples into the
program gives another means of evaluating the analytical
performance of a laboratory. Although the results of the
blind duplicates have not appeared in the QAP reports to
date, a section containing this data will appear in the
statistical summary report for calendar year 1979. A preview
of the results obtained from duplicate analyses of a soil
sample is shown in Table IV.

Although the statistical evaluation of the QAP results
is relatively simple, it has provided OES with hard data with
which to judge the quality of their contractor laboratories.
Generally, there has been an improvement in both the accuracy
and precision of the reported results indicating that the
quality assurance program is accomplishing its goal of
providing a degree of confidence in environmental measurement
data.

REFERENCES

1. J. H. Harley, "Experience with an Analytical Quality
    Control Program," Effluent and Environmental Radiation
    Surveillance, ASTM STP 698, J. J. Kelly, Ed., American
    Society for Testing and Materials, to be published (1980).

2. G. A. Welford, and J. H. Harley, HASL Participation in
    IAEA Intercomparisons, U. S. Energy Research and Develop-
    ment Administration Report HASL-322, New York (1977).

TABLE IV

BLIND DUPLICATE ANALYSES OF A QAP SOIL SAMPLE

Nuclide	Laboratory	Mean (pCi/g)	% Standard Deviation
^{239}Pu	EML	3.7	3
	2	2.7	22
	3	1.8	47
	4	4.0	8
	5	3.0	34
	6	3.3	36
	7	2.8	8
	8	2.9	23
	9	2.6	13
^{238}Pu	EML	0.09	4
	2	0.05	16
	3	0.02	225
	4	0.07	22
	5	0.06	29
	6	0.05	14
	7	0.06	28
	8	0.05	32
^{137}Cs	EML	0.48	1
	2	0.49	10
	3	0.45	6
	4	1.7	125
	5	0.50	4
	6	0.68	15
^{90}Sr	EML	0.25	6
	2	0.24	3
	3	0.28	6
	4	0.54	60
	5	0.27	8
	6	0.46	43
	7	0.26	8

VALIDATION OF ANALYTICAL LABORATORY DATA
FOR THE CHARACTERIZATION OF HANFORD
DEFENSE HIGH-LEVEL CHEMICAL WASTES

A. G. Miller. Chemical Sciences Group, Research
Department, Rockwell Hanford Operations, Richland,
Washington, USA.

ABSTRACT

An application of a proposed comprehensive data validation program within Rockwell Hanford Operations is being
developed for the characterization of stored, nuclear defense
wastes. It will encompass sample preparation, sample analysis, and automation. The benefits expected from this validation activity are improved results in terms of precision and
accuracy, information on significant deviations in analytical
method and/or analyst performance in near real time, identification of needed improvements in analytical methodology,
information on long-term precision and accuracy of analytical
determinations, and the several automation benefits.

INTRODUCTION

The characterization of Hanford high-level wastes stored
in underground tanks is required to support the Hanford Long-
Term Waste Management Program (HLWM). Specifically, information is required for evaluation of alternative courses of
action for disposing of radioactive wastes. The information
will assist in design decisions for proposed waste processing
plants and in evaluating environmental impact of the various
toxic and radioactive constituents for the storage alternative. The quality of the desired characterization data is
determined by the requirements of site-specific, Hanford waste
management alternatives. These requirements have led to the
development of various analytical precision and accuracy specifications for the characterization of Hanford defense waste.[1]

Information generated on chemical and physical characteristics of stored waste originates with analytical measurements
on samples prepared from solid cores taken from the stored
wastes. Reliable analyses, which ensure that the Waste Characterization activity meets HLWM specifications, are of para-
mount importance to the program. Therefore, the ability of
the analytical laboratory (AL) to meet customer specifications
must be demonstrated and proven in conjunction with the analysis of waste tank samples. This implies that a laboratory
data validation program (DVP) consistent with the goals of
the Waste Characterization Activity must be implemented.
Therefore, this document describes and defines a DVP to be
applied to the analysis of Waste Characterization samples.

A DVP which will enable AL to meet Waste Characterization specifications, basically consists of two parts:

● A quality control program, which ensures a product
meets defined criteria, and may include provision to determine the degree to which the criteria are met. Information
derived is useful in making necessary improvements to the
processes producing the product.

● An automation program, which is a planned approach
to using computers for the control and operation of equipment
involved in production.

In the circumstances we consider here, the product is
analytical results derived from chemical determinations on
samples of stored nuclear chemical waste, the equipment is
analytical instrumentation, and production includes collecting raw data, calculation of results, and producing, distri-
buting, and storing the final analytical reports. The
Quality Control Program (QCP), with automation, will provide
the real time control not currently provided.

THE QUALITY CONTROL PROGRAM

The QCP to be implemented shall use working reference
materials (WRMs) that simulate actual samples as nearly as
possible, contain known quantities of constituents for which
quality control is desired, and have a verified or characterized chemical composition. These WRMs shall be known, single-
and double-blind standards defined as follows:

● Known Standard - a reference material for which the
analyst knows the chemical content.

● Single-Blind Standard - a reference material for
which the analyst does not know the chemical content, but does
know that it is a WRM.

● Double-Blind Standard - a reference material for which the analyst does not know the chemical content and does not know that it is a WRM.

These standards will be used in the following ways:

● To prove that each analytical method/analyst is capable of performing to HLWM specifications.

● To demonstrate adequate performance in sample analysis.

Working Reference Materials

Validation of analytical information generated for these activities requires development and quality control usage of chemical WRMs. The purpose of such WRMs is to simulate the sample materials such that analytical methods can be logically expected to perform in a similar manner on both standard and sample materials. Therefore, such WRMs can be used to judge performance of analytical methods and develop accuracy and precision data relevant to the actual samples.

Since there are various types of stored, nuclear-chemical waste which originated from several chemical processes and chemical differentiation during storage, it is necessary to have WRMs which simulate the major waste types (water-soluble and water-insoluble). Water-soluble waste is to be simulated by a number (such as three) preparations. Their compositions will vary around that typical of the samples. The following list is a proposed analytical standard for dissolved, water-soluble waste:

$0.00020\underline{M}$ Bi^{+++}	$0.20\underline{M}$ CO$_3^{2-}$	$0.05\underline{M}$ OH
$0.000040\underline{M}$ Cd^{++}	$0.13\underline{M}$ C	$3.0\underline{M}$ Na^{+}
$0.00020\underline{M}$ Ce^{+++}	$0.10\underline{M}$ SO$_4^{=}$	500 μCi/ℓ ^{137}Cs
$0.00030\underline{M}$ Hg^{+++}	$0.10\underline{M}$ Cl^{-}	3.0 μCi/ℓ ^{106}Ru
$0.00020\underline{M}$ Ni^{++}	$0.030\underline{M}$ PO$_4^{3-}$	0.020 μCi/ℓ ^{99}Tc
$0.000010\underline{M}$ Pb^{++}	$0.010\underline{M}$ CN^{-}	1.0 μCi/ℓ ^{125}Sb
$0.00010\underline{M}$ UO$_2^{++}$	$0.20\underline{M}$ NO$_2^{-}$	10 μCi/ℓ ^{90}Sr
$0.00020\underline{M}$ Zr	$2.00\underline{M}$ NO$_3^{-}$	0.010 μCi/ℓ ^{63}Ni

The composition of water-insoluble waste is indicative of the chemical processes. Simulations of KOH fusion-dissolved water-insoluble wastes have compositions such as:

$1 \times 10^{-3} \underline{M}$ Al	$7 \times 10^{-3} \underline{M}$ Na	90 μCi/ℓ ^{137}Cs
$3 \times 10^{-3} \underline{M}$ Bi	$1 \times 10^{-3} \underline{M}$ Ni	15 μCi/ℓ ^{63}Ni
$4 \times 10^{-4} \underline{M}$ Cd	$1 \times 10^{-3} \underline{M}$ NO$_3^-$	10 μCi/ℓ ^{125}Sb
$4 \times 10^{-4} \underline{M}$ Ce	$0.01\underline{M}$ PO$_4^{3-}$	200 μCi/ℓ ^{90}Sr
$1 \times 10^{-3} \underline{M}$ Cr	$4 \times 10^{-3} \underline{M}$ Si	0.1 μCi/ℓ ^{79}Se
$0.01\underline{M}$ F$^-$	$3 \times 10^{-3} \underline{M}$ SO$_4^-$	20 μCi/ℓ ^{241}Am
$0.06\underline{M}$ Fe	$1 \times 10^{-3} \underline{M}$ Sr	7 μCi/ℓ ^{239}Pu
$5 \times 10^{-6} \underline{M}$ Hg	$1 \times 10^{-4} \underline{M}$ U	0.4 μCi/ℓ ^{99}Tc
$2 \times 10^{-3} \underline{M}$ La	$8 \times 10^{-5} \underline{M}$ W	Matrix - 6$\underline{M}$ HCl
$3 \times 10^{-3} \underline{M}$ Mn	$1 \times 10^{-4} \underline{M}$ Zr	

Solid WRMs are also required to monitor the sample preparation process. Solid, synthetic salt cake standards (water-soluble fraction) have been prepared.[2] Solid, synthetic sludge standards (water-insoluble fraction) are planned.

For a substance to function as a WRM, its composition, or more exactly, the concentration of the constituents in which it is to be standard, must be known to an adequate degree of confidence. This information is to be derived from the preparation and/or analytical determinations constituting either verification of the preparation values by chemical analysis or generation of characterized values from chemical analyses by more than one method.

Analytical Qualification

The ability of the analytical methods and analysts to perform chemical analyses such that Waste Characterization specifications can be met must be demonstrated prior to analyzing samples on which required specifications are to be applied. Specific aspects of this qualification of the analytical method/analyst are:

- Analysis of known standards for training, verification of calibration, etc.

- Analysis of two single-blind WRMs in triplicate.

- The pooled standard deviation of an individual determination calculated from the results must be within the Waste Characterization specification for that analyte.

• The arithmetic average of the results obtained on each single-blind WRM must agree with the standard value to within the Waste Characterization specification for that analyte.

• Failure of any of the three results to meet the Waste Characterization specification shall void the qualification effort.

Analysis of Waste Characterization Samples

The continual demonstration of the capability for meeting the Waste Characterization specification is necessary to assure that the specification is being met. Many factors, which may vary at any time, can enter into an analytical procedure and cause a loss in capability. Therefore, the analytical capability of AL must be routinely demonstrated with the analysis of the Waste Characterization samples. This will be accomplished by the following requirements:

• Analysis of a single-blind WRM at the minimum frequency of 1/day or 1/batch of samples, whichever is lesser.

• Analysis of double-blind WRMs submitted as samples by the sample preparation laboratory at the proposed frequency of 1/batch of samples.

• Analysis of samples prepared from solid WRMs and submitted to AL as samples. Such materials are primarily for the purpose of controlling the sample preparation process, but analytical data obtained from their analysis can be used in a subordinate role for evaluating analytical control.

Loss of Analytical Control

Unsatisfactory analytical results (do not meet Waste Characterization specifications) on a blind standard shall require satisfactory performance on an independent blind standard to validate the sample analyses. The independent blind standard (single- or double-blind) may be analyzed either concurrently with the samples or during the same calendar day by the same analyst. Failure to obtain satisfactory analyses on two consecutive blind standards shall constitute loss of analytical control, requiring AL to take corrective action.

Validation of Sample Preparation

     Control of the sample preparation process can only be
shown by analysis of WRMs processed with the samples and in
the same manner. The solid WRMs are primarily for this purpose. Analyses of two consecutive samples prepared from solid
WRMs for which results on one or more constituents do not meet
the specification, but are validated by primary control standards (solution WRMs), shall constitute "loss of sample-
processing control". Samples processed during the period in
question shall be flagged as questionable. Each loss of
sample-processing control condition shall be handled on an
individual basis by representatives of the Rockwell components
involved. Resolution/correction of the sample processing difficulty to the satisfaction of those functional representatives will permit resumption of sample processing. Satisfactory results on the next solid WRM processed officially
terminate the loss of sample-processing-control condition.
Some or all of the analyses on samples processed during the
period in question may be validated by the investigating
committee.

STATISTICAL ANALYSES

     An important product of the QCP is statistical evaluation
of the total analytical results provided by the Rockwell
Statistics Group. Analytical uncertainties are to be based on
the analyses of known, single-, and double-blind standards.
The analyses of the solid WRMs will provide information on
sample preparation error. The statistical analyses, to
include current and long-term recoveries and standard deviations at the 95% confidence level, are to be reported quarterly or on request.

QUALITY CONTROL PROGRAM COMPUTER

     The fully programmed QCP computer will:

     1. Receive and store standard values, identifications
of WRMs, sample identification numbers and analytical results.

     2. Evaluate qualification attempts and report results to
the analyst.

     3. Evaluate and report results on blind WRMs.

     4. Identify sample-processing loss of control conditions
and flag samples affected.

5. Store all pertinent WRM and sample analytical data
for calculation of statistical results on WRM and sample
analyses.

6. Tabulate all analytical results and current method
performance data for each sample into selected format for
reporting to the appropriate Rockwell component.

CURRENT STATUS OF DVP DEVELOPMENT

The DVP described is the fully developed program planned
for Waste Characterization samples at Hanford. Currently,
Waste Characterization is in preliminary stages, where some
waste tanks are being cored, WRMs are being developed, and
software/automation is either partially in place or absent.
Most aspects of this DVP are expected to be put in place
during FY 1980.

REFERENCES

1. G. W. Grimes, U.S. DOE Document, RHO-CD-573, "Hanford
    Defense High-Level Waste Sampling and Characterization
    Plan," (1978).

2. A. E. Schilling and A. G. Miller, "Synthetic Salt Cake
    Standards for Analytical Laboratory Quality Control,"
    23rd Oak Ridge National Conference on Analytical
    Chemistry in Energy Technology, Gatlinburg, Tennessee,
    October, 1979.

SYNTHETIC SALT CAKE STANDARDS FOR ANALYTICAL LABORATORY
QUALITY CONTROL

A. E. Schilling and A. G. Miller, Rockwell Hanford
Operations, Richland, Washington, USA.

ABSTRACT

The validation of analytical results in the characterization of Hanford Nuclear Defense Waste requires the preparation
of synthetic waste for standard reference materials. Two
independent synthetic salt cake standards have been prepared
to monitor laboratory quality control for the chemical characterization of high-level salt cake and sludge waste in support
of Rockwell Hanford Operations' High-Level Waste Management
Program. Each synthetic salt cake standard contains 15 characterized chemical species and was subjected to an extensive
verification/characterization program in two phases. Phase I
consisted of an initial verification of each analyte in salt
cake form in order to determine the current analytical capability for chemical analysis. Phase II consisted of a final
characterization of those chemical species in solution form
where conflicting verification data were observed. The 95
percent confidence interval on the mean for the following
analytes within each standard is provided: sodium, nitrate,
nitrite, phosphate, carbonate, sulfate, hydroxide, chromate,
chloride, fluoride, aluminum, plutonium-239/240, strontium-90,
cesium-137, and water.

INTRODUCTION

The High-Level Waste Management Program at Rockwell
Hanford Operations is currently responsible for both the present and future management of all high-level defense waste
stored on the Hanford Reservation. This waste, a by-product
of the reprocessing of spent reactor fuel, has been accumulated
and stored in various forms on the Hanford Reservation since
the 1940's. The defense waste has now been converted into
salt cake and sludge components. Of particular interest to

Rockwell's High-Level Waste Management Program is the future
management of these components.

In order to assure that the future management of this
salt cake and sludge waste is accomplished in a safe and
secure manner, a complete chemical characterization of both
salt cake and sludge waste is being undertaken by the High-
Level Waste Management Program at Rockwell Hanford Operations.
To ensure adequate laboratory quality control for this characterization effort, two synthetic salt cake standards have been
prepared by Rockwell Hanford Operations' Research Department
in cooperation with the Analytical Laboratories Department.
In general, this article briefly outlines the preparation of
these standards and the verification/characterization program
used to assign the overall uncertainty to each analyte within
each standard. The concepts of data validation outlining the
general use of these standards to demonstrate laboratory quality control are discussed in Reference 1.

SYNTHETIC SALT CAKE STANDARD PREPARATION

Since the preparation of Synthetic Salt Cake Standards I
and II is similar, this section shall deal specifically with
the preparation of the more complex synthetic salt cake standard, Standard II. While the independent chemical stock
sources for both standards are provided in this section, the
use of these stock sources to prepare Standard I will not be
discussed; however, the relationship between these stock
sources and the preparation of Standards I and II will be
briefly outlined in the conclusions at the end of this article.

STANDARD II PREPARATION

Synthetic Salt Cake Standard II was prepared from five
independent chemical stock sources: solid core for Standard
II; an evaporator bottoms liquor (EBL solutions); an
americium-241 spike solution (AmSoL), and two basic salt solutions, Solutions A and B. Reagent grade or ultrex grade
chemicals were used throughout the preparation of the standard.
Super "Q" water (greater than 10 megaohms resistance), calibrated pipets, and calibrated volumetric flasks were used in
the preparation of all solutions. The weight of all chemicals
and/or solutions was determined by difference using calibrated
balances with accuracy sufficient to ensure five significant
figures (i.e., 1 kg ± 0.1 g or 1 g ± 0.01 mg); however, no
correction for buoyancy was made. The atomic weights for all
elements were obtained from Reference 2. The chemical makeup
for each of these five independent stock sources is provided
in Tables I through IV.

TABLE I

SOLID CORE FOR STANDARDS I AND II

| Chemical Species | Quantity | |
	Standard I (Grams)	Standard II (Grams)
$NaNO_3$	2,100.0	1,500.0
Na_2CO_3	70.0	280.22
$Na_3PO_4 \cdot 12H_2O$	152.39	594.87
Na_2SO_4	15.001	180.03
$NaNO_2$	20.327	15.063
Total H_2O[a]	78.945	325.36

[a]Based on an experimental measurement for the percent water in $Na_3PO_4 \cdot 12H_2O$.

TABLE II

AMERICIUM–241 SPIKE SOLUTION

Chemical Species	Final Concentration
NaOH	2.6754$\underline{M}$
$NaNO_3$	1.6328$\underline{M}$
^{241}Am	898.90 $\mu ci/\ell$
H_2O	79.259 percent

TABLE III

BASIC SALT SOLUTIONS A AND B

| Chemical Species | Final Concentration | |
	Solution A	Solution B
NaOH	9.1454$\underline{M}$	5.1591$\underline{M}$
$NaNO_3$	1.5971$\underline{M}$	2.5004$\underline{M}$
$NaNO_2$	---	1.8301$\underline{M}$
H_2O	63.328%	59.994%

TABLE IV

EVAPORATOR BOTTOMS LIQUOR[a]

Chemical Species	Final Concentration
$Na_4N_2C_{10}H_{12}O_8$ (EDTA)	$0.02546\underline{M}$
$Na_3N_2C_{10}H_{15}O_7$ (HEDTA)	$0.02642\underline{M}$
$Na_3C_6H_9O_9$	$0.02600\underline{M}$
$^{239/240}Pu$	36.650 µci/ℓ
^{137}Cs	4,180.0 "
^{90}Sr	241.36 "
^{241}Am	2.9080 "
$NaNO_2$	$1.5997\underline{M}$
Na_2SO_4	$0.05001\underline{M}$
$Na_3PO_4 \cdot 12H_2O$	$0.01000\underline{M}$
$Na_2Cr_2O_7 \cdot 2H_2O$	$0.05000\underline{M}$
$NaCl$	$0.30000\underline{M}$
$Na_2SiO_2 \cdot 9H_2O$	$0.00500\underline{M}$
NaF	$0.00403\underline{M}$
$NaAlO_2$	$1.2007\underline{M}$
$NaNO_3$	$2.4015\underline{M}$
Na_2CO_3	$0.08004\underline{M}$
Free OH^-	$3.0691\underline{M}$
Total H_2O[b]	$56.850\underline{\%}$

[a]EBL solution is the combination fo three solutions: a complexant solution, a caustic salt solution, and a caustic aluminum solution.

[b]Based on theoretical values for percent water in $Na_3PO_4 \cdot 12H_2O$; $Na_2SiO_3 \cdot 9H_2O$, and $Na_2Cr_2O_7 \cdot 2H_2O$.

The following sequence of steps was followed to prepare synthetic Salt Cake Standard II.

1. Solid core material for Standard II (2,570.2 grams) and EBL solution (282.36 grams) were added by weight to a two-gallon (7.6-liter) Sterling Mixer, Model LDM-2. This material was mixed for 0.33 hours at speed number two after which 34.440 grams of the AmSoL solution were added.

2. The beaters and sides of the mixer were then scraped periodically during 2.92 hours of mixing. After mixing, samples were taken in duplicate for gamma-ray analyses to determine the distribution of cesium-137 throughout this standard.

3. Solution A (119.52 grams) was then added by weight. This
    material was then mixed for 15 minutes. Solution B
    (252.69 grams) was then added by weight.

4. This material was again mixed and the beaters and side
    walls of the mixer scraped periodically for 8.32 hours.
    Throughout this period, the material was yellowish and
    lumpy at the start and yellowish and creamy-sticky at the
    end.

VERIFICATION/CHARACTERIZATION PROGRAM

Each synthetic salt cake standard was subjected to an
extensive verification/characterization program in two phases.
Phase I consisted of an initial verification of each analyte
in salt cake form in order to determine the current analytical
capability for chemical analysis of this material. Phase II
consisted of a final characterization of those analytes in
solution form when conflicting verification data were observed
(i.e., two independent analytical methods for the same analyte
resulting in two statistically different mean values or such
mean values different from the preparation value). All data
from both phases were submitted to the Statistics Unit of
Rockwell Hanford Operations' Research Department for analysis.

Results of this statistical analysis, summarized in
Table V below, indicated that at least two independent averages (i.e., averages from two independent methods) of good
precision are available for the following analytes: sodium,
nitrate, sulfate, aluminum, chromate, nitrite, cesium-137, and
plutonium-239/240. Since none of the recoveries for these
analytes showed a significant bias, they were combined into
high- and low-variance groups. The low-variance group includes sodium, nitrate, sulfate, aluminum, and plutonium-239/
240. The high-variance group includes chromate, nitrite, and
cesium-137. All remaining analytes (i.e., phosphate, carbonate, hydroxide, chloride, fluoride, water, and strontium-90)
had one method for chemical analysis with significantly better
precision.

As indicated earlier in the preparation of Standard II
(see Step 2, page 4), a test to determine the degree of homogeneity for this standard was made. Specifically, 18 two-dram
aliquots were taken at random from the sides, top, and bottom
of the mixer during the preparation of both standards. These
two-dram aliquots were analyzed by gamma-ray spectroscopy for
cesium-137. Again, all data were submitted to Rockwell
Hanford Operations' Statistics Unit for analysis. The results
of this statistical analysis indicated no significant variation at the 95 percent confidence interval between the

measured cesium-137 activity and the position from which the samples were taken. From these data, it is apparent, barring any unknown selective chemical interaction, that both synthetic standards are homogeneous.

TABLE V

COMBINED STATISTICAL RESULTS
BASED ON MAKEUP VALUES

Analyte	Standard I		Standard II	
	Percent Average Recovery	95 Percent[a] Limit	Percent Average Recovery	95 Percent[a] Limit
Na^+	100.77	1.55	100.77	1.55
NO_3^-	100.99	1.55	100.99	1.55
$SO_4^=$	100.99	1.55	100.99	1.55
Al^{+3}	100.99	1.55	100.99	1.55
$^{239/240}Pu$	100.67	1.55	100.70	1.55
$CrO_4^=$	101.6	30	97.3	6.5
NO_2^-	101.6	30	97.3	6.5
^{137}Cs	101.6	30	97.3	6.5
PO_4^{-3}	92.58	1.54	98.78	1.54
$CO_3^=$	101.89	0.48	100.46	0.45
OH^-	100.58	2.39	97.35	2.39
F^-	126.01	11.74	112.18	1.96
Ce^-	116.82	0.49	90.59	0.12
^{90}Sr	82.81	7.59	93.01	7.59
H_2O	85.54	1.31	97.44	0.29

[a]Ninety-five percent confidence interval on the mean.

CONCLUSIONS

Table VI lists the standard value assigned to each analyte for both synthetic standards as a result of this verification/characterization program. The uncertainty for each analyte has been provided at the 95 percent confidence interval on the mean.

The total weight percent uncertainty for each standard at the 95 percent confidence interval on the mean is ±1.05 percent for Standard I and ±0.69 percent for Standard II. A total of 2,737 grams of Standard I and 3,259 grams of Standard II was prepared.

TABLE VI

STANDARD VALUES/SYNTHETIC
SALT CAKE STANDARDS I AND II

Chemical Species	Standard I (mg/gstd)	95 Percent Limit	Standard II (mg/gstd)	95 Percent Limit
Na^+	256.8	1.55	251.12	1.55
NO_3^-	580.2	1.55	361.03	1.55
NO_2^-	11.96	30.0	12.071	6.5
$PO_4^{\equiv}$	14.473	1.54	47.377	1.54
$CO_3^=$	15.206	0.48	49.200	0.45
$SO_4^=$	4.1926	1.55	38.026	1.55
OH^-	5.1850	2.39	12.446	2.39
$CrO_3^=$	1.0913	30.0	0.69530	6.5
Cl^-	1.1487	0.49	0.59343	0.12
$SiO_3^=$	0.03523	----[a]	0.02344	----[a]
F^-	0.00894	11.74	0.00530	1.96
Al^{3+}	3.0299	1.55	2.0157	1.55
$^{239/240}Pu$	3.323 nCi/gstd	1.55	2.212 nCi/gstd	1.55
^{241}Am	6.616 nCi/gstd	----[a]	8.195 nCi/gstd	----[a]
^{90}Sr	18.51 nCi/gstd	7.59	18.830 nCi/gstd	7.59
^{137}Cs	393.3 nCi/gstd	30.0	250.6 nCi/gstd	6.5
H_2O	93.673	1.31	221.37	0.29

[a] Concentration below detectable limits.

The major difference between the preparation of Standard
I and Standard II is the lack of Solutions A and B in the preparation of Standard I. Specifically, Standard I was prepared
by adding fractions of the EBL solution and the AmSoL solution
to a solid core material (Solid Core I), in a similar fashion
as described in the preparation for Standard II.

Each of these standards was used independently as blind
and double-blind standards during the characterization of salt
cake waste by Rockwell Hanford Operations' Analytical Laboratories Department. Data obtained on the quality control for
this laboratory characterization of salt cake waste, using
these standards, will be published at the conclusion of this
characterization effort.

REFERENCES

1. A. G. Miller, <u>Validation of Analytical Laboratory Data for Characterization of Hanford High-Level Waste</u>, RHO-SA-133, Rockwell Hanford Operations, Richland, Washington (1979).

2. <u>Chart of the Nuclides</u>, Knolls Atomic Power Laboratory, Naval Reactors, USAEC, 9th Edition (1966).